Lecture Notes in Mathematics

Edited by A. Dold and B. Eckmann

551

Algebraic K-Theory

Proceedings of the Conference Held at Northwestern University Evanston, January 12–16, 1976

Edited by Michael R. Stein

Springer-Verlag
Berlin · Heidelberg · New York 1976

Editor
Michael R. Stein
Department of Mathematics
Northwestern University
Evanston, Il. 60201/USA

Library of Congress Cataloging in Publication Data
Main entry under title:

Algebraic K-theory.

(Lecture notes in mathematics ; 551)
Bibliography: p.
Includes index.
1. K-theory--Congresses. 2. Homology theory--Congresses. 3. Rings (Algebra)--Congresses. I. Stein, Michael R., 1943- II. Series: Lecture notes in mathematics (Berlin) ; 551.
QA3.L28 no. 551 [QA612.33] 510'.8s [514'.23] 76-49894
ISBN

AMS Subject Classifications (1970): 13D15, 14C99, 14F15, 16A54, 18F25, 18H10, 20C10, 20G05, 20G35, 55E10, 57A70

ISBN 3-540-07996-3 Springer-Verlag Berlin · Heidelberg · New York
ISBN 0-387-07996-3 Springer-Verlag New York · Heidelberg · Berlin

Printed in Germany
Printing and binding: Beltz Offsetdruck, Hemsbach/Bergstr.

Introduction

A conference on algebraic K-theory, jointly supported by the National Science Foundation and Northwestern University, was held at Northwestern University January 12-16, 1976. These proceedings contain papers presented at that conference, survey articles on certain subspecialities represented at the conference, and related papers, some by mathematicians who did not attend the conference.

The diversity of mathematical interests subsumed under the title "algebraic K-theory" is by now well-known; a glance at the contents of this volume will confirm this. To deal with this diversity, a large block of time was left free for participants to organize themselves into seminars on topics of their choice. A list of these seminar talks has been included following the list of lectures given to the full group. I have also listed the names of conference participants and of authors of articles along with their addresses as of January 1976.

It seems appropriate to mention here that shortly after the end of this conference, Quillen and Suslin, working independently, found a positive solution to Serre's problem,which motivated some of the earliest research in the "classical" period of algebraic K-theory. Quillen's solution is to appear in Inventiones Math.

On behalf of the participants, I would like to thank the NSF and Northwestern University's College of Arts and Sciences for their financial support. I would also like to thank Madalyn Kuharick and Georgette Savino of Northwestern's Mathematics Department for their excellent administrative and secretarial help.

Michael R. Stein

Evanston, July, 1976

Algebraic K-Theory Conference - List of Talks

LECTURES

Monday, January 12, 1976

Bloch: K-theory of group schemes
Roberts: Reducible curves
Hatcher: Some new algebraic K-theories

Tuesday, January 13, 1976

van der Kallen: Injective stability for K_2
Dennis: Algebraic K-theory and Hochschild homology
Grayson: $+ = Q$

Wednesday, January 14, 1976

Krusemeyer: Serre's problem
Wagoner: Continuous cohomology and K-theory
Szczarba: $K_3(\mathbb{Z})$

Thursday, January 15, 1976

Giffen: Algebraic K_2 and K_3 invariants of Hermitian forms
Loday: Stable homotopy and higher Whitehead groups
Bass: Projective modules over infinite groups

Friday, January 16, 1976

Quinn: A new surgery obstraction group
Pardon: Localization in L-theory
Hausmann: Homology spheres in algebraic K-theory

SEMINARS

Algebraic geometry and K-theory

Bloch: Some examples in the theory of algebraic cycles
Murthy: Cancellation theorems for projective modules on affine surfaces
Kazhdan: Some strange groups

Cohomology of groups

Soulé: Cohomology of $SL_3(\mathbb{Z})$
Brown: Tate cohomology of infinite groups
Alperin: Stability for $H_2(SU_n)$
Fiedorowicz: Homology of classical groups over finite fields
Wagoner: Stability for $H_*(GL_n(A))$, A a local ring
Evens: Chern classes of induced representations

K_0 and K_1

Dayton: $SK_1(CX)$
Martin: $NK_1(\mathbb{Z}\pi)$, π finitely generated abelian
Magurn: SK_1 of dihedral groups
Kuku: SK_n of orders
Wright: K-theory of the category of invertible algebras

K_2

Dunwoody: K_2 of a Euclidean ring
Geller: KG^V theory
Green: K_2 of a division ring
van der Kallen: K_2 of a regular local noetherian ring of dimension 2 injects into K_2 of its field of fractions
Krusemeyer: $K_2(F[x,y])$ and related computations
Strooker: The fundamental group of GL_n

L-theory

Ranicki: Algebraic theory of surgery

Topology and K-theory

Hatcher: The simple homotopy space Wh(X)
May: Remarks on Brauer lifting, Frobenius and $KO_*(\mathbb{Z})$
Waldhausen: Computation of pseudo-isotopies by non-additive Q construction
Browder: Complete intersections, fixed-point free involutions and the Kervaire invariant
Giffen: Segal K-theory
Browder: K-theory and stable homotopy
Edwards: Steenrod homotopy

LIST OF PARTICIPANTS AND AUTHORS

Professor Roger Alperin
Department of Mathematics
Brown University
Providence, Rhode Island 02912

Mr. David F. Anderson
Department of Mathematics
University of Chicago
Chicago, Illinois 60637

Professor Anthony Bak
Fakultät für Mathematik
Universität Bielefeld
48 Bielefeld
Federal Republic of Germany

Professor Hyman Bass
Department of Mathematics
Columbia University
New York, New York 10027

Professor Spencer Bloch
IHES
91440 Bures-sur-Yvette
France

Professor William Browder
Department of Mathematics
Princeton University
Princeton, New Jersey 08540

Professor Kenneth S. Brown
Department of Mathematics
Cornell University
Ithaca, New York 14853

Ms. Ruth Charney
Department of Mathematics
Princeton University
Princeton, New Jersey 08540

Dr. Barry Dayton
Department of Mathematics
Northeastern Illinois University
Bryn Mawr at St. Louis
Chicago, Illinois 60625

Professor R. K. Dennis
Department of Mathematics
Cornell University
Ithaca, New York 14853

Professor Andreas Dress
Fakultät für Mathematik
Universität Bielefeld
48 Bielefeld
Federal Republic of Germany

Professor M. J. Dunwoody
Mathematics Division
University of Sussex
Falmer, Brighton BN1 9QF
England

Professor David A. Edwards
Department of Mathematics
State University of New York
Binghamton, New York 13901

Professor Helmut Epp
Department of Mathematics
De Paul University
25 East Jackson Boulevard
Chicago, Illinois 60604

Professor Leonard Evens
Department of Mathematics
Northwestern University
Evanston, Illinois 60201

Professor Zbigniew Fiedorowicz
Department of Mathematics
University of Michigan
Ann Arbor, Michigan 48104

Dr. Edward Formanek
Department of Mathematics
University of Chicago
Chicago, Illinois 60637

Professor Eric Friedlander
Department of Mathematics
Northwestern University
Evanston, Illinois 60201

Professor S. Geller
Department of Mathematics
Purdue University
West Lafayette, Indiana 47907

Professor C. H. Giffen
Department of Mathematics
University of Virginia
Charlottesville, Virginia 22901

Mr. Daniel R. Grayson
MIT
Room 2-087
Cambridge, Massachusetts 02139

Professor Sherry M. Green
Department of Mathematics
University of Utah
Salt Lake City, Utah 84112

Professor Bruno Harris
Department of Mathematics
Brown University
Providence, Rhode Island 02912

Professor Allen Hatcher
School of Mathematics
Institute for Advanced Study
Princeton, New Jersey 08540

Dr. J. C. Hausmann
School of Mathematics
Institute for Advanced Study
Princeton, New Jersey 08540

Dr. Peter T. Johnstone
Department of Mathematics
University of Chicago
Chicago, Illinois 60637

Professor Wilberd van der Kallen
Department of Mathematics
Northwestern University
Evanston, Illinois 60201

Dr. T. Kambayashi
Department of Mathematics
Northern Illinois University
DeKalb, Illinois 60115

Professor David Kazhdan
Department of Mathematics
Harvard University
Cambridge, Massachusetts 02138

Professor Michel Kervaire
Department of Mathématiques
Université de Genève
2-4 rue du Lièvre
Genève 24, Switzerland

Professor Mark Krusemeyer
Department of Mathematics
Columbia University
New York, New York 10027

Professor A. O. Kuku
Department of Mathematics
University of Ibadan
Ibadan
Nigeria

Professor R. Lee
Department of Mathematics
Yale University
New Haven, Connecticut 06520

Mr. H. W. Lenstra, Jr.
Mathematisch Instituut
Roeterstraat 15
Amsterdam -C
The Netherlands

Dr. Jean-Louis Loday
Institut de Recherche
Mathematique Avancée
7, rue Réné Descartes
67084 Strasbourg, France

Mr. Bruce Magurn
Department of Mathematics
Northwestern University
Evanston, Illinois 60201

Professor Robert Martin
Department of Mathematics - Box 1093
Hunter College
695 Park Avenue
New York, New York 10021

Professor Peter May
Department of Mathematics
University of Chicago
Chicago, Illinois 60637

Professor M. P. Murthy
Department of Mathematics
University of Chicago
Chicago, Illinois 60637

Professor W. Pardon
Department of Mathematics
Columbia University
New York, New York 10027

Mr. Barton Plumstead
Department of Mathematics
University of Chicago
Chicago, Illinois 60637

Professor Stewart Priddy
Department of Mathematics
Northwestern University
Evanston, Illinois 60201

Professor Daniel G. Quillen
Department of Mathematics
Massachusetts Institute of Technology
Cambridge, Massachusetts 02139

Professor Frank Quinn
Department of Mathematics
Yale University
New Haven, Connecticut 06520

Dr. Andrew Ranicki
Trinity College
Cambridge
England

Dr. Ulf Rehmann
Fakultät für Mathematik
Universität Bielefeld
48 Bielefeld
Federal Republic of Germany

Professor I. Reiner
Department of Mathematics
University of Illinois
Urbana, Illinois 61801

Professor L. G. Roberts
Department of Mathematics
Queen's University
Kingston, Ontario
Canada K7L 3N6

Mr. Paul Selick
Department of Mathematics
Princeton University
Princeton, New Jersey 08540

Dr. Jack M. Shapiro
Department of Mathematics
Washington University
St. Louis, Missouri 63130

M. Christophe Soulé
Université Paris VII
2, Place Jussieu
75221 Paris CEDEX 05
France

Mr. Ross Staffeldt
Department of Mathematics
University of California
Berkeley, California 94720

Professor Michael R. Stein
Department of Mathematics
Northwestern University
Evanston, Illinois 60201

Professor Jan R. Strooker
Mathematisch Institut RU
Budapestlaan, de Uithof
Utrecht
The Netherlands

Professor R. Szczarba
Department of Mathematics
Yale University
New Haven, Connecticut 06520

Mr. R. Thomason
Department of Mathematics
Princeton University
Princeton, New Jersey 08540

Professor J. B. Wagoner
Department of Mathematics
University of California
Berkeley, California 94720

Professor F. Waldhausen
Fakultät für Mathematik
Universität Bielefeld
48 Bielefeld
Federal Republic of Germany

Mr. Charles A. Weibel
Department of Mathematics
University of Chicago
Chicago, Illinois 60637

Professor Julian S. Williams
Mathematics Department
University of Wisconsin - Parkside
Kenosha, Wisconsin 53140

Dr. David Wright
Department of Mathematics
Washington University
St. Louis, Missouri 63130

Table of Contents

S. BLOCH, An Example in the Theory of Algebraic Cycles........................ 1

B. DAYTON, SK_1 of Commutative Normed Algebras................................ 30

L. ROBERTS, The K-Theory of Some Reducible Affine Curves: A Combinatorial Approach.. 44

A. O. KUKU, SK_n of orders and G_n of finite rings.............................. 60

H. W. LENSTRA, Jr., K_2 of a global field consists of symbols................. 69

S.M. GREEN, Generators and relations for K_2 of a division ring............... 74

W. VAN DER KALLEN, Injective stability for K_2.................................. 77

J. L. LODAY, Les matrices monomiales et le groupe de Whitehead Wh_2............ 155

U. REHMANN & C. SOULÉ, Finitely presented groups of matrices.................. 164

J.-C. HAUSMANN, Homology sphere bordism and Quillen plus construction......... 170

D. QUILLEN, Letter from Quillen to Milnor on $\mathrm{Im}(\pi_i O \xrightarrow{J} \pi_i^s \longrightarrow K_i\mathbb{Z})$...... 182

D. QUILLEN, Characteristic classes of representations........................ 189

D. GRAYSON, Higher algebraic K-theory: II (after D. Quillen)................. 217

J. B. WAGONER, Continuous Cohomology and p-Adic K-Theory...................... 241

K. S. BROWN, Cohomology of groups (Summary of talks).......................... 249

J. M. SHAPIRO, On the homology and cohomology of the orthogonal and symplectic groups over a finite field of odd characteristic.............. 260

Z. FIEDOROWICZ & S. PRIDDY, Homology of Classical Groups over a Finite Field... 269

B. HARRIS, Group cohomology classes with differential form coefficients....... 278

R. ALPERIN, Stability for $H_2(SU_n)$.. 283

E. M. FRIEDLANDER, Homological stability for classical groups over finite fields.. 290

W. PARDON, Hermitian K-theory in topology: A survey of some recent results... 303

J. L. LODAY, Higher Witt groups: A survey...................................... 311

W. PARDON, The exact sequence of a localization for Witt groups............... 336

A. W. M. DRESS, Orthogonal Representations on Positive Definite Lattices...... 380

A. BAK, The computation of surgery groups of finite groups with abelian 2-hyperelementary subgroups.................................. 384

An Example in the Theory of Algebraic Cycles

Spencer Bloch*

Once upon a time, algebraic K-theory meant $K_0(X)$ for X a smooth, quasi-projective variety. Despite all the recent progress in the subject, the structure of this original object remains a mystery. To see the point, filter $K_0(X)$ by codimension of support, so $\text{filt}^i K_0(X)$ is generated by $[F]$ for F a coherent sheaf on X with codim. supp $F \geq i$. The successive quotients $gr^i K_0(X) = \text{filt}^i/\text{filt}^{i+1}$ are isomorphic to the Chow groups $CH^i(X)$ of codimension i cycles modulo rational equivalence. Another way to describe $CH^i(X)$ is to take the free abelian group $Z^i(X)$ generated by codimension i cycles and factor out the subgroup $B^i(X)$ generated by classes $z_0 - z_\infty$, where z_0 and z_∞ denote the fibre of a cycle $Z \in Z^i(X \times \mathbf{P}^1)$ over 0 and infinite respectively. Let $A^i(X) \subset CH^i(X)$ denote the subgroup generated by cycles $z_a - z_b$ where again z_a and z_b are fibres of a cycle $Z \in Z^i(X \times C)$, but this time C is a smooth curve; $a,b \in C$. $A^i(X)$ should be thought of as the "continuous part" of $CH^i(X)$.

The particular mystery which fascinates me involves $A^i(X)$. $A^0(X) = (0)$, and $A^1(X)$ forms in a natural way (assuming, say, X smooth, projective over an algebraically closed field) the set of points of an algebraic variety ($A^1(X)$ is said to be representable). Naively speaking, one can draw a picture of $A^1(X)$. In fact, $A^1(X)$ can be shown to be an abelian variety (= complex torus which admits a projective embedding) so the picture looks like

$A^1(X) =$ 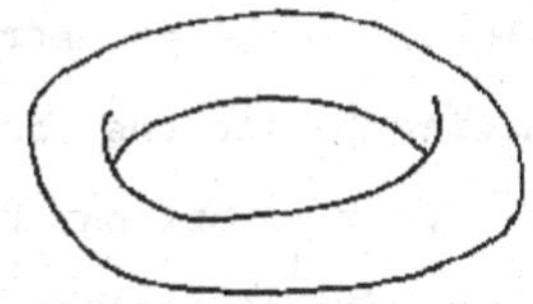

*Supported by a NATO fellowship and the C. N. R. S.
The author also gratefully acknowledges the hospitality of the I.H.E.S.

The extraordinary fact is that $A^i(X)$ for $i \geq 2$ is in general not representable. Actually, algebraic geometers have dealt with non-representable objects before (stacks, algebraic spaces) but these have always been in some sense very close to algebraic varieties. With the cycle groups, one encounters for the first time objects which are geometric in content and yet joyously non-representable.

Two questions arise.

I. What sort of geoemtric structure do the $A^i(X)$ have?

II. Granted that the $A^i(X)$ are not in general representable, are there any hypotheses on X which will insure that they are?

In what follows I will be primarily concerned with $A^2(X)$. One conjectures in this case that the vanishing of $H^2(X, O_X)$ and $H^3(X, O_X)$ will imply representability for $A^2(X)$. The main result will be a proof of this fact for complete intersection 3-folds (the argument works also for all, or at least all known, Fano 3-folds.) Actually, all these varieties are unirational (covered by varieties with rational function fields) except possibly the quartic 3-fold (hypersurface of degree 4 in projective 4-space). In the unirational case, the argument is straightforward. The proof for the quartic is somewhat longer, but is completely naive and geometric. There is a brief discussion of the more difficult structural question I at the end.

I am aware that this material is perhaps somewhat foreign to the spirit of the conference and to my talk at the conference, which was more algebraic and concerned with the higher K-groups. On the other hand, the novelty and interest in the higher K's is quite precisely the algebraic analogue of the geometric novelty of the higher cycle groups. It is amusing to see the concrete geometry interact with the abstract algebra. Every effort has been made to make the arguments comprehensible to the non-spacialist. I hope to publish jointly with J. Murre a more detailed account of the structure of cycles on

Fano 3-folds. Finally, I am endebted to D. Mumford for bringing to my attention the problem of cycles on the quartic threefold, and for giving me the proof of Lemma (3.2).

§1. Some sorites for cycles

Let X be a smooth projective variety over an algebraically closed field k, and let $A^n(X)$ (resp. $A_n(X)$) denote the group of codimension n (resp. dimension n) cycles algebraically equivalent to zero on X, modulo rational equivalence. If Y is another such variety, and T is a cycle class of codimension n on $Y \times X$ ($T \in CH^n(Y \times X)$) we get a correspondence

$$T_*: A_0(Y) \to A^n(X)$$

$$T_*(y) = \pi_{X*}(T \cdot \pi_Y^*(y))$$

where π_Y, π_X denote the projection maps and $\cdot$ means to intersect the cycles. (I should have mentioned that the groups CH, A, have covariant and contravariant functioriality properties, obtained by "pushing forward" and "pulling back" cycles, as well as multiplicative structure obtained by intersection [16].) When Y is a curve, the group $A_0(Y) = A^1(Y) = Pic^0(Y)$, the group of divisors of degree zero on Y modulo rational equivalence. This group has a natural structure of abelian variety, the Jacobian variety $J(Y)$.

Definition (1.1). $A^n(X)$ is representable if there exists a curve Y and a cycle $T \in CH^n(Y \times X)$ such that $T_*: J(Y) \to A^n(X)$ is surjective and $\mathrm{Ker}\, T_* \subset J(Y)$ is a closed algebraic subgroup (i.e. $\mathrm{Ker}\, T_* = \coprod_{\text{finite}}$ closed subvariety of $J(Y)$).

Note that $A^n(X) \cong J(Y)/\mathrm{Ker}\, T_*$ inherits a structure of abelian variety in the representable case. It is possible to show that this structure does not depend on the choice of Y and T. When the ground field k is the complex numbers, Griffiths has defined a complex torus $J^n(X)$ (the n-th intermediate jacobian), and there is a natural map (defined quite generally without any hypothesis of representability)

$$\Theta: A^n(X) \to J^n(X).$$

It is always true that Image Θ is an abelian variety, but Θ is in general neither injective nor surjective.

Suppose, for example, that we are interested in the case of curves on a three-fold, i.e., $A^2(X)$ for $\dim X = 3$. For $V \subset H^3(X,\mathbb{C})$ a complex subvectorspace, there is a natural map $H_3(X,\mathbb{Z}) \to V^* = \mathrm{Hom}(V,\mathbb{C})$ obtained by integration, $\gamma \to \int_\gamma$. Taking for V the subspace of differentials of type (3,0) plus (2,1)

$$V = H^0(X,\Omega^3_X) + H^1(X,\Omega^2_X)$$

the quotient $V^*/H_3(X,\mathbb{Z}) \underset{\text{dfn.}}{=} J^2(X)$. If γ denotes a codimension 2 algebraic cycle algebraically equivalent to zero on X, then one can choose a triangulation of X and a chain C (of dimension 3) such that $\partial C = \gamma$. It turns out that for $\omega \in V$, the expression $\int_C \omega$ depends only on γ, at least up to $\int_\tau \omega$ for $\tau \in H_3(X,\mathbb{Z})$. The map

$$\Theta: A^2(X) \to J^2(X)$$

is defined by $\gamma \to \int_C$.

<u>Definition</u> (1.2). Assuming $k = \mathbb{C}$, $A^n(X)$ is said to be isogenous to the jacobian if $\Theta: A^n(X) \to J^n(X)$ is surjective with finite kernel.

<u>Proposition</u> (1.3). If $A^n(X)$ is isogenous to the jacobian, it is representable.

<u>Proof</u>. I claim first that $A^n(X)$ is divisible. In fact, by definition

of algebraic equivalence, there exists a family of curves C_i, $i \in I$ and correspondences T_{i*} such that $\coprod_{i \in I} J(C_i) \xrightarrow{\coprod T_{i*}} A^n(X)$ is surjective. Since $J(C_i)$ is divisible, $A^n(X)$ is as well. We now obtain a diagram of surjective arrows for some $N \geq 1$

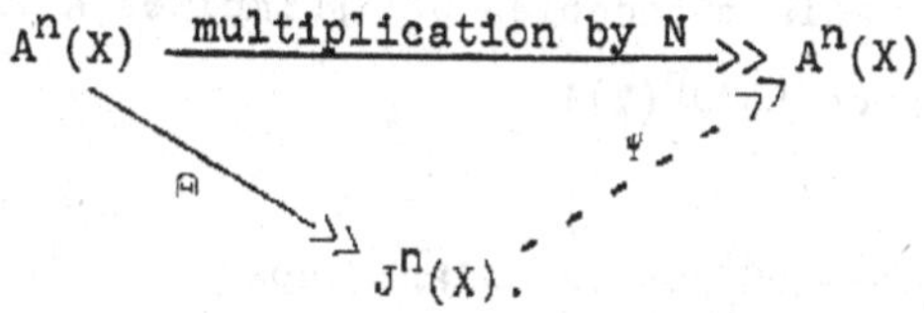

For $I' \subset I$ a finite subset, the image of the composition $\coprod_{i \in I'} J(C_i) \to A^n(X) \xrightarrow{\Theta} J^n(X)$ will be a subtorus of $J^n(X)$, which will necessarily be all of $J^n(X)$ for I' sufficiently large. Let $C \subset \Pi C_i$ be a smooth linear space section of dimension 1. The map $J(C) \to \coprod_{i \in I'} J(C_i)$ will be <u>surjective</u>. (This is a general fact about linear space sections of varieties. The reader who is uncomfortable with this sort of thing may skip any discussion of intermediate jacobians, as they appear only as an adjoint to the final results.)

We now have a diagram (defining Σ and T_*)

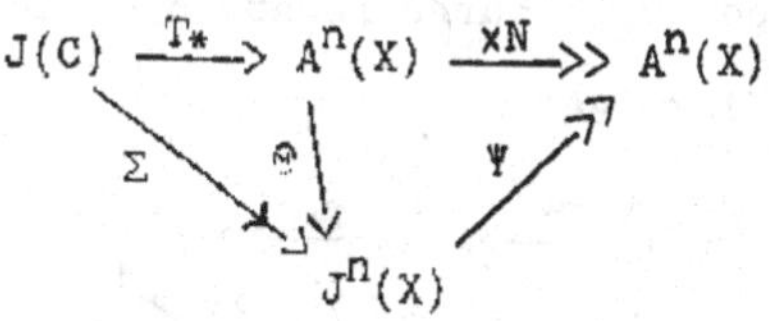

where Σ is a morphism of complex tori, and hence Ker Σ is an algebraic subgroup of $J(C)$. But Ker Σ/Ker T_* is N-torsion, so Ker T_* is algebraic as well. It is clear from the diagram that T_* is surjective, so we are done.

Q. E. D.

A proper morphism of varieties $f: X \to Y$ is said to be of finite degree d if $[k(X), k(Y)] = d$.

<u>Proposition</u> (1.4). Let X and Y be smooth, projective varieties over an algebraically closed field k, and let $f: X \to Y$ be a morphism of degree d. Then $A^n(X)$ representable (resp. $k = \mathbb{C}$ and $A^n(X)$ isogenous to the intermediate jacobian $J^n(X)$) implies $A^n(Y)$ representable (resp. $A^n(Y)$ isogenous to $J^n(Y)$).

<u>Proof</u>. There are co- and contravariant maps

$$A^n(Y) \xrightarrow{f^*} A^n(X) \xrightarrow{f_*} A^n(Y)$$

and the composition is multiplication by d. The key point is that $A^n(X)$ representable implies $f^*(A^n(Y)) \subseteq A^n(X)$ is a sub-abelian variety. In characteristic 0, when we dispose of an intermediate jacobian $J^n(X)$ isogenous to $A^n(X)$ this can be proved by considering correspondences as in (1.3).

$$\coprod J(C_i) \to A^n(Y) \xrightarrow{f^*} A^n(X) \to J^n(X),$$

noting that the image is always a subtorus. In general, the proof requires rather more algebraic geometry. In particular one needs a more precise definition of representability than I have given. Details will appear in the paper with Murre referred to in the introduction.

Writing $A = f^*A(Y)$ we get

$$A^n(Y) \xrightarrow{f^*} A \xrightarrow{f_*} A^n(Y).$$
$$\text{(composite: } d\text{)}$$

It follows that $\mathrm{Ker}(f_*: A \to A^n(Y)) \subset \mathrm{Ker}(A \xrightarrow{d} A)$, whence $A^n(Y) \cong A/(\text{finite group})$. Using this, the reader can verify (reasonable exercise) that $A^n(Y)$ is representable.

Suppose now that we have intermediate jacobians $J^n(X)$ and $J^n(Y)$. Functoriality gives commutative squares

$$\begin{array}{ccc} A^n(X) & \xrightarrow{\Theta_X} & J^n(X) \\ {\scriptstyle f^*}\uparrow & & \uparrow{\scriptstyle f^*} \\ A^n(Y) & \xrightarrow{\Theta_Y} & J^n(Y) \end{array} \qquad \begin{array}{ccc} A^n(X) & \xrightarrow{\Theta_X} & J^n(X) \\ {\scriptstyle f_*}\downarrow & & \downarrow{\scriptstyle f_*} \\ A^n(Y) & \xrightarrow{\Theta_Y} & J^n(Y) \end{array}$$

and f_*f^* = multiplication by d on $J^n(Y)$. Assuming Θ_X an isogeny, we see by (1.3) and the above that $A^n(Y)$ is representable. From the square on the left, one finds Ker $\Theta_Y \subseteq A^n(Y)$(N-torsion) for some integer N. From the square on the right one gets Θ_Y surjective. Hence Θ_Y is an isogeny. Q. E. D.

In the sequel, we will frequently have to deal with rational maps of smooth varieties, i.e. maps such as $f: X \to Y$ below

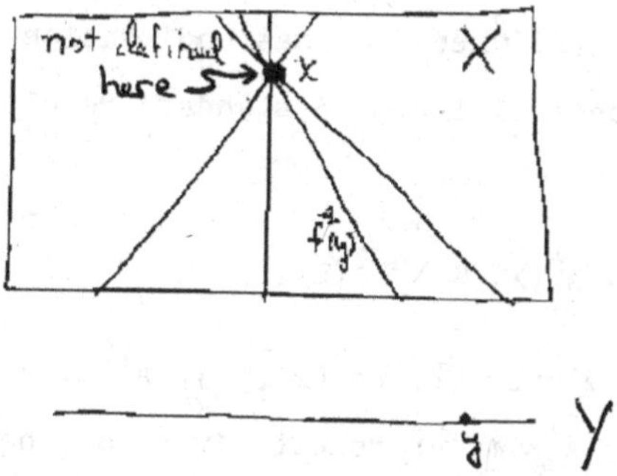

where f is not defined at x. "Blowing up" at x gives one a variety X' and a morphism $\pi: X' \to X$. π is an isomorphism away from x, and $\pi^{-1}(x) = E$ is a divisor in X' which is isomorphic to projective space of dimension = dim $X - 1$. Frequently the map f will lift to an everywhere defined map $f': X' \to Y$

$$\begin{array}{ccc} & X' & \\ {\scriptstyle \pi}\swarrow & & \searrow{\scriptstyle f'} \\ X & \overset{f}{\dashrightarrow} & Y \end{array}$$

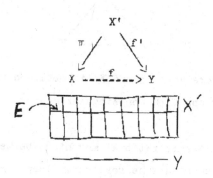

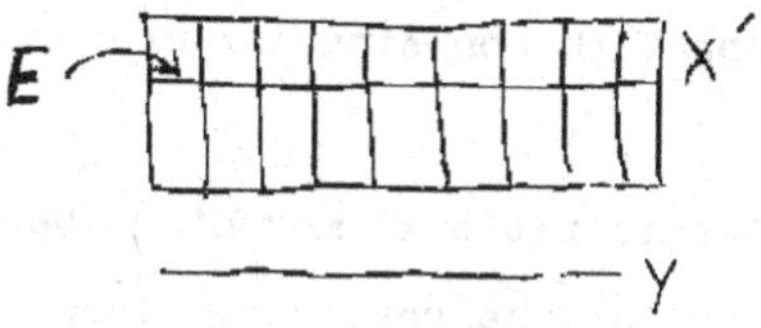

More generally, given $Z \subset X$ a smooth closed subvariety, one can construct $X' = BL_Z(X) \xrightarrow{\pi} X$ such that π is an isomorphism of Z and $\pi^{-1}(z) \cong P^{\dim X - \dim Z - 1}$ for all $z \in Z$. X' is called the <u>blowing up</u> of X along <u>center</u> Z.

<u>Lemma</u> (1.5). Let $\pi\colon X' \to X$ be obtained by a sequence of blowing ups with non-singular centers. Then $A^2(X)$ is represented (resp. isogenous to $J^2(X)$) if and only if $A^2(X')$ is representable, (resp. isogenous to $J^2(X)$).

<u>Proof</u>. It suffices to consider the case when π is obtained by a single blowing up, say $X' = BL_Z(X)$ where Z has codimension r in X. Let $E \subset X'$ be the exceptional divisor. A standard result about algebraic cycles [ ] implies

$$A^n(X') \cong A^n(X) \oplus A^{n-1}(E).$$

Note E is a projective bundle over Z, so by [ ], $A^1(E) \cong A^1(Z)$. Hence $A^2(X') \cong A^2(X) \oplus A^1(Z)$. A similar result gives for the jacobians $J^2(X') \cong J^2(X) \oplus J^1(Z)$. Since $J^1(Z) \cong A^1(Z)$ = Picard variety, we are done. Q. E. D.

Let me say that a rational morphism $f\colon X \dashrightarrow Y$ can be <u>resolved</u> if there exists a diagram

(*)

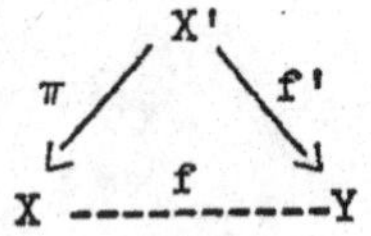

where f' is everywhere defined and X' is obtained from X by a sequence of blowing ups with non-singular centers.

<u>Proposition</u> (1.6). Assume rational morphisms between varieties of dimension = dim X can always be resolved. Then the question of

representability (resp. isogeny with the intermediate jacobian) for $A^2(X)$ is birational in X.

Proof. We consider diagram (*), with dim X = dim Y. Using (1.4)-(1.5),

$A^2(X)$ representable (resp. isogenous to $J^2(X)$)

implies

$A^2(X')$ representable (resp. isogenous to $J^2(X')$)

implies

$A^2(Y)$ representable (resp. isogenous to $J^2(Y)$).

Assuming f is birational, there will be an inverse rational map f^{-1}: $Y \to X$ and we can reverse the implications. Q. E. D.

Remark. The cycle group $A^2(X)$, itself, is not a birational invariant. For example if $X = \mathbb{P}^n$, $Z \subset X$ a smooth curve, $X' = BL_Z(X)$, we have $X' \xrightarrow[\text{birational}]{} X$ but $A^2(X) = (0)$, $A^2(X') =$ Jacobian (C).

A variety X of dimension n is said to be unirational if there exists a rational map f: $\mathbb{P}^n$ ----> X such that $[k(\mathbb{P}^n): k(X)] < \infty$ (Geometrically, f is defined on some open $U \subset \mathbb{P}^n$ and we require that $f(U)$ be dense in X.)

Proposition (1.7). Assume rational maps between varieties of dimension n are resolvable, and let X be a unirational variety of dim n. Then $A^2(X)$ is representable (resp. isogenous to the intermediate jacobian).

Proof. Construct a diagram

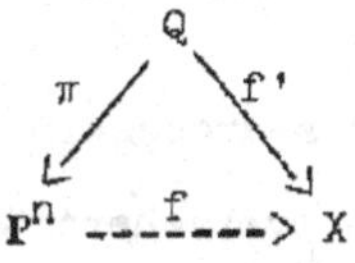

We have $A^2(\mathbb{P}^n)$ representable (= (0)) implies $A^2(Q)$ representable implies $A^2(X)$ representable. Isogeny with the intermediate jacobian is handled similarly.

The following definition is not quite standard, but it is convenient for our purposes.

<u>Definition</u> (1.8). A <u>conic bundle</u> over a variety S is a variety U together with a rational map $f\colon U \dashrightarrow S$ such that the <u>geometric generic fibre</u> of f is isomorphic to $\mathbb{P}^1$. In other words, there exists a surjective map of finite degree $g\colon S' \to S$ such that the pullback $U \times_S S'$ is birational with $\mathbb{P}^1 \times S'$.

As a final example which will be important in the sequel, we have:

<u>Proposition</u> (1.9). Let S be a surface ($\dim S = 2$) such that $A^2(S)$ is representable (resp. isogenous to $J^2(S)$), and let $\pi\colon U \dashrightarrow S$ be a conic bundle. Then $A^2(U)$ is representable (resp. isogenous to $J^2(U)$).

<u>Proof</u>. By hypothesis we can find a surface S', a morphism $f\colon S' \to S$ of finite degree, and a rational map of finite degree $\psi\colon S' \times \mathbb{P}^1 \dashrightarrow U$ covering f. Rational maps of varieties of dimension 3 can be resolved so we can find a variety V and a diagram

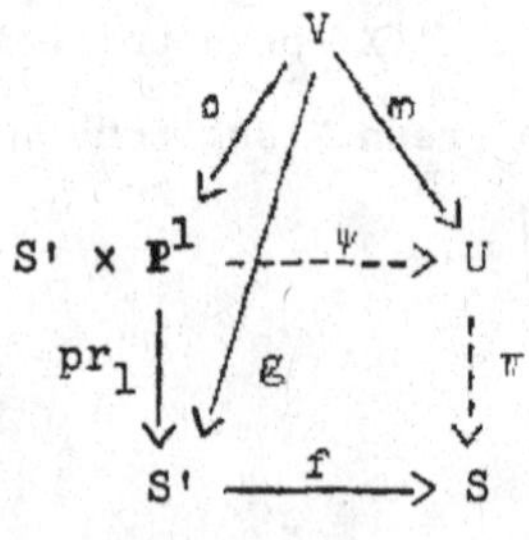

such that ρ is obtained by a succession of blowings up with non-singular centers, and ϕ is everywhere defined of finite degree. The difficulty is that $A^2(S') = A_0(S')$ is not necessarily representable.

Recall that for any variety X one has defined an abelian variety $\mathrm{Alb}(X)$ ($\mathrm{Alb}(X) = J_0(X)$ for $X/\mathbb{C}$) together with a map $\phi: X \to \mathrm{Alb}(X)$ such that

(a) ϕ is well defined up to the choice of a base point $x_0 \in X$ with $\phi(x_0) = 0$, and $\phi(X)$ generates $\mathrm{Alb}(X)$.

(b) Given an abelian variety B and a map $\emptyset: X \to B$ with $\emptyset(x_0) = 0$, there exists a unique homomorphism $\Theta: \mathrm{Alb}(X) \to B$ such that $\emptyset = \Theta\phi$.

(c) The assignment $(x) \to \phi(x)$ defines a surjection (independent of choice of base point x_0) $A_0(X) \twoheadrightarrow \mathrm{Alb}(X)$.

Using these properties, it is not hard to show that $A_0(X)$ representable implies $A_0(X) \cong \mathrm{Alb}(X)$. Define, in general, $T(X) = \mathrm{Ker}\ (A_0(X) \to \mathrm{Alb}(X))$. We have in our situation

$$\begin{array}{ccccccccc} 0 & \to & T(S') & \to & A_0(S') & \to & \mathrm{Alb}(S') & \to & 0 \\ & & & & {\scriptstyle f_*}\downarrow & & \downarrow & & \\ & & & & A_0(S) & \cong & \mathrm{Alb}(S) & & \end{array}$$

whence $f_*(T(S')) = (0)$. I claim $\emptyset_* g^*(T(S')) = (0)$. Indeed, the assertion in (1.9) is birational in U by (1.6), so we may assume π is everywhere defined. We now use

<u>Lemma</u> (1.10). Suppose given a diagram of varieties

$$\begin{array}{ccc} V & \xrightarrow{\emptyset} & U \\ {\scriptstyle g}\downarrow & & \downarrow{\scriptstyle \pi} \\ S' & \xrightarrow{f} & S \end{array}$$

such that V is birational to $U \underset{S}{\times} S'$. Then there exists a non-empty open set $S'_0 \subset S'$ such that for any $s \in S'_0$, $\phi_* g^*(s) = \pi^* f_*(s)$ as cycles (i.e. algebraic sets with multiplicities) on U.

<u>Proof of Lemma</u>. Since U and V are varieties, there exists a non-empty open set $S'_1 \subset S'$ such that the fibres V_s and $U_{f(s)}$ are reduced and irreducible for $s \in S'_1$. Since V is birational with $S' \underset{S}{\times} U$ we may restrict further to some $S'_0 \subset S'_1$ and assume that $\phi_s \colon V_s \to U_{f(s)}$ is birational For $s \in S'_0$, $\pi^* f_*(s) = \phi_* g^*(s) = U_{f(s)}$ with multiplicity 1.

Q. E. D.

Returning to the proof of (1.9), we see from the lemma that $\phi_* g^*(T(S')) = \pi^* f_*(T(S')) = (0)$. (In fact a given class $z \in T(S')$ can by general position arguments be represented by a cycle $\tilde{z}$ supported on the open set S'_0 of (1.10).) It follows that we have maps

$$\underbrace{A^2(U) \xrightarrow{\phi^*} A^2(V)/g^*T(S') \xrightarrow{\phi_*} A^2(U)}_{\text{mult. by deg. } \phi}.$$

Now $A^2(S' \times \mathbb{P}^1)/pr_1^* T(S') \cong Alb(S')$ is representable (resp. isogenous to $J^2(S' \times \mathbb{P}^1)$), so by repeating the argument in (1.6) we find $A^2(V)/g^*T(S')$ is representable (resp...). Arguing with ϕ^* and ϕ_* as in (1.4), we conclude $A^2(U)$ is representable (resp...) as well.

Q. E. D.

§2

Complete intersections of dimension 3 and Hodge level 1

Let $V_n(a_1,\ldots,a_d)$ denote a non-singular complete intersection of d hypersurfaces of degrees $a_1,\ldots,a_d$, in $\mathbb{P}_k^{n+d}$. For $V = V_3(a_1,\ldots,a_d)$, the condition $H^2(V,\mathcal{O}_V) = (0)$ is automatic. The condition $H^3(V,\mathcal{O}_V) = (0)$ holds only in the following cases [ ]

$$V_3(2,2),\ V_3(2,2,2),\ V_3(3),\ V_3(2,3),\ V_3(4).$$

Our objective is to show $A^2(V)$ is representable in these cases.

Proposition (2.1). $V_3(2,2)$, $V_3(2,2,2)$, $V_3(3)$, and $V_3(2,3)$ are unirational.

For details the reader is referred to [18]. By way of example, let me sketch the argument given in [12] for $V = V_3(3)$. Let $T \to V$ denote the tangent bundle, and let $\mathbb{P}(T) \to V$ be the projectivization of T. A point $x \in \mathbb{P}(T)$ corresponds to a point $y \in V$ together with a line $\ell \subset \mathbb{P}^4$ tangent to V at y.

One has a rational map $f\colon \mathbb{P}(T) \to V$ obtained by sending $x = (y,\ell)$ to the "third point of intersection" of ℓ with V (wite ℓ intersects V in three points counting multiplicities, and at least two of these points are centered at y). If $L \subset V \subset \mathbb{P}^4$ is a general line on V, one shows that f restricts to a rational map of finite degree $f|_L\colon \mathbb{P}(T)|_L \to V$. Since $\mathbb{P}(T)|_L$ is locally isomorphic to affine three-space, we conclude V is unirational.

Combining (1.7) and (2.1), we get:

Corollary (2.2). Let V denote a variety of one of the four types $V_3(2,2)$, $V_3(2,2,2)$, $V_3(3)$, $V_3(2,3)$. Then $A^2(V)$ is representable. When the ground field k is the complex numbers, $A^2(V)$ is isogenous to the intermediate jacobian of V.

§3

The quartic three-fold

In this section $V \subset \mathbb{P}^4$ will denote a non-singular hypersurface of degree 4.

Theorem (3.1). $A^2(V)$ is representable (resp. isogenous to the intermediate jacobian when $k = \mathbb{C}$).

Proof. We will construct a conic bundle (U,S,f) in the sense of (1.8) such that the hypotheses of (1.9) are satisfied, together with a rational map of finite degree $h: U \dashrightarrow V$. It will follow from (1.4), (1.5), and (1.9) that $A^2(V)$ is representable as claimed.

Lemma (3.2). There exists a non-empty Zariski open set $V^0 \subset V$ such that for all $x \in V^0$ we have

(i) The intersection of the tangent hyperplane H_x to V at x with V, $H_x \cap V$, has a single ordinary double point at x and no other singularities.

(ii) There exists no line $\ell \subset \mathbb{P}^4$ supported on V and passing through x.

Proof. (i) is a consequence of standard Lefschetz theory. For an explicit argument, see [5]. The following proof of (ii) was suggested by Mumford.

Quite generally, the collection of subvarieties of a projective variety V forms a scheme, the Hilbert scheme of V. If $W \subset V$ is a subvariety defined by an ideal $\mathcal{I} \subset O_V$ = sheaf of functions on V, then the Zariski tangent space to the point on Hilbert scheme V corresponding to W is given by the vector space

$$\underset{\sim\sim}{\mathrm{Hom}}_{O_W}(\mathscr{I}/\mathscr{I}^2, O_W)$$

(the best reference for this is [7]). Suppose now that W is locally a complete intersection in V (this is always the case when V and W are non-singular). Then $\mathscr{I}/\mathscr{I}^2$ is a locally free O_W-module whose dual is the <u>normal bundle</u> of W in V

$$N_{W/V} = \underset{\sim\sim}{\mathrm{Hom}}(\mathscr{I}/\mathscr{I}^2, O_W).$$

Thus tangent space to Hilb V at $W \cong \Gamma(W, N_{V/W})$.

Suppose now that V and W are smooth. Thinking differential-geometrically(assume for a moment the ground field is $\mathbf{C}$), we can put a metric on $N_{W/V}$ and take B = ball bundle of vectors v with $\|v\| \leq 1$. B can be identified with some tubular neighborhood of W in V

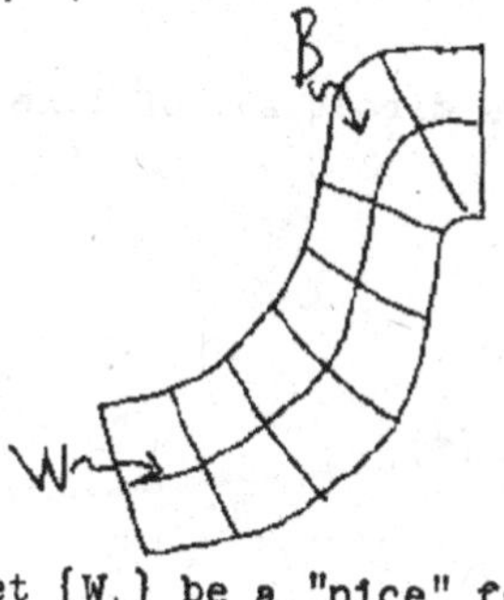

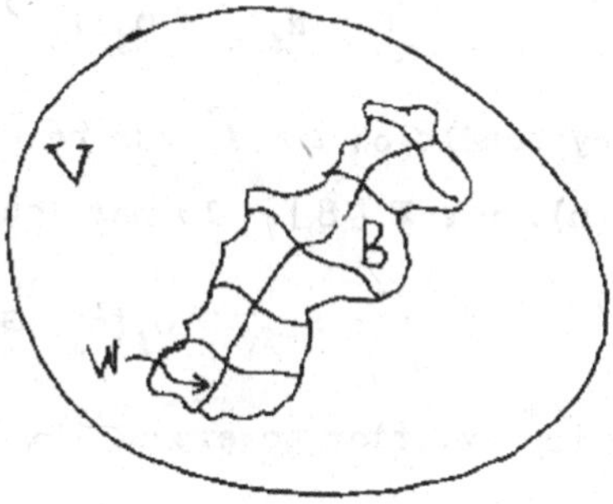

Let $\{W_t\}$ be a "nice" family of subvarieties of V, with $W_0 = W$. Then for t near 0, W_t gives a section of B and hence of $N_{W/V}$. In other words, small displacements of W are parameterized by sections of $N_{W/V}$. In particular, if displacements of W "fill out" V, the bundle $N_{W/V}$ will be generated by its global sections.

Consider now the case $V = V_3(4)$, $W = \ell$ = line in V. If a general point of V lies on a line supported in V, it follows from the above discussion that $N_{\ell/V}$ is generated by global sections. To see this is not the case, consider the exact sequence of bundles

$$0 \to N_{\ell/V} \to N_{\ell/\mathbf{P}^4} \to N_{V/\mathbf{P}^4}\big|_{\ell} \to 0.$$

Note that the normal bundle of a hypersurface H of degree d is given by $N_{H/\mathbb{P}^4} \cong \mathcal{O}_{\mathbb{P}^4}(d)|_H$. Also, if V_1 and V_2 are smooth subvarieties of V intersecting transversally we get

$$N_{V_1 \cap V_2/V} \cong N_{V_1/V}\big|_{V_1 \cap V_2} \oplus N_{V_2/V}\big|_{V_1 \cap V_2}.$$

Using these facts, we get

$$N_{V/\mathbb{P}^4} \cong \mathcal{O}_V(4)$$

$$N_{\ell/\mathbb{P}^4} \cong \mathcal{O}_\ell(1)^{\oplus 3},$$

so the above exact sequence becomes

$$(*) \qquad 0 \to N_{\ell/V} \to \mathcal{O}_\ell(1)^{\oplus 3} \to \mathcal{O}_\ell(4) \to 0.$$

Now any bundle on $\ell \cong \mathbb{P}^1$ can be written as a direct sum of line bundles $\mathcal{O}_\ell(n)$, $n \in \mathbb{Z}$ [8]. In particular,

$$N_{\ell/V} \cong \mathcal{O}_\ell(n_1) \oplus \mathcal{O}_\ell(n_2).$$

Taking the top exterior powers of the bundles in (*)

$$\wedge^3 \mathcal{O}_\ell(1)^{\oplus 3} \cong (\wedge^2 N_{\ell/V}) \otimes \mathcal{O}_\ell(4)$$

$$\mathcal{O}_\ell(3) \cong \mathcal{O}_\ell(4 + n_1 + n_2).$$

In particular, one of the $n_i < 0$, so $N_{\ell/V}$ is not generated by its global sections.

Q. E. D.

We turn now to the proof of (3.1). For $x \in V^0$, let H_x denote the tangent hyperplane to V at x. We have $H_x \cong \mathbb{P}^3$, and $V \cap H_x$ is a hypersurface of degree 4 inside H_x with an ordinary double point at x and no other singularities. Let $Q_x \subset H_x$ be the <u>tangent</u> <u>cone</u> to x on $H_x \cap V$. Set theoretically, Q_x is the union of all lines $\mathbb{P}^1 \subset \mathbb{P}^4$ which are (at least) triply tangent to V at x. In terms of equations, one can choose homogeneous forms $X_0, \ldots, X_4$ on $\mathbb{P}^4$ such that

$x = (1,0,0,0,0)$ and H_*: $X_4 = 0$. The fact that x is an ordinary double point on $V \cap H_x$ means that the equation for V has the form

$$X_0^2 \cdot q_2(X_1,X_2,X_3) + X_0 \cdot r_3(X_1,\ldots,X_4) + s_4(X_1,\ldots,X_4) + X_4 \cdot t_3(X_0,\ldots,X_4)$$

where q_2, r_3, s_4, t_3 denote homogeneous polynomials of degrees 2,3,4, and 3 respectively. Moreover q_2 is non-degenerate (i.e. equivalent after change of coordinates to the quadric $X_1^2 + X_2X_3$), and
Q_x: $X_4 = q_2(X_1,X_2,X_3) = 0$. In particular, Q_x is a cone over the smooth rational curve D_x defined by

$$D_x: X_0 = X_4 = q_2(X_1,X_2,X_3) = 0.$$

There is a conic bundle over V, π: D ----> V defined with $\pi^{-1}(x) = D_x$ for $v \in V^0$.

There is a rational morphism ρ: D ----> V defined as follows: for $x \in V^0$ and $d \in D_x$, let $\ell_d \subset Q_x$ be the line (ruling) passing through d and x. ℓ_d meets V in 4 points, at least three of which coincide with x. Define $\rho(d)$ = "fourth point of intersection of ℓ_d and V."

Lemma (3.3). Let $x \in V^0$ and define $C_x = \rho(D_x)$. Then $C_x = Q_x \cap V$ is a reduced, irreducible rational curve of degree 8. If $x' \in V^0$, $x' \neq x$, then $C_{x'} \neq C_x$.

Proof of Lemma: $Q_x \cap V = Q_x \cap (V \cap H_x)$ is the intersection of two distinct irreducible hypersurfaces in H_x, and hence has pure dimension 1. In particular, $\{x\} \in Q_x \cap V$ is not an isolated component. It is clear, set-theoretically, that $\rho(D_x) \subset Q_x \cap V$ and $Q_x \cap V - \{x\} \subset \rho(D_x)$, so we have $\rho(D_x) = Q_x \cap V$.

The degrees of Q_x and V are 2 and 4 respectively, so $\deg C_x = 8$. Also $C_x = \rho(D_x)$ so C_x is an irreducible rational curve. To show that the intersection $Q_x \cap V$ has multiplicity 1, choose two general points $p_1, p_2 \in C_x$ and let $\ell_1, \ell_2 \subset Q_x$ be the corresponding lines. There exists a hyperplane $L \subset \mathbf{P}^4$ such that $L \cap Q_x = \ell_1 \cup \ell_2$. Note that $(\ell_1 \cup \ell_2) \cap V$ contains p_1, p_2 with multiplicity one, because $\ell_1 \cap V$ contains x with multiplicity 3. On the other hand

$$(\ell_1 \cup \ell_2) \cap V = L \cap Q_x \cap V = L \cap C_x .$$

If the intersection $Q_x \cap V$ were not smooth at p_i, the multiplicity of p_i on $L \cap Q_x \cap V$ would be > 1.

It remains to show that $x' \in V^0$, $x' \neq x$, implies $C_{x'} \neq C_x$. Note that $C_x \subset H_x$, $C_{x'} \subset H_{x'}$, $H_x \neq H_{x'}$ by (3.2)(i). If $C_x = C_{x'}$, we would have $C_x \subset Q_x \cap H_{x'}$ = curve of degree 2. This is impossible (even set-theoretically) by reason of degrees. Q. E. D.

Returning to the proof of (3.1), let $\pi^0 \colon D^0 \to V^0$ denote the restriction of π to $D^0 = \pi^{-1}(V^0)$. π^0 is a regular morphism, as is $\rho^0 \colon D^0 \to V$. The idea now is to construct a surface S and a conic bundle U over S by a "bootstrap" technique. Let $\Gamma \subset V$ be a reduced, irreducible curve. We assume $\Gamma^0 = \Gamma \cap V^0 \neq \emptyset$, and that for some $x \in \Gamma^0$ $\Gamma \neq C_x$. It follows from these hypotheses that $S^0 = \pi^{0-1}(\Gamma^0)$ is an irreducible quasi-projective surface, and $\rho^0(S^0) \subset V$ has dimension 2. Note that $\Gamma^0 \subset \rho^0(S^0)$ (because $x \in C_x = \rho(D_x)$) so $\rho^0(S^0) \cap V^0 \neq \emptyset$. Choose a projective desingularization and completion S of S^0 such that $\rho^0|_{S^0}$ extends to a morphism a morphism $\psi\colon S \to V$. Define U^0 by the fibre square

$$\begin{array}{ccccc} & & h^0 \ (U^0 \to V) & & \\ U^0 & \longrightarrow & D^0 & \xrightarrow{\rho^0} & V \\ \downarrow{\scriptstyle f^0} & & \downarrow{\scriptstyle \pi^0} & & \\ S & \xrightarrow{\psi} & V & & \end{array}$$

$U^0 \neq \emptyset$, and we have by composition a regular morphism $h^0 \colon U^0 \to V$.

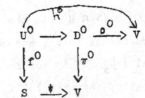

S is generically fibred over Γ with rational fibres, so it is easy to see $A^2(S) \cong \mathrm{Alb}(S)$. Also $f^{0-1}(s) \cong \mathbf{P}^1$ for $s \in S$ a general point. The hypotheses of (1.9) are thus verified (taking U = suitable completion of U^0) and (3.1) will follow if we show $h^0(U^0)$ is dense in V.

Suppose $h^0(U^0)$ is not dense in V. This can only happen if for all $x \in \phi(S) \cap V^0$, $C_x \subset \phi(S)$. Take another curve Γ' analogous to Γ such that $\Gamma' \not\subset \phi(S)$ but $\Gamma' \cap \phi(S) \cap V^0 \neq \emptyset$. Now repeat the above construction replacing Γ by Γ'!! We get ϕ': $S' \to V$. Every surface in V is a hypersurface section by Lefschetz theory, so $\phi(S) \cap \phi(S)$ = finite union of curves. Let $E = V^0 \cap \phi'(S') \cap \phi(S)$. E is non-empty (it meets Γ') and open in $\phi(S) \cap \phi'(S')$, hence E is an infinite set.

Assume now that the triple (U',S',f') is no good either, i.e. $\dim h'(U'^0) < 3$. Then for every $x \in E$, we must have $C_x \subset \phi(S) \cap \phi'(S')$. This intersection is only a finite union of curves, so the infinite set E must contain points $x \neq x'$ with $C_x = C_{x'}$. This contradicts (3.3), so we conclude one of the maps h: U ---> V, h': U' ----> V must be of finite degree. This completes the proof of (3.1). Q. E. D.

§4

Conjectures

What can one say about the structure of $CH^2(X)$ in cases when it is not representable? I want to discuss a conjecture, but first it is important to endow $CH^2(X)$ with more structure than just that of abstract group. Let $\mathscr{V}$ denote the category of non-singular quasi-projective k-varieties. Define a functor

$$CH^2_X\colon \mathscr{V} \to \text{(abelian groups)}$$

by

$$CH^2_X(S) \underset{\text{dfn.}}{=\!=} \varinjlim_{\substack{U\subset S\\ \text{open, dense}}} CH^2(X \times U).$$

Alternately, we can describe $CH^2_X(S)$:

$$CH^2_X(S) = CH^2(X \times S)/\left\{\begin{array}{l}\text{cycles supported on } X \times Z \text{ for}\\ \text{some } Z \subsetneq S \text{ nowhere dense}\end{array}\right\}.$$

If, say, S is irreducible with generic point η (i.e. $k(\eta)$ = field of rational functions on S), then

$$CH^2_X(S) = CH^2(X \underset{\text{Sp } k}{\times} \text{Sp } k(\eta)).$$

Intuitively, one should think of $CH^2_X(S)$ as being equivalence classes of families of cycles parameterized by S. Note that $CH^2(X) = CH^2_X(\text{Sp } k)$, and that $CH^2(X)$ not representable means in particular that CH^2_X is not the functor of points of some group scheme over k.

For any functor $F\colon \mathscr{V} \to$ (ab. groups), we can consider a functor

$$\underline{\text{Hom}}(F, CH^2_X)\colon \mathscr{V} \to \text{(ab. groups)}$$

defined by

$$\underset{\sim}{\mathrm{Hom}}(F,CH_X^2)(S) \underset{\text{dfn.}}{=\!=} \mathrm{Hom}_{\text{functor}}(F|\mathscr{V}_S, CH_X^2|\mathscr{V}_S)$$

where $\mathscr{V}_S$ denotes the category of objects "over S" (i.e. objects $T \in \mathrm{Ob}\mathscr{V}$ plus a morphism $T \to S$) and $F|\mathscr{V}_S, CH_X^2|\mathscr{V}_S$ are defined e.g.

$F|\mathscr{V}_S(T \to S) = F(T)$.

This is all a bit abstract, but consider as a special case, the case $F = A$ = abelian variety. In other words, let A be an abelian variety over k, and view A as a functor $\mathscr{V} \to$ (abelian groups)

$$A(S) \underset{\text{dfn.}}{=\!=} \mathrm{Morph}(S,A).$$

Note $A \in \mathrm{Ob}\mathscr{V}$ (i.e. A is representable), and there is a canonical element $\mathrm{id} \in A(A)$. A morphism of functors ϕ: $A \to CH_X^2$ gives an element $\phi(\mathrm{id}) \in CH_X^2(A)$. The addition law and the two projections give three maps $A \times A \overset{p_1,\ \mu,\ p_2}{\longrightarrow} A$, and one checks that ϕ is a homomorphism of group-valued functors if and only if

$$p_1^*(\phi(\mathrm{id})) + p_2^*(\phi(\mathrm{id})) = \mu^*(\phi(\mathrm{id})) \in CH_X^2(A \times A).$$

In this way one can show

$$\underset{\sim}{\mathrm{Hom}}(A,CH_X^2)(S) = \{z \in CH_X^2(S \times A) \mid p_1^*z + p_2^*z = \mu^*z \in CH_X^2(S \times A \times A)\}.$$

Conjecture (4.1). $\underset{\sim}{\mathrm{Hom}}(A,CH_X^2)$ is a representable functor, represented by an extension of a discrete group by an abelian variety.

The key case to consider is when $A = J(C)$ = jacobian of a curve C. A codimension 2 cycle Γ on $C \times X$ gives rise to a correspondence

points on $C \to$ codimension 2 cycles on X

and hence a morphism of functors

$$J(C) = \frac{\text{0-cycles of degree 0 on } C}{\text{rational equivalence}} \to CH^2_X$$

Such a Γ is trivial if

$$\text{Supp } \Gamma \subset [(\text{finite set of points on } C) \times X] \cup [C \times (\text{cod. 2 cycle on } X)]$$

and we define the group of codimension 2 correspondences

$$Cor^2(C,X) \underset{\text{dfn.}}{=\!=} CH^2(C \times X)/(\text{trivial cycles})$$

$$= CH^2_X(C)/\{(\text{cycles supported on } C \times (\text{cod. 2 cycle on } X)\}$$

One checks that

$$Cor^2(C,X) \cong \underset{\sim}{Hom}(J(C), CH^2_X)(\text{Sp } k)$$

(One can also define a functor $Cor^2_{C,X} = \underset{\sim}{Hom}(J(C), CH^2_X)$, but let's not.)

Conjecture (4.2). $Cor^2(C,X) \cong$ group of k-points of an extension of a discrete group by an abelian variety $Cor^2(C,X)^0$, where $Cor^2(C,X)^0 = \text{Image}(A^2(C \times X) \to Cor^2(C \times X))$.

Example (4.3). Suppose $X = C \times C'$ is a product of curves, and consider $Cor^2(C,X)^0$. Any point on X is the intersection of a vertical and a horizontal divisor, so we can write $CH^2(X)$ as a quotient of $\text{Pic } C \otimes_{\mathbb{Z}} \text{Pic } C'$. If we pick base points on C and C', we can identify

$$\text{Pic}(C) \cong J(C) \oplus \mathbb{Z}, \quad \text{Pic}(C') \cong J(C') \oplus \mathbb{Z}$$

so $CH^2(C)$ is a quotient of

$$\mathbb{Z} \oplus J(C) \oplus J(C') \oplus (J(C) \otimes J(C'))$$

The map

$$J(C) \otimes J(C') \to CH^2(C \times C')$$

is non-trivial (this follows from the techniques of [11]), so we get a map

$$J(C') \to \mathrm{Hom}(J(C),\ CH^2(C \times C')).$$

Note that the group J(C) is divisible, so torsion in J(C') must die in the Hom. However, it is possible that the map of functors

$$J(C') \to \underset{\sim}{\mathrm{Hom}}(J(C), CH^2_{C\times C'})$$

is an injection.

Recall earlier we conjectured that $A^2(X)$ was representable when $H^2(O_X) = H^3(O_X) = (0)$. Actually, it is conceivable that representability is controlled by $H^2(O_X)$, and that the vanishing of $H^3(O_X)$ serves to insure that the abelian variety varies nicely with parameters. Notice for C a curve we have

$$H^2(C \times X, O_{C\times X}) \cong [H^1(O_C) \underset{k}{\otimes} H^1(O_X)] \oplus H^2(O_X)$$

Since $H^1(O_C)$ and $H^1(O_X)$ are the tangent spaces to the respective picard varieties, it is natural to associate $H^1(O_C) \otimes H^1(O_X)$ with the image

$$A^1(C) \underset{\mathbb{Z}}{\otimes} A^1(X) \xrightarrow{\text{intersection}} A^2(C \times X).$$

Similarly, $H^2(O_X) \subset H^2(O_{X\times C})$ should be associated to the pullback $A^2(X) \subset A^2(C \times X)$. These two subgroups generate the subgroup of trivial cycles in $A^2(C \times X)$, so, intuitively at least, the obstruction to representing

$$\mathrm{Cor}^2(C,X)^0 = A^2(C \times X)/\{\text{trivial cycles}\}$$

vanishes.

I mentioned earlier that the geometry of cycles was analogous to the algebra of the higher K-groups of Quillen. In particular, CH^2 is related to K_2. Broadly speaking, (4.1) suggests

Hom(representable functor, functor assoc. to K_2) = representable functor.

Writing $\mathbb{G}_m$ for the multiplicative group of units, viewed as a representable functor on the category of commutative rings ($\mathbb{G}_m(R)$ = mult. grp. of invertible elements in R), I gave a proof in my talk at the conference of the algebraic result

$$\underset{\sim}{\mathrm{Hom}}(\mathbb{G}_m, K_n) \cong K_{n-1}.$$

In particular, since K_1(local ring R) = $\mathbb{G}_m(R)$, the Zariski sheaf associated to $\underset{\sim}{\mathrm{Hom}}(\mathbb{G}_m, K_2)$ is $\mathbb{G}_m$. Broadly speaking, one can envision a table

	algebraic object	geometric object
0	$\mathbb{Z}$	discrete group functor
1	$\mathbb{G}_m, K_1$	picard group, picard variety, abelian variety $H^1(X, \mathbb{G}_m)$
2	K_2	$CH^2(X)$, $H^2(X, K_2)$
3	K_3	$CH^3(X)$, $H^3(X, K_3)$
⋮	⋮	⋮

For $n \geq m$ integers, one expects roughly

$\underset{\sim}{\mathrm{Hom}}$(object m-th row, object in n-th row) = object in (n-m)th row.

For example, it is known that A, B abelian varieties implies

Hom(A,B) = finitely generated discrete group.

Here is a final bit of motivation for (4.1). Suppose we are

given abelian varieties A_1, A_2, A_3 and a morphism of functors

$$(*) \qquad A_1 \times A_2 \times A_3 \to CH^2_X$$

which is tri-linear (i.e. linear in each A_i separately). This gives a bilinear morphism

$$A_1 \times A_2 \to \underset{\sim}{\mathrm{Hom}}(A_3, CH^2_X)$$

Assuming the right hand side representable (4.1), such a morphism is necessarily trivial by rigidity ([9], p. 10), so the morphism (*) is zero also.

A typical example of a trilinear morphism like (*) is obtained by taking curves C_1, C_2, C_3 and a correspondence Γ on $C_1 \times C_2 \times C_3 \times X$. We take the composition

$$(**) \quad J(C_1) \times J(C_2) \times J(C_3) \xrightarrow{\text{intersection}} CH^3_{C_1 \times C_2 \times C_3} \xrightarrow{\Gamma} CH^2_X .$$

This leads to

<u>Conjecture</u> (4.3). Let Γ be a codimension 2 cycle on $C_1 \times C_2 \times C_3 \times X$, so we have

$$\Gamma_*: CH_0(C_1 \times C_2 \times C_3) \to CH^2(X).$$

Let $c_i^{(1)}, c_i^{(2)} \in C_i$ be points. Then

$$\sum_{1 \le i,j,k \le 2} (-1)^{i+j+k} \Gamma_*(c_1^{(i)}, c_2^{(j)}, c_3^{(k)}) = 0 \text{ in } CH^2(X).$$

To see what is involved here, suppose the ground field $k = \mathbb{C}$, the complex numbers. We have the cohomological interpretation [13]

$$CH^2(X) \cong H^2(X, K_2(O_X))$$

where $K_2(O_X)$ denotes the sheaf in the Zariski topology associated to the functor K_2, but we also have the cohomology of X with constant coefficients in the classical (analytic) topology denoted X^{an}. For example we can consider $H^2(X^{an},K_2(\mathbb{C}))$, where $K_2(\mathbb{C})$ denotes the constant sheaf. On the K_1 level, one can consider groups $H^1(X^{an},\mathbb{C}^*)$ and $H^1(X,O_X^*) = \mathrm{Pic}(X)$. Here the situation is rather nicer, because one has also a sheaf of analytic units $O^*_{X^{an}}$, and one knows

$$H^1(X,O_X^*) \cong H^1(X^{an},O^*_{X^{an}}).$$

$H^1(X^{an},\mathbb{C}^*)$ is the group of flat line bundles on X^{an}, and the natural map $\mathbb{C}^* \to O^*_{X^{an}}$ induces a map

$$H^1(X^{an},\mathbb{C}^*) \to H^1(X,O_X^*)$$

whose image is the group of line bundles $\mathscr{L}$ such that some tensor multiple $\mathscr{L}^{\otimes N}$ is algebraically equivalent to 0.

Conjecture (4.4). There exists a natural homomorphism $H^2(X^{an},K_2(\mathbb{C})) \to CH^2(X)$.

The word natural here should mean first that given varieties X, Y and a cycle Γ on $Y \times X$ with $\dim \Gamma = \dim X$, one gets a commutative diagram

$$\begin{array}{ccc} H^2(Y^{an},K_2(\mathbb{C})) & \xrightarrow{\Gamma_{\mathbb{Z}} \otimes K_2(\mathbb{C})} & H^2(X^{an},K_2(\mathbb{C})) \\ \downarrow & & \downarrow \\ CH^2(Y) & \xrightarrow{\Gamma_*} & CH^2(X) \end{array}$$

(Here $\Gamma_{\mathbb{Z}}$ denotes the correspondence on singular cohomology with integer coefficients induced by the cycle Γ. Note $K_2(\mathbb{C})$ is a uniquely divisible group, so $H^2(X^{an},K_2(\mathbb{C})) \cong H^2(X^{an},\mathbb{Z}) \otimes_{\mathbb{Z}} K_2(\mathbb{C})$.)

Moreover, we assume that the symbol pairing $\mathbb{C}^* \times \mathbb{C}^* \to K_2(\mathbb{C})$ gives rise to a commutative diagram

$$\begin{array}{ccc} H^1(X^{an},\mathbb{C}^*) \times H^1(X^{an},\mathbb{C}^*) & \xrightarrow{\text{symbol}} & H^2(X^{an},K_2(\mathbb{C})) \\ \downarrow & & \downarrow \\ H^1(X,O_X^*) \times H^1(X,O_X^*) & \xrightarrow{\text{intersection}} & CH^2(X). \end{array}$$

One can define a sheaf $K_2(O_X^{an})$, but the map $CH^2(X) \to H^2(X^{an},K_2(O_{X^{an}}))$ is probably not an isomorphism, though it may be one modulo torsion. If so, (4.4) could be weakened to give only a map $H^2(X^{an},K_2(\mathbb{C})) \to CH^2(X) \otimes_{\mathbb{Z}} \mathbb{Q}$.

Proposition (4.5). Conjecture (4.4) implies conjecture (4.3) (for varieties over $\mathbb{C}$).

Proof. Let Γ be a codimension two cycle on $C_1 \times C_2 \times C_3 \times X$, and view Γ as a family $\{\Gamma_t\}_{t\in C_3}$ of codimension two cycles on $C_1 \times C_2 \times X$. We are given points $c_i^{(1)}, c_i^{(2)} \in C_i$, and we can choose a flat structure $\mathscr{L}_i \in H^1(C_i^{an}, \mathbb{C}^*)$ on the line bundle L_i with divisor $c_i^{(2)} - c_i^{(1)}$. Let $\{\mathscr{L}_1,\mathscr{L}_2\} \in H^2(C_1^{an} \times C_2^{an},K_2(\mathbb{C}))$ denote the image of the pair $(\mathscr{L}_1,\mathscr{L}_2)$ under the symbol map. The reader can easily check that the cycle

$$\sum_{1\le i,j,k\le 2} (-1)^{i+j+k}\Gamma(c_1^{(i)},c_2^{(j)},c_3^{(k)}) \in CH^2(X)$$

is the image of the class

$$(\Gamma_{c_3^{(2)},\mathbb{Z}} \otimes K_2(\mathbb{C}))\{\mathscr{L}_1,\mathscr{L}_2\}$$

$$- (\Gamma_{c_3^{(1)},\mathbb{Z}} \otimes K_2(\mathbb{C}))\{\mathscr{L}_1,\mathscr{L}_2\} \in H^2(X^{an},K_2(\mathbb{C})).$$

But the correspondence $\Gamma_{t,\mathbb{Z}}$ depends only on the homology class of Γ_t, and hence is independent of t. This proves the desired vanishing.

Q. E. D.

References

1. M. Artin, D. Mumford, Some elementary examples of unirational varieties which are not rational, Proc. London Math. Soc (3) 25 (1972) 75-95.
2. S. Bloch, Some elementary theorems about algebraic cycles on abelian varieities, Inventiones Math, (to appear).
3. S. Bloch, K_2 of artinian Q-algebras, with application to algebraic cycles, Communications in Algebra 3(1975) 405-428.
4. S. Bloch, D. Lieberman, A. Kas, 0-cycles on algebraic surfaces with Pg = 0, Compositio Math. (to appear).
5. C. H. Clemens and P. A. Griffiths, The intermediate jacobian of the cubic three-fold, Ann. Math 95(1972) 281-356.
6. P. A. Griffiths, Periods of integrals on algebraic manifolds; summary of main results and discussion of open problems, Bull. Amer. Math. Soc. (2) 76(1970).
7. A. Grothendieck, FGA, Secretariat Math., 11 Rue Pierre Curie, Paris 5e (1962).
8. A. Grothendieck, Sur la classification des fibrés holomorphes sur la sphère de Riemann, Amer. J. Math. 79(1957)121-138.
9. S. Lang, Abelian Varieties, New York, Interscience, (1959).
10. V. A. Iskovskikh, Ju. I. Manin, Three-dimensional quartics and counterexamples to the Lüroth problem, Mat. Sb. 86(1971)140-66.
11. D. Mumford, Rational equivalence of 0-cycles on surfaces, J. Math. Kyoto Univ. 9(1969)195-204.
12. J. P. Murre, Algebraic equivalence modulo rational equivalence on a cubic three-fold, Compositio Math. Vol. 25, Fasc. 2, (1972) pp. 161-206.
13. D. Quillen, Higher Algebraic K-theory I, Algebraic K-theory I, Lecture notes in Math. 341, Springer Verlag (1973).

14. A. A. Roitman, On Γ-equivalence of zero-dimensional cycles, Math. USSR Sbornik 15(1971)555-567.

15. A. A. Roitman, Rational equivalence of 0-cycles Math. USSR Sbornik 18(1972)571-588.

16. P. Samuel, Relations d'équivalence en géometrie algébrique, Proc. Int. Cong. Math. Edinburgh (1958) 470-487.

17. A. N. Tyurin, Five Lectures on three dimensional varieties, Russian Math. Surveys 27(1972).

18. L. Roth, Algebraic Threefolds with Special Regard to Problems of Rationality, Ergebnisse der Math. 6, Springer-Verlag, Berlin and New York, 1955.

SK_1 of Commutative Normed Algebras

Barry H. Dayton

Banach algebras have often been used as examples in algebraic K-theory since information can be obtained by relatively simple topological methods. In this paper we show that many of these methods apply to a larger class of normed algebras, the "special" normed algebras. One advantage of this larger class is that every normed algebra has a localization which is special. We show that this localization has the same effect on SK_1 as does completion of the normed algebra.

In some cases this localization or completion gives an isomorphism of SK_1 but it is in general neither injective or surjective. We give examples to show that what happens seems to depend on both algebraic properties and the choice of a norm. In the last section we use our results to describe a map from algebraic K-theory to topological K-theory for the higher K-theory functors K_n, $n \geq 2$.

1. Normed Algebras

By normed algebra we shall mean a commutative, associative and unitary real (R) algebra A with a function $\|\cdot\|: A \longrightarrow R$ satisfying

i) $\|a\| \geq 0$, $a = 0$ if and only if $a = 0$, $a \in A$

ii) $\|a + b\| \leq \|a\| + \|b\|$, $\|ab\| \leq \|a\|\,\|b\|$ $a, b \in A$

iii) $\|ra\| = |r|\,\|a\|$ for $r \in R$, $a \in A$, $\|1\| = 1$

We shall denote by $\hat{A}$ the completion of A with respect to its norm. $\hat{A}$ is, of course, a Banach Algebra with A as a subalgebra. $A^{\cdot}$ will denote the group of units of A. We state the following [Bourbaki, IX, § 3.7]:

Proposition 1.1: Let A be a normed algebra, $a \in A$. If $\|a\| < 1$ and $1 - a$ is invertible then $\|(1 - a)^{-1}\| \leq (1 - \|a\|)^{-1}$, in particular the function $A^{\cdot} \longrightarrow A^{\cdot}$ given by $a \longmapsto a^{-1}$ is continuous. Furthermore if A is complete then $\|a\| < 1$ implies $1 - a$ is invertible in A.

We will call a normed algebra special if $\|a\| < 1$ implies $1 - a$ is invertible. In particular every Banach algebra is special. Given a normed algebra A, let $S = \{ a \in A | a \in (\hat{A})^{\cdot}\}$. S is a multiplicatively closed set and we write $\bar{\bar{A}} = A_S$. Then $A \subset \bar{\bar{A}} \subset \hat{A}$ and $\bar{\bar{A}}$ is special.

By a morphism of normed algebras we mean a continuous morphism of unitary real algebras. If $f:A \longrightarrow B$ is a morphism of normed algebras it is in fact uniformly continuous and so induces a morphism $f:\hat{A} \longrightarrow \hat{B}$ which, when restricted, gives a morphism $\bar{\bar{f}}:\bar{\bar{A}} \longrightarrow \bar{\bar{B}}$.

If A is a normed algebra and X is a compact Hausdorff space we let $A^X = \{f:X \longrightarrow A | f \text{ is continuous}\}$ and define a norm on A^X by $\|f\| = \sup\{\|f(x)\| \,|\, x \in X\}$ Of particular intrest will be the case where A is R or ¢ (complex numbers). If $A = R$ or ¢ and B is dense in A^X then $\hat{B} = A^X$ and $\bar{\bar{B}} = B_S$ where $S = \{f \in B | f(x)$ does not vanish on X } .

Proposition 1.2: Let A be a normed algebra, B a subalgebra of R^X. Then $B \otimes_R A$ is cannonically isomorphic to a subalgebra of A^X. Furthermore if B is dense in R^X then $B \otimes_R A$ (viewed as a subalgebra of A^X) is dense in A^X.

Proof: The map $\xi:B \times A \longrightarrow A^X$ given by $\xi(f,a)(x) = f(x)a$ is bilinear and factors through $B \otimes_R A$. It is not difficult to show that this gives a monomorphism of real algebras. The second part is a restatement of [Bourbaki, X, §4.4 Proposition 5].

In particular $R^X \otimes_R A$ is dense in A^X. If A,B are actually algebras over ¢ and B is dense in $¢^X$ then $B \otimes_{¢} A$ is dense in $¢^X \otimes_{¢} A$ which in turn is dense in A^X since $¢^X \otimes_{¢} A$ contains a copy of $R^X \otimes_R A$. This gives

Proposition 1.3: If A,B are normed algebras over ¢ and B is a dense subalgebra of $¢^X$ then $B \otimes_{¢} A$ is cannonically isomorphic to a dense subalgebra of A^X.

A similar argument gives

Proposition 1.4: If D is a dense subalgebra of A then D^X is a dense subalgebra of A^X.

For a normed algebra A we will always view $A[x] = A \otimes_R R[x]$ as a dense subalgebra of A^I where I denotes the unit interval. We will view $\mathbb{C}[z,z^{-1}]$ as a dense subalgebra of $\mathbb{C}^{S^1}$ where S^1 is imbedded as the unit circle in $\mathbb{C}$.

2. SK_1 of Special Normed Algebras

For definitions and notation in this section the reader is referred to [Milnor]. If A is a normed algebra we define for an n x n matrix (α_{ij}) with entries in A, $\|(\alpha_{ij})\| = \max_{i,j} \|\alpha_{ij}\|$. With the topology induced by this norm SL(n,A), GL(n,A) become topological groups (Using Prop. 1.1 and continuity of the determinant).

Thanks to Prop. 1.1 and the definition of special normed algebra the proof of [Milnor, Lemma 7.4] gives:

Proposition 2.1: Let A be a special normed algebra. If $\alpha \in SL(n,A)$ such that $\|\alpha - I\| < 1/(n-1)$ then $\alpha \in E(n,A)$. In particular E(n,A) is open in SL(n,A).

E(n,A) is path connected so we get

Corollary 2.2: If A is a special normed algebra E(n,A) is the component and path component of I in SL(n,A).

We note that Corollary 2.2 implies that E(n,A) is always a normal subgroup of SL(n,A). We shall identify $SL(n,A^X)$ with $SL(n,A)^X$ and hence $SL(A^X) = \lim SL(n,A^X)$ with $SL(A)^X = \lim SL(n,A)^X$ where $SL(A^X)$ and SL(A) are given the direct limit topology. Then identifying homotopy classes in $SL(A)^X$ with path components of $SL(A^X)$ we get:

Theorem 2.3: If A is a special normed algebra and [X,Y] denotes the set of free homotopy classes of maps $X \longrightarrow Y$ then there is an isomorphism $SK_1(A^X) \longrightarrow [X, SL(A)]$.

In the category of special normed algebras SK_1 is a homotopy functor in the following sense:

Proposition 2.4: Let $f,g: A \longrightarrow B$ be morphisms of normed algebras with B special.

If there is a continuous function $h:A \longrightarrow B^I$ with $\varepsilon_0 h = f$ and $\varepsilon_1 h = g$ where $\varepsilon_t:B^I \longrightarrow B$ is evaluation at t then $SK_1(f) = SK_1(g)$.

Proof: If $\alpha \in SL(A)$ then $h(\alpha)$ is a path from $f(\alpha) = \varepsilon_0 h(\alpha)$ to $g(\alpha) = \varepsilon_1 h(\alpha)$. Thus $f(\alpha)$, $g(\alpha)$ differ by an element of $E(B)$ and so are in the same class in $SK_1(B)$.

From either 2.3 or 2.4 we get that if I^n is in the unit n-cube then the inclusion $A \longrightarrow A^{I^n}$ viewing A as the subalgebra of constant maps induces an isomorphism $SK_1(A) \longrightarrow SK_1(A^{I^n})$ whenever A is a special normed algebra.

We need the following two lemmas before proving our main theorem.

Lemma 2.5: Let A be a dense subalgebra of B. Then $E(n,A)$ is dense in $E(n,B)$.

Proof: Each $e_{ij}^b \in E(n,B)$ can be approximated arbitrarily closely by elementary matrices $e_{ij}^a \in E(n,A)$. It follows from continuity of multiplication that elements of $E(n,B)$ can be approximated by elements of $E(n,A)$.

Lemma 2.6 Let A, B be special normed algebras with A dense in B. Then if $a \in A$ is invertible in B it is invertible in A.

Proof: Let $b \in B$ an inverse to a and let $c \in A$ be so close to b that $\| 1 - ac \| < 1$. Then ac is invertible in A, hence a is.

Theorem 2.7: Let A, B be special normed algebras with A dense in B. Then inclusion induces an isomorphism $SK_1(A) \longrightarrow SK_1(B)$.

Proof: Let $\alpha \in SL(n,A)$ such that $\alpha \in E(n,B)$. It follows from 2.5 that $\alpha \in \overline{E(n,A)}$. But since, by 2.2, $E(n,A)$ is open in $SL(n,A)$ it is closed and so $\alpha \in E(n,A)$. It follows that $SK_1(A) \longrightarrow SK_1(B)$ is injective.

Now let $\beta \in SL(n,B)$. Since $B^\cdot$ is open $GL(n,B)$ is open in the normed space of n x n matricies so we can choose an ε-ball about β contained in $GL(n,B)$. Pick a matrix α with entries in A contained in this ε-ball. Then if we set $\gamma(t) = t\alpha + (1 - t)\beta$ the image of γ, for $t \in I$ lies in this ε-ball so γ is a path from β to α in $GL(n,B)$. Let $\delta(t)$ be the diagonal matrix with $\delta(t)_{11} = (\det \gamma(t))^{-1}$ and $\delta(t)_{ii} = 1$ for $i \neq 1$. Now $\det \gamma(1) \in A$ and is invertible in B so that by 2.6 is invertible in A, thus $\delta(1) \in A$. Then $\delta\gamma$ is a path in $SL(n,B)$ from β to $\delta(1)\alpha$ so by 2.2 β is congruent to $\delta(1)\alpha$ modulo $E(n,B)$.

But $\delta(1)\alpha \in SL(n,A)$ so it follows that $SK_1(A) \longrightarrow SK_1(B)$ is surjective.

In particular this theorem says that for any normed algebra A, $SK_1(\bar{\bar{A}}) \longrightarrow SK_1(\hat{A})$ is an isomorphism. Thus for SK_1 the completion problem is equivalent to a localization problem. We now describe the kernel and cokernel of the map $SK_1(A) \longrightarrow SK_1(\bar{\bar{A}})$.

<u>Proposition 2.8:</u> Let A be a normed algebra, $\Theta: SK_1(A) \longrightarrow SK_1(\bar{\bar{A}})$ the morphism induced by inclusion, Then

$$\ker \theta = \lim(\overline{E(n,A)}/E(n,A))$$

$$\operatorname{coker} \theta = \lim(SL(n,\bar{\bar{A}})/\overline{SL(n,A)})$$

where $\overline{E(n,A)}$ denotes the closure of $E(n,A)$ in $SL(n,A)$ and $\overline{SL(n,A)}$ denotes the closure of $SL(n,A)$ in $SL(n,\bar{\bar{A}})$.

<u>Proof:</u> For the first part we note that the kernel consists of those classes in $SK_1(A)$ represented by $\alpha \in SL(n,A)$ for some n such that $\alpha \in E(n,\bar{\bar{A}})$. If $\alpha \in SL(n,A)$ and $\alpha \in E(n,\bar{\bar{A}})$ then by 2.5 $\alpha \in \overline{E(n,A)}$. On the other hand if $\alpha \in \overline{E(n,A)}$ then since $E(n,\bar{\bar{A}})$ is closed $\alpha \in E(n,\bar{\bar{A}})$.

We next note that since $E(n,A)$ is dense in $E(n,\bar{\bar{A}})$ we have $E(n,\bar{\bar{A}}) \subset \overline{E(n,A)} \subset \overline{SL(n,A)}$ where here $\overline{E(n,A)}$ denotes the closure of $E(n,A)$ in $SL(n,\bar{\bar{A}})$. Since $E(n,\bar{\bar{A}})$ is closed in $SL(n,\bar{\bar{A}})$ cosets are closed so $SL(n,A)E(n,\bar{\bar{A}})/E(n,\bar{\bar{A}}) = \overline{SL(n,A)}/E(n,\bar{\bar{A}})$. Thus $\operatorname{Im} \theta = \lim(\overline{SL(n,A)}/E(n,\bar{\bar{A}}))$ and so $\operatorname{coker} \theta = \lim(SL(n,\bar{\bar{A}})/E(n,\bar{\bar{A}}) / \overline{SL(n,A)}/E(n,\bar{\bar{A}})) = \lim(SL(n,\bar{\bar{A}})/\overline{SL(n,A)})$.

3. <u>Examples</u>

Theorems 2.3 and 2.7 give a powerful method of computing $SK_1(A)$ when A is a special normed algebra such that $\hat{A}$ is a function algebra. In this section we give some examples of this and also examine the behavior of the homomorphism $SK_1(A) \longrightarrow SK_1(\bar{\bar{A}}) \approx SK_1(\hat{A})$ when A is a normed algebra.

<u>Lemma 3.1:</u> Let X be a compact Hausdorff space. Then there is a split exact sequence

$$(1) \qquad 0 \longrightarrow [X,S^1] \longrightarrow \tilde{K}^{-1}(X) \longrightarrow SK_1(\mathbb{C}^X) \longrightarrow 0$$

and whenever X is connected an isomorphism $\widetilde{KO}^{-1}(X) \longrightarrow SK_1(R^X)$ where $\tilde{K}^{-1}$, $\widetilde{KO}^{-1}$ denote the complex and real topological reduced K-theory functors.

Proof: By [Atiyah, p 75] $\tilde{K}^{-1}(X) \approx [X, GL(¢)]$. Now GL(¢) is algebraically a semidirect product of SL(¢) with ¢˙ and topologically a product. Since multiplication is commutative in [X, GL(¢)] we have [X, GL(¢)] = [X, SL(¢)] ⊕ [X, ¢˙] Since ¢˙ has the homotopy type of S^1 sequence (1) follows from 2.3.

Likewise $\widetilde{KO}^{-1}(X) \approx [X,O]_0$ where $O = \lim O(n)$ is the infinite orthogonal group and $[X,O]_0$ denotes the group of homotopy classes of base point preserving maps. If X is connected the image of such a map lies in SO, the component of the identity. But SO is a deformation retract of SL(R) so the isomorphism follows from theorem 2.3.

Since $\tilde{K}^{-1}(X) = \tilde{K}^0(SX)$ and $\widetilde{KO}^{-1}(X) = \widetilde{KO}(SX)$ where SX denotes the reduced suspension of X [Atiyah, p 67], for the n-sphere S^n using $S^{n+1} \approx SS^n$ we can recover $\tilde{K}^{-1}(S^n)$ and $\widetilde{KO}^{-1}(S^n)$ from well known computations of $\tilde{K}^0(S^n)$ and $\widetilde{KO}(S^n)$. For instance the groups $\tilde{K}^0(S^n)$ are computed in [Atiyah] and the groups $\widetilde{KO}(S^n)$ are computed in [Atiyah, Bott, Shapiro], the results of which we summarize in the following table:

n ≡ (mod 8)	0	1	2	3	4	5	6	7
$\tilde{K}^{-1}(S^n)$	0	Z	0	Z	0	Z	0	Z
$\widetilde{KO}^{-1}(S^n)$	Z_2	Z_2	0	Z	0	0	0	Z

Example 3.2: Let $A_n = R[x_1, x_2, \dots x_{n+1}]/(x_1^2 + \dots + x_{n+1}^2 - 1)$ and $B_n = A_n \otimes_R ¢ = ¢[x_1, \dots, x_{n+1}]/(x_1^2 + \dots + x_{n+1}^2 - 1)$. A_n, B_n are dense subalgebras of real, complex valued continuous functions on the n-sphere S^n. Since S^n is connected for $n > 1$ $SK_1(\bar{\bar{A}}_n) \approx SK_1(\hat{A}_n) = SK_1(R^{S^n}) \approx KO^{-1}(S^n)$. For $n \geq 2$ $[S^n, S^1] = 0$ so $SK_1(\bar{\bar{B}}_n) \approx SK_1(\hat{B}_n) = SK_1(¢^{S^n}) \approx \tilde{K}^{-1}(S^n)$. For $n = 1$ $[S^1, S^1] = Z$ and since the sequence (1) of 3.1 splits $SK_1(\bar{\bar{B}}_1) = 0$.

It is well known [Bass, p 714] that $SK_1(A_1) \longrightarrow SK_1(\hat{A}_1) \approx Z_2$ and $SK_1(B_1) \longrightarrow SK_1(\hat{B}_1) = 0$ are isomorphisms. Using the description of KO^{-n}(point) given in [Atiyah, Bott, Shapiro] one can show that $SK_1(A_n) \longrightarrow SK_1(\hat{A}_n)$ is always surjective.

Example 3.3: Let A be a noetherian regular special normed algebra (Eg. R, ¢, the algebras $\bar{A}_n$, $\bar{B}_n$ of example 3.2). Then the inclusion $A \to A[x_1,x_2\ldots,x_n]$ induces an isomorphism of SK_1. Thus from the comment after prop. 2.4 $SK_1(A[x_1,\ldots,x_n]) \longrightarrow SK_1(\overline{A[x_1,\ldots,x_n]}) \approx SK_1(A^{I^n})$ is an isomorphism.

We have just seen examples of noetherian regular rings where the localization $SK_1(A) \longrightarrow SK_1(\bar{A})$ is an isomorphism. We now show, for these same rings, that if the norm is not properly chosen this localization need be neither injective nor surjective.

Example 3.4: Let $A = R[x,y]$ be normed as a subalgebra of the algebra of continuous real valued functions on the annulus $X = \{(x,y) \in R^2 \mid 1 \leqslant x^2 + y^2 \leqslant 2\}$. Then $SK_1(A) = 0$ but $SK_1(\bar{A}) = SK_1(\hat{A}) = [X, SL(R)]$. Since X has the homotopy type of S^1 this latter group is Z_2. Here $SK_1(A) \longrightarrow SK_1(\bar{A})$ is not surjective. See also proposition 3.5 and example 4.5.

Let $B = R[x,y]/(x^2 + y^2 - 1)$ viewed as a subalgebra of the algebra of continuous real valued functions on the arc $S^1_+ = \{(x,y) \in R^2 \mid x^2 + y^2 = 1, x \geqslant 0\}$. Then $SK_1(B) = Z_2$ but $SK_1(\bar{\bar{B}}) \approx SK_1(\hat{B}) \approx [S^1_+, SL(R)] \approx [I, SL(R)] = 0$. Here $SK_1(B) \longrightarrow SK_1(\bar{B})$ is not injective. See also example 5.2.

A more general result on localizations of $R[x_1,\ldots,x_n]$ is

Proposition 3.5: Let $A = R[x_1,x_2,\ldots,x_n]$, X a connected compact subset of R^n and $S = \{f \in A \mid f \text{ does not vanish on } X\}$. Then $SK_1(A_S) \approx \widetilde{KO}^{-1}(X)$.

Proof: Let J be the ideal of A_S consisting of all elements that vanish identically on X. Then A_S/J may be viewed as a dense subalgebra of R^X with the property that all non-vanishing functions are invertible, i.e. A_S/J is special. Thus from 2.7 and 3.1 $SK_1(A_S/J) \approx SK_1(R^X) \approx \widetilde{KO}^{-1}(X)$. Now for each $x \in X$, $M_x = \{f \in A \mid f(x) = 0\}$ is a maximal ideal of A disjoint from S. Suppose I is an ideal of A not contained some M_x, $x \in X$. Then since X is compact we can find $f_1, f_2,\ldots,f_k$ in I so that for any x at least one of the f_j's does not vanish at x, hence $f_1^2 + f_2^2 + \cdots + f_k^2 \in I \cap S$. Thus any ideal of A disjoint from S is contained in some M_x so the maximal ideals of A_S are the ideals $(M_x)_S$, $x \in X$. Since J is the intersection of these ideals, J is the Jacobson radical of A_S. It follows from[Bass, Chap. IX 1.3, 3.9, 3.10] that $SK_1(A_S) \to SK_1(A_S/J)$ is an isomorphism.

4. Laurent Polynomials.

It has been observed that there is a close connection between the fundamental theorem of algebraic K-theory and the Bott periodicity theorem of complex topological K-theory. Swan actually gives a proof of the periodicity theorem using the fundamental theorem. The content of his key lemma [Swan, 17.4], in the terminology of this paper, is basically that $SK_1(B[z,z^{-1}]) \longrightarrow SK_1(\widehat{B[z,z^{-1}]})$ is surjective when $B = \mathbb{C}^X$. Here we will use both the periodicity theorem and the fundamental theorem to compute the kernel and cokernel of this completion for certain subalgebras B of $\mathbb{C}^X$.

In this section X will be a compact Hausdorff space and B will be a dense, complex subalgebra of $\mathbb{C}^X$. If we let $A = \mathbb{C}[z,z^{-1}]$ since A is dense in $\mathbb{C}^{S^1}$ by proposition 1.3 $B \otimes_{\mathbb{C}} A$ is dense in A^X. By 1.4 A^X is dense in $(\mathbb{C}^{S^1})^X \approx \mathbb{C}^{X \times S^1}$ so we can view $B[z,z^{-1}] \approx B \otimes_{\mathbb{C}} A$ as a dense subalgebra of $\mathbb{C}^{X \times S^1}$.

From the fundamental theorem [Bass, Chapter XII, §7] or [Swan, 16.4] there is a short sequence

$$(1) \qquad 0 \to K_1(B) \longrightarrow K_1(B[z,z^{-1}]) \longrightarrow K_0(B) \longrightarrow 0$$

exact at $K_1(B)$, $K_0(B)$ but not at $K_1(B[z,z^{-1}])$ in general. Sequence (1) is exact when B is noetherian regular.

The inclusion $B \subset \mathbb{C}^X$ gives a map $GL(B) \subset GL(\hat{B}) = GL(\mathbb{C}^X) = GL(\mathbb{C})^X \longrightarrow [X, GL(\mathbb{C})] = \tilde{K}^{-1}(X)$ which induces a map $\theta_1 \colon K_1(B) \to \tilde{K}^{-1}(X)$. In the same way there is a map $\theta \colon K_1(B[z,z^{-1}]) \longrightarrow \tilde{K}^{-1}(X \times S^1)$. It follows from [Atiyah, 2.4.8] and the Bott periodicity theorem that

$$(2) \qquad 0 \to \tilde{K}^{-1}(X) \longrightarrow \tilde{K}^{-1}(X \times S^1) \longrightarrow K^0(X) \longrightarrow 0$$

is split exact. This is also shown directly in [Swan, Chapter 17] from where it follows that the diagram

$$(3) \qquad \begin{array}{ccccccccc} 0 & \to & K_1(B) & \longrightarrow & K_1(B[z,z^{-1}]) & \longrightarrow & K_0(B) & \longrightarrow & 0 \\ & & \downarrow{\scriptstyle\theta_1} & & \downarrow{\scriptstyle\theta} & & \downarrow{\scriptstyle\theta_0} & & \\ 0 & \to & \tilde{K}^{-1}(X) & \longrightarrow & \tilde{K}^{-1}(X \times S^1) & \longrightarrow & K^0(X) & \longrightarrow & 0 \end{array}$$

commutes where θ_0 is $K_0(B) \longrightarrow K_0(\hat{B}) = K_0(\mathbb{C}^X)$ followed by the isomorphism $K_0(\mathbb{C}^X) \longrightarrow K^0(X)$. If we now insist that B is special it follows from 2.6 and [Evans, Theorem 1] that θ_0 is a monomorphism.

Proposition 4.1: Let B be a noetherian regular special normed algebra. Then the restriction $\theta': SK_1(B[z,z^{-1}]) \longrightarrow SK_1(\widehat{B[z,z^{-1}]})$ of θ is a monomorphism.

Proof: Let $[\alpha] \in \ker \theta'$. Then since θ_0 is injective and (1) is exact, $[\alpha]$ is in the image of $K_1(B) \longrightarrow K_1(B[z,z^{-1}])$, i.e. there exists $\beta \in GL(B)$ with $[\beta] = [\alpha]$. Thus for large n, α and β differ by an element of $E(n, B[z,z^{-1}]) \subset SL(B[z,z^{-1}])$. So $[\beta] \in SK_1(B)$. Now $[\beta]$ is in the kernel of $SK_1(B) \longrightarrow SK_1(\hat{B}) \subset \tilde{K}^{-1}(X)$ (See 3.1) which, since B is special, is trivial by 2.7. Thus $[\alpha] = 0$ in $SK_1(B[z,z^{-1}])$ showing θ' is injective.

Before computing coker θ' we need a lemma.

Lemma 4.2: Let X be connected and let $r: \mathbb{C}^{\cdot} \longrightarrow S^1$ to be the retraction $r(z) = z/|z|$. Then if B is special the map $B[z,z^{-1}]^{\cdot} \longrightarrow [X \times S^1, S^1]$ given by $b \mapsto r\circ b$ is surjective.

Proof: It follows from elementary considerations of S^1 that two maps from a space into S^1 are homotopic if they are close together. Let $f: X \times S^1 \longrightarrow S^1$ be given. Then letting $i_x: S^1 \longrightarrow X \times S^1$ be the slice $i_x(z) = (x,z)$ it follows that the map $X \longrightarrow Z$ given by $x \mapsto \deg(f\circ i_x)$ is continuous where Z has the discrete topology. Since X is connected $\deg(f\circ i_x)$ is constant, say $\deg(f\circ i_x) = n$ all x. Let $g: X \times S^1 \longrightarrow S^1$ be given by $g(x,z) = f(x,1)z^n$. Then $(fg^{-1})(x,1) = 1$ and $(fg^{-1})\circ i_x$ is nullhomotopic for each x. Thus for each x there is a unique $\phi_x: S^1 \longrightarrow R$ such that $(fg^{-1})(x,z) = e^{i\phi_x(z)}$ and $\phi_x(1) = 0$.

One can then show that the map $F: X \times S^1 \times I \longrightarrow S^1$ given by $F(x,z,t) = e^{it\phi_x(z)}$ is continuous and so is a homotopy from the constant map to (fg^{-1}).

Thus f is homotopic to g. Now let $j: X \longrightarrow X \times S^1$ be the slice $j(x) = (x,1)$. Since B is dense in $\mathbb{C}^X$ there is a $h \in B$ such that h is close to $f\circ j$. Since $(\mathbb{C}^X)^{\cdot}$ is open $h \in B^{\cdot}$ by 2.6 since B is special. Now $r\circ h$ is close to $f\circ j$ so $r\circ(hz^n) = (r\circ h)z^n$ is close to $(f\circ j)z^n = g$ and so is homotopic to g. Thus $hz^n \in B[z,z^{-1}]^{\cdot}$ maps to f in $[X \times S^1, S^1]$.

We remark that 4.2 holds without the hypothesis X is connected if instead we assume each idempotent in $\hat{B}$ is contained in B (for example $B = ¢^X$).

Proposition 4.3: If X is connected and B is special then coker $\theta' =$ coker θ_0 where θ' is the map of 4.1 and θ_0 is the map $K_0(B) \longrightarrow K^0(X)$.

Proof: From 4.2 and 3.1 we get the diagram (4). A chase of

$$(4) \quad \begin{array}{ccccccccc} 0 & \longrightarrow & B[z,z^{-1}] & \longrightarrow & K_1(B[z,z^{-1}]) & \longrightarrow & SK_1(B[z,z^{-1}]) & \longrightarrow & 0 \\ & & \downarrow & & \downarrow \theta & & \downarrow \theta' & & \\ 0 & \longrightarrow & [X \times S, S^1] & \longrightarrow & \tilde{K}^{-1}(X \times S^1) & \longrightarrow & SK_1(\widehat{B[z,z^{-1}]}) & \longrightarrow & 0 \end{array}$$

(4) yields an isomorphism coker $\theta' \approx$ coker θ. As in the last part of the proof of 4.2 since B is special and dense in $¢^X$ $B^{\cdot} \rightarrow [X,S^1]$ is surjective. By 2.7 $SK_1(B) \rightarrow SK_1(\hat{B})$ is surjective so it follows from 3.1 that $K_1(B) \rightarrow \tilde{K}^{-1}(X)$ is surjective. A diagram chase of (3) (exactness is not needed at $K_1(B[z,z^{-1}])$) yields an isomorphism coker $\theta \approx$ coker θ_0.

Corollary 4.4: Let X be a compact connected Hausdorff space and B a noetherian regular special algebra dense in $¢^X$. Let $\theta_0 : K_0(B) \longrightarrow K^0(X)$ be induced by the inclusion $B \subset \hat{B}$. Then

$$(5) \qquad 0 \rightarrow SK_1(B[z,z^{-1}]) \rightarrow SK_1(\widehat{B[z,z^{-1}]}) \rightarrow \text{coker } \theta_0 \rightarrow 0$$

is exact.

Example 4.5: Define D_n by $D_0 = ¢$ and $D_{n+1} = D_n[z_{n+1}, z_{n+1}^{-1}]$. It follows from 1.3 and 1.4 that D_n is dense in $¢^{T^n}$ where $T^n = S^1 x \cdots x S^1$ is the n-torus. By [Swan, 5.4, 6.3 and 5.15] D_n is noetherian regular, $K_0(D_n) = 0$ and so $K_0(\bar{\bar{D}}_n) = 0$

On the other hand, using [Atiyah, 2.4.8] and the fact that $\tilde{K}^0(S^1) = 0$ and $\tilde{K}^{-1}(S^1) = Z$ we have $\tilde{K}^0(X \times S^1) = \tilde{K}^{-1}(X) \oplus \tilde{K}^0(X)$ and $\tilde{K}^{-1}(X \times S^1) = \tilde{K}^{-2}(X) \oplus \tilde{K}^{-1}(X) \oplus Z = \tilde{K}^0(X) \oplus \tilde{K}^{-1}(X) \oplus Z$ where the last equality follows from the periodicity theorem. Thus starting with $T^1 = S^1$ we find by induction that $\tilde{K}^0(T^n)$ is free of rank $2^{n-1}-1$ for $n \geq 1$.

From 4.4, 2.7 and the above for each n we have an exact sequence

$$(6) \qquad 0 \longrightarrow SK_1(\bar{\bar{D}}_n[z_{n+1}, z_{n+1}^{-1}]) \longrightarrow SK_1(\bar{\bar{D}}_{n+1}) \longrightarrow K^0(T^n) \longrightarrow 0$$

Now $\bar{\bar{D}}_n$ is an integral domain so $\bar{\bar{D}}_n[z,z^{-1}]^{\cdot} \approx \bar{\bar{D}}_n^{\cdot} \times Z$ from which it follows that that $SK_1(\bar{\bar{D}}_n[z,z^{-1}]) \approx SK_1(\bar{\bar{D}}_n)$ using the fundamental theorem $K_1(\bar{\bar{D}}_n[z,z^{-1}]) \approx K_0(\bar{\bar{D}}_n) \oplus K_1(\bar{\bar{D}}_n)$ and the previous computation $K_0(\bar{\bar{D}}_n) = 0$ which implies $K_0(\bar{\bar{D}}_n) = Z$. (The same argument shows $SK_1(D_{n+1}) \approx SK_1(D_n)$ and so $SK_1(D_n) = 0$ all n.) Since $\tilde{K}^0(T^n)$ is free, (6) splits and so by induction $SK_1(\bar{\bar{D}}_n)$ is free of rank $2^{n-1} - n$ for $n \geqslant 1$. Thus both localizations $SK_1(\bar{\bar{D}}_n[z_{n+1}, z_{n+1}]) \to SK_1(\bar{D}_{n+1})$ and $SK_1(D_n) \longrightarrow SK_1(\bar{\bar{D}}_n)$ are injective but the first is surjective only when $n = 0,1$ and the second only when $n = 0, 1, 2$.

5. Higher K-theory of Normed Algebras

We will write K_n for the Quillen K-theory functors and K_n^h for the Karoubi-Villamayor K-theory functors. See [Gersten] for a discussion of these functors.

For a normed algebra A $\Omega^n A = (x_1 \cdots x_n(x_1-1) \cdots (x_n-1))A[x_1, \dots, x_n]$ can be viewed as the ideal of all polynomial functions $R^n \to A$ which vanish on ∂I^n, the boundary of the unit n-cube. We can thus regard $\Sigma_A^n = A[x_1, x_2, \dots, x_n]/\Omega^n A$ as a subalgebra of $A^{\partial I^n}$. Let $\varepsilon: \Sigma_A^n \longrightarrow A$ be evaluation at the point $(0,0,\dots,0) \in \partial I^n$.

Proposition 5.1: Let A be a noetherian regular normed algebra. Then for $n \geqslant 2$ there is a split exact sequence

$$(1) \qquad 0 \longrightarrow SK_1(A) \longrightarrow SK_1(\Sigma_A^n) \longrightarrow K_n(A) \longrightarrow 0.$$

Proof: Since A is noetherian regular by [Gersten, 3.14, 3.9] $K_n(A) \approx K_n^h(A)$ $K_0(\Omega^n A)$ where here $K_0(\Omega^n A) = \ker (K_0(\Omega^n A^+) \longrightarrow K_0(Z))$ where $\Omega^n A^+$ is the unitary ring obtained from $\Omega^n A$. By excision [Bass, p 483] $K_0(\Omega^n A) \approx K_0(A[x_1, \dots, x_n], \Omega^n A)$ so the K_0, K_1 sequence for an ideal gives the exact sequence

$$(2) \qquad \begin{array}{ccccc} & K_1(A[x_1, \dots, x_n]) & \longrightarrow & K_1(\Sigma_A^n) & \longrightarrow \\ K_0(\Omega^n A) & \longrightarrow & K_0(A[x_1, \dots, x_n]) & \longrightarrow & K_0(\Sigma_A^n). \end{array}$$

Now the diagram

$$\begin{array}{ccc} A & \xrightarrow{\kappa} & A[x_1, \dots, x_n] \\ \downarrow{\scriptstyle 1} & & \downarrow \\ A & \xrightarrow{\varepsilon} & \Sigma_A^n \end{array}$$

commutes where κ, p are the inclusion and projection. Since A is noetherian regular $K_i(\kappa)$ is an isomorphism for $i = 0,1$ so $K_i(\kappa\varepsilon)$ is a left inverse for $K_i(p)$. Thus (2) gives the split exact

$$(3) \qquad 0 \longrightarrow K_1(A) \xrightarrow{K_1(p\kappa)} K_1(\Sigma_A^n) \longrightarrow K_0(\Omega^n A) \longrightarrow 0$$

Let $\sigma_j^i = \{(x_1,\ldots,x_n) \in I^n | x_j = i\}$ $j = 1,2,\ldots,n$; $i = 0,1$ be a face of ∂I^n. Now the algebra of polynomial functions on σ_j^i is isomorphic to $A[t_1,\ldots,t_{n-1}]$ so restriction to σ_j^i gives a homomorphism $f_j^i : \Sigma_A^n \longrightarrow A[t_1,\ldots,t_{n-1}]$. Now if u is a unit of Σ_A^n then $f_j^i(u)$ is a unit of $A[t_1,\ldots,t_{n-1}]$ and so, since A is noetherian regular, must be constant. Thus units of Σ_A^n are constant on each face of ∂I^n and so, since ∂I^n is connected for $n \geqslant 2$, must be constant, i.e. come from a unit in A. Hence $K_1(p\kappa)$ maps $A^{\cdot}$ onto $(\Sigma_A^n)^{\cdot}$ so sequence (1) follows from (3).

It should be noted that 5.1 holds if A is any noetherian regular commutative ring. The proof is the same except for a slight modification to show that units of Σ_A^n come from A. For n=2 5.1 is a special case of results in [Roberts].

Example 5.2: Let A = R. Then $\Sigma_R^2 = R[x,y]/(x^2-x)(y^2-y)$ is a dense subalgebra of $R^{\partial I^2}$. Since ∂I^2 is topologically S^1 $SK_1(\widehat{\Sigma_R^2}) \approx [\partial I^2, SL(R)] = Z_2$. By 5.1 since $SK_1(R) = 0$ then $SK_1(\Sigma_R^2) \approx K_2(R)$. As $K_2(R)$ is uncountable the localization $SK_1(\Sigma_R^2) \longrightarrow SK_1(\overline{\overline{\Sigma_R^2}}) \approx SK_1(\widehat{\Sigma_R^2})$ has a large kernel. One can show directly that $SK_1(\Sigma_R^2) \longrightarrow SK(\overline{\overline{\Sigma_R^2}})$ is surjective.

More generally Σ_R^n is a dense subalgebra of $R^{\partial I^n} \approx R^{S^{n-1}}$ and so also $\Sigma_{¢}^n \approx \Sigma_R^n \otimes_R ¢$ is by 1.2 a dense subalgebra of $¢^{\partial I^n} \approx ¢^{S^{n-1}}$. Thus by 3.1 $SK_1(\widehat{\Sigma_R^n}) \approx \widetilde{KO}^{-1}(S^{n-1}) \approx \widetilde{KO}(S^n) = KO^{-n}(\text{point})$ for $n \geqslant 2$. Also for $n \geqslant 3$ $[S^{n-1}, S^1] = 0$ so $SK_1(\widehat{\Sigma_{¢}^n}) \approx \tilde{K}^{-1}(S^{n-1}) \approx \tilde{K}^0(S^n) = K^{-n}(\text{point})$. Now $SK_1(R) = SK_1(¢) = 0$ so from 5.1 and 2.7 we get

Proposition 5.3: There are maps

$$K_n(R) \longrightarrow KO^{-n}(\text{point}) \qquad n \geqslant 2$$
$$K_n(¢) \longrightarrow K^{-n}(\text{point}) \qquad n \geqslant 3$$

whose kernels and cokernels are exactly those of the localizations $SK_1(\Sigma_R^n) \longrightarrow SK_1(\overline{\overline{\Sigma_R^n}})$ and $SK_1(\Sigma_{¢}^n) \longrightarrow SK_1(\overline{\overline{\Sigma_{¢}^n}})$ respectively.

For a special normed algebra A we can interpert the quoteint $SK_1(\overline{\overline{\Sigma}}{}^n_A)/SK_1(A)$ geometrically as follows:

<u>Lemma 5.4:</u> Let A be a special normed algebra. Then there is a split exact sequence

$$(4) \qquad 0 \longrightarrow SK_1(A) \longrightarrow SK_1(\overline{\Sigma}{}^n_A) \longrightarrow \pi_{n-1}(SL(A)) \longrightarrow 0$$

<u>Proof:</u> Since $\Sigma^n_A = \Sigma^n_R \otimes_R A$ by 1.2 Σ^n_A is dense in $A^{\partial I^n} \approx A^{S^{n-1}}$. Now since A is special so is $A^{S^{n-1}}$ so we can view $\overline{\overline{\Sigma}}{}^{\pi}_A$ as a special dense subalgebra of $A^{S^{n-1}}$. Thus by theorems 2.7 and 2.3 $SK_1(\overline{\overline{\Sigma}}{}^n_A) \approx SK_1(A^{S^{n-1}}) \approx [S^{n-1}, SL(A)]$. But as A is special $SK_1(A)$ may be viewed as the group of path components of SL(A) whose image in $[S^{n-1}, SL(A)]$ is the subgroup of classes of constant maps. The lemma easily follows from this.

For a commutative Banach algebra there is a well known surjection $K_2(A) \longrightarrow \pi_1(SL(A))$ [Milnor, 7.6]. As in 2.1 the proof for a Banach algebra generalizes to the case where A is a special normed algebra. If, however, A is a regular special normed algebra from 5.1 and 5.4 the localization $SK_1(\Sigma^n_A) \longrightarrow SK_1(\overline{\overline{\Sigma}}{}^n_1)$ induces a map $K_n(A) \longrightarrow \pi_{n-1}(SL(A))$ by the diagram

$$\begin{array}{ccccccccc}
0 & \longrightarrow & SK_1(A) & \longrightarrow & SK_1(\Sigma^n_A) & \longrightarrow & K_n(A) & \longrightarrow & 0 \\
 & & \| & & \downarrow & & \vdots \downarrow & & \\
0 & \longrightarrow & SK_1(A) & \longrightarrow & SK_1(\overline{\overline{\Sigma}}{}^n_A) & \longrightarrow & \pi_{n-1}(SL(A)) & \longrightarrow & 0
\end{array}$$

giving

<u>Proposition 5.5:</u> Let A be a noetherian regular special normed algebra. Then there is a map $K_n(A) \longrightarrow \pi_{n-1}(SL(A))$ for $n \geqslant 2$ whose kernel and cokernel is the same as that of the localization $SK_1(\Sigma^n_A) \longrightarrow SK_1(\overline{\overline{\Sigma}}{}^n_A)$.

References

M. F. Atiyah, K-theory, Benjamin, 1967.

M. F. Atiyah, R. Bott, and A. Shapiro, Clifford Modules, Topology 3 (1964) pp. 3 - 38.

H. Bass, Algebraic K-theory, Benjamin, 1968.

Bourbaki, General Topology (English edition), Hermann, 1966.

E. G. Evans Jr., Projective modules as Fiber Bundles, Proc. AMS Vol. 27 No. 3 (1971) pp. 623 - 626.

S. M. Gersten, Higher K-theory of Rings, Algebraic K-Theory I, Lecture Notes in Math. #341, Springer 1973.

J. Milnor, Introduction to Algebraic K-theory, Annals of Math. Studies #72, 1971.

L. G. Roberts, The K-Theory of Some Reducible Affine Varieties, Journal of Algebra Vol 35 Nos 1-3 (1975).

R. G. Swan, Algebraic K-Theory, Lecture Notes in Math. #76, Springer, 1968.

Northeastern Illinois University
Chicago, Illinois 60625

The K-Theory of some Reducible Affine Curves: A Combinatorial Approach

by

Leslie G. Roberts

1. Introduction

The K-theory of the reducible curves considered in [12] is essentially combinatorial in nature. Here I would like to pursue this idea further. I originally considered only K_0 and K_1. However at this conference I had several valuable conversations with Barry Dayton, in which we concluded that my method applied also to Karoubi - Villamayor K-theory K_i^h (in the notation of [5]). In this paper I will study only curves over a field k, although as in [12] more general varieties and ground rings could have been considered. However, unlike [12], I will allow points in the same component to be identified. By using known relations between K_i and K_i^h we can obtain some results about K_2 of the curves under consideration. This was not possible using only the Mayer Vietoris sequence for K_2, K_1, K_0 since that sequence stops with K_2.

Let $X = \text{Spec } A$ be a reduced affine curve over a field k with irreducible components $X_i = \text{Spec } A_i$ $(1 \le i \le n)$. Let $\bar{A}_i$ be the integral closure of A_i. The idea is to consider the Mayer-Vietoris sequence of the Cartesian square

$$\text{(C1)} \qquad \begin{array}{ccc} A & \to & \bar{A} \\ \downarrow & & \downarrow \\ A/I & \to & \bar{A}/I \end{array}$$

where I is the conductor of A in $\bar{A} = \Pi_{i=1}^{n} \bar{A}_i$. If $\bar{A}/I$ (hence also A/I) is reduced the results are purely combinatorial (ie depend only on the incidence relations among the components). In [14] it is shown that if $A = k[X,Y]/(Y-\alpha_1X)(Y-\alpha_2X)(Y-\alpha_3X)(Y-\alpha_4X)$ (α_i distinct) then $SK_1(A)$ depends on the slopes α_i, so the result in this case is not entirely combinatorial.

Throughout, Z = integers.

2. The Category $\underline{C}$

First we assume

(1) The non-regular points of X are k-rational.

(2) Each point of $\overline{X} = \operatorname{Spec} \overline{A}$ lying over a non-regular point of X is k-rational.

Let $m(X)$ be the set of non-regular points of X and $M(X)$ the points in $\overline{X}$ lying over $m(X)$. We have an inclusion $A \to \overline{A} = \prod_{i=1}^n \overline{A}_i$. Let B be the subring of all $\prod_{i=1}^n f_i$ $(f_i \in \overline{A}_i)$ such that $f_i(P_1) = f_j(P_2)$ whenever $P_1 \in \operatorname{Spec} \overline{A}_i$ and $P_2 \in \operatorname{Spec} \overline{A}_j$ lie over the same point $Q \in m(X)$ ($i=j$ is possible). There is an inclusion $\delta : A \to B$. The varieties defined by identification in section 2 of [12] are those for which δ is an isomorphism (except that now points in the same irreducible component can be identified). Pedrini has a similar construction in [9]. I got this idea originally from chapter IV of [15].

<u>Theorem 1</u> The following are equivalent

[$\underline{1}$] δ is an isomorphism

[$\underline{2}$] $\overline{A}/I$ is reduced

[$\underline{3}$] At each $Q \in m(X)$ we have smooth branches with independent tangent directions.

<u>Proof</u> [$\underline{1}$] $\Rightarrow$ [$\underline{2}$] is clear. Let J be the conductor of B in $\overline{B} = \overline{A}$. Then $I \subset J$ so we have a morphism of Cartesian squares

$$\begin{array}{ccccccc} A & \to & \overline{A} & & B & \to & \overline{B} \\ \downarrow & & \downarrow & \overset{\psi}{\to} & \downarrow & & \downarrow \\ A/I & \to & \overline{A}/I & & B/J & \to & \overline{B}/J \end{array}$$

Then $A = B$ because the other three corners are mapped isomorphically under ψ, so $[\underline{2}] \Rightarrow [\underline{1}]$. $[\underline{2}] \Leftrightarrow [\underline{3}]$ by Theorem 2 of [2]. Under these conditions $A/I = \prod_{i=1}^{m} k$ and $\overline{A}/I = \prod_{i=1}^{M} k$. I will not try to describe what $[\underline{3}]$ means, however $[\underline{3}]$ is satisfied by the curves W considered in §6.

Now assume

(3) $\overline{A}_i = k[t_i]$ (t_i an indeterminate).

Curves $X = \operatorname{Spec} A$ such that $\overline{A}/I$ is reduced and satisfying (1), (2), (3) will be referred to as Type II. If (3) is replaced by (3'): $A_i = k[t_i]$ the curve will be referred to as type I. Then type I $\subset$ type II.

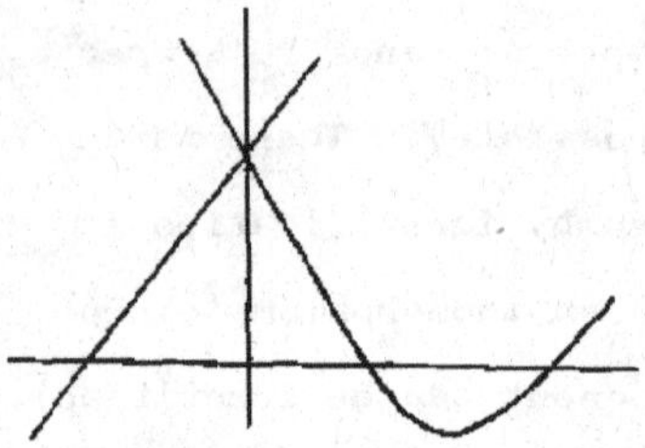

type I

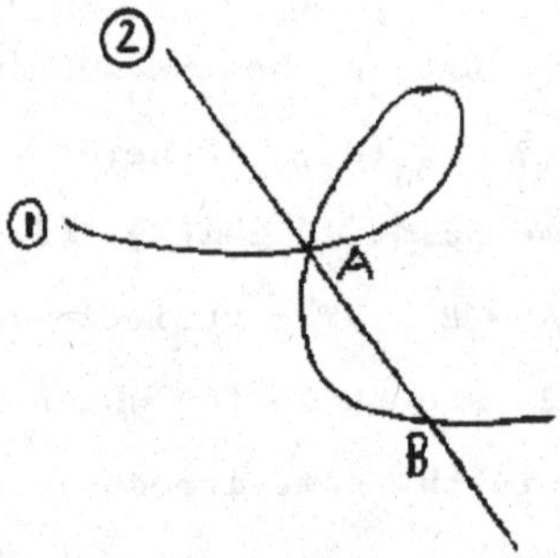

type II

The examples are curves in 3-space with independent tangent directions at the triple intersection points.

Let $\underline{C}$ denote the category whose objects are the affine curves /k of type II. For objects T, U in $\underline{C}$ there is a morphism $T \to U$ in $\underline{C}$ if (a) the irreducible components of T are a subset of those of U (b) if X_i, X_j are irreducible components of T ($i=j$ is possible) and $P_1 \in X_i$ has been identified with $P_2 \in X_j$ in T, then P_1 and P_2 have been identified in U. For any object S in $\underline{C}$ denote its co-ordinate ring by $k[S]$. If the components of U are $X_1,\ldots,X_r$ then $k[U] \subset \Pi_{i=1}^{r} k[t_i]$ and projection onto the factors corresponding to components of T maps $k[U]$ into $k[T]$. Thus a morphism in $\underline{C}$ gives rise to a morphism of schemes. (More general morphisms could have been considered, but the above suffice for later applications.) Let $n(T)$ denote the components of T. Clearly $n(T)$, $m(T)$, and $M(T)$ are give functors from $\underline{C}$ to finite sets. Hence the corresponding free abelian groups (denoted $\underline{n}(T)$, $\underline{m}(T)$, $\underline{M}(T)$) will be functors too.

Let F be a product preserving functor from rings with unit to abelian groups such that

(H) For K a field, t an indeterminate, $F(K) \to FK[t]$ is an isomorphism.

For $T \in \mathrm{obj}\ \underline{C}$ choose a closed k-rational point from each connected component, and let $\overline{F}(T)$ be the kernel of the resulting split surjection $F(T) \to \Pi F(k)$. I claim $\overline{F}(T)$ is independent of the choice. To prove this it suffices to assume T connected and that the two points are in the same irreducible component, say X_1. Let $f_1 : k[T] \to k$ and $f_2 : k[T] \to k$ be induced by the two closed points. Then f_1 and f_2 both factor

$$k[T] \to k[t_1] \xrightarrow{\tilde{f}_i} k$$

and by (H), $F(\tilde{f}_1) = F(\tilde{f}_2)$. Thus $F(f_1) = F(f_2)$ so their kernels are equal. Hence we have a functorial direct sum decomposition $F(T) = \overline{F}(T) \oplus (\underline{n}(T) \otimes_Z F(k))$.

3. Matroids and the Mayer-Vietoris Sequence

Let F_1 and F_0 be two product preserving functors from rings (with unit) to abelian groups which satisfy (H) and such that (F_1,F_0) forms a Mayer-Victoris pair (as defined on page 674 of [1]) for squares of the type Cl . I will work first in the category $\underline{C}$ and then at the end of the paper sketch some results in the non-rational intersection case.

Our functors are product preserving so we can without loss of generality take $X = \operatorname{Spec} A$ to be a connected object in $\underline{C}$. The Mayer-Vietoris sequence for Cl yields an exact sequence

$$
(\text{M.V.})\qquad
\begin{array}{ccccccc}
F_1(\bar{A}) \oplus F_1(A/I) & \xrightarrow{\alpha_1} & F_1(\bar{A}/I) & \rightarrow & F_0(A) & \xrightarrow{d} & F_0(\bar{A}) \oplus F_0(A/I) \\
\| & & \| & & \| & & \| \\
nF_1(k) \oplus mF_1(k) & & MF_1(k) & & F_0(k) \oplus \bar{F}_0(A) & & nF_0(k) \oplus mF_0(k)
\end{array}
$$

where n = number of elements in $n(X)$, similarly for m, M . Then $d\bar{F}_0(A) = 0$ and d maps $F_0(k)$ injectively. Therefore $\bar{F}_0(A) = \operatorname{coker} \alpha_1$. Finally α_1 is given by an $M\times(n+m)$ matrix $\alpha = (\beta,\gamma)$ with coeficients 0,1. The rows of α correspond to $M(X)$, the columns of β to $n(X)$, and the columns of γ to $m(X)$, with a 1 if the row and column are incident, 0 otherwise. It is easy to check that ${}^t\alpha : \underline{M}(X) \rightarrow (\underline{m} \oplus \underline{n})(X)$ is a morphism of functors, hence $\alpha : (\underline{m} \oplus \underline{n})(X) \rightarrow \underline{M}(X)$ is a morphism of contravariant functors. Because of the assumed naturality of the Mayer-Vietoris sequence we have an equivalence of functors $\bar{F}_0(A) = (\operatorname{coker} \alpha) \otimes_Z F_1(k)$.

In order to interpret coker α geometrically the language of matroids [17] is helpful. Let M be an $r\times s$ matrix over a field F. Then a circuit of the row matroid of M is a set of rows which is dependent but such that each subset is independent. The relation of linear dependence among the rows of a circuit is unique up to a non-zero constant multiple, and has only non-zero coefficients. If M is the incidence matrix of a graph (rows corresponding to edges and columns to vertices) and $F = Z/2Z$ then circuits of the row matroid of M correspond to circuits of the graph. (Graph theoretic terminology seems to vary somewhat. Circuits of the graph are as defined on p.27 of [18]). Each row of M has two non-zero entries. Regard M as a matrix over the rationals Q, and change one of the entries in each row from $+1$ to -1. Denote the resulting matrix by N. Then it is not hard to show that the row matroids of N (over Q) and M (over $Z/2Z$) are the same. Furthermore in each circuit of N the coefficients can be taken as ± 1. (N can be thought of as the incidence matrix of a digraph obtained by orienting each edge of our original graph.)

Now observe that given an object T in $\underline{C}$, the matrix α is the incidence matrix of the bipartite graph G whose vertices are $n(T)$ and $m(T)$ and whose edges are $M(T)$. Replacing β by $-\beta$ does not change coker α or the row matroid of α, so the above considerations can be applied to α. Now let α correspond to our variety X. It was observed in [12] p520 that the only relation of dependence among the columns of α is sum of columns of β = sum of columns of γ. Therefore α is of rank $m+n-1$. Arrange the rows of α so that the first $m+n-1$ rows are independent. By section 3.7 of [16] the $(m+n-1)\times(m+n-1)$ matrix in the upper left hand corner of α has det ± 1. (The proof, by induction, is easy.) Thus the image of α is a direct summand of $\underline{M}(X)$, and coker α is free abelian of rank $M-m-n+1$. (This result is equivalent to the lemma on page 70 of [19].) Each of the last $M-m-n+1$ rows of α is a linear combination of the first $m+n-1$ rows. This set of circuits is called a fundamental system of circuits (relative to the chosen basis of the row space). For any object T in $\underline{C}$ we have $(\text{coker }\alpha)^* = \ker {}^t\alpha$. The latter is called the group of cycles $c(T)$, and is a covariant functor. In particular $c(X)$ is free abelian of rank $M-m-n+1$ with the fundamental system of circuits as basis (more precisely, the relation of dependence with coefficients ± 1 corresponding to each circuit).

Now I will realize coker α in a more concrete manner.

Let $C_1, C_2, \ldots C_{M-m-n+1}$ be the above system of fundamental circuits of X. Then each C_i corresponds to a circuit $\overline{C}_i$ of the bipartite graph G associated to X. The circuit $\overline{C}_i$ is of the form $Q_{i1}X_{i1}Q_{i2}X_{i2} \ldots Q_{ir}X_{ir}Q_{i1}$ where $Q_{ij} \in m(X)$, $X_{ij} \in n(X)$ are distinct and adjacent Q's and X's are incident. The case $r=1$ corresponds to a loop of G or a node in X. (The expression $Q_{i1}X_{i1} \ldots Q_{i1}$ can be ambiguous for a type II curve. Strictly speaking I should include the edge $P_{ij} \in M(X) \cap X_{ij}$ (or $P_{i,j-1} \in M(X) \cap X_{i,j-1}$) joining Q_{ij} and X_{ij} (or $X_{i,j-1}$) but this this would clutter up the notation too much). Now construct an object Y_i in $\underline{C}$ with components X_{ij} $(1 \leq j \leq r)$ with $P_{ij} \in X_{ij}$ identified with $P_{i,j-1} \in X_{i,j-1}$ (as in [12] section 2). Y_i will be called the geometric realization of C_i. Furthermore $c(Y_i) = Z$ and the image of $1 \in c(Y_i)$ under the homomorphism $c(Y_i) \to c(X)$ is C_i. Thus $\bigoplus_{i=1}^{M-m-n+1} c(Y_i) \to c(X)$ is an isomorphism. Let α correspond to X and α_i to Y_i. Then taking duals we have

<u>Theorem 3</u> Let Y_i $(1 \leq i \leq M-m-n+1)$ be the geometric realizations of a fundamental system of circuits of X. Then coker $\alpha_i = Z$ and the morphisms $Y_i \to X$ induce an isomorphism coker $\alpha = \Pi_i Z$.

4. An example.

Let X be the example of a type II curve given earlier. Then α and the bipartite graph are

$$\alpha = \begin{array}{c} \\ P_1 \\ P_2 \\ P_3 \\ P_4 \\ P_5 \end{array} \begin{array}{c} \text{①} \ \text{②} \ A \ B \\ \begin{pmatrix} 1 & 0 & 1 & 0 \\ 0 & 1 & 1 & 0 \\ 1 & 0 & 0 & 1 \\ 1 & 0 & 1 & 0 \\ 0 & 1 & 0 & 1 \end{pmatrix} \end{array}$$

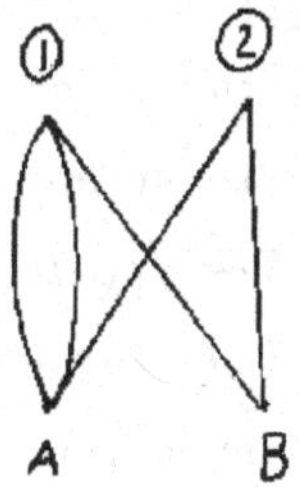

Here $m(X) = \{A,B\}$, $M(X) = \{P_i\}$, $M = 5$, $n = 2$, $m = 2$ and $M-m-n+1 = 2$. The first three rows of α are independent and the fundamental system of circuits is A1A (or P_1-P_4) and A1B2A (or $P_1-P_2-P_3+P_5$).

5. Applications

Here I give some examples where the calculation in §3 applies. Let $X = \operatorname{Spec} A$ be any object in $\underline{C}$ (not necessarily connected).

(1) Let $F_1 = K_1$ and $F_0 = K_0$. Then by [8] p28 the sequence (M.V.) holds for any X of type II . $\overline{F}_0 = \operatorname{Pic}$, $F_1(k) = k^*$ (units of k) so we have $\operatorname{Pic} A = (\operatorname{coker} \alpha) \otimes_Z k^*$.

(2) Let $F_1 = K_2$ and $F_0 = K_1$. Then by [4] p26, M.V. holds for X of type I . $\overline{F}_0 = SK_1$ so we have $SK_1(A) = (\operatorname{coker} \alpha) \otimes_Z K_2(k)$.

(3) Let $F_1 = K_0$ and $F_0 = K_{-1}$. Here M.V. holds for any X of type II, by [1] p676 . Thus $\overline{K}_{-1}(A) = K_{-1}(A) = \operatorname{coker} \alpha$. Similarly $K_{-i}(A) = 0$, $i \geq 2$. Here we have not used (1), (2) or (3) and by proposition 10.1 p685 of [1] we don't even need $\overline{A}/I$ reduced. For X irreducible this result was obtained in [11].

(4) Let $F_i = \ker(K_iB[t_1,\ldots,t_n] \to K_iB)$ $(i=0,1)$ for any commutative ring B . Then as on page 861 of [10] M.V. holds whenever it holds for K_i . Also $F_1(k) = 0$. Thus curves X of type II are K_0-regular (i.e. $K_0(A) \xrightarrow{\cong} K_0A[t_1,\ldots,t_n]$). This of course is well known [1] p691 . Similarly curves of type I are K_1-regular.

(5) Let $F_1 = K^h_{i+1}$, $F_0 = K^h_i$, $i \geq 0$, (Karoubi-Villamayor K-theory in the notation of [5]). Then $\overline{A}/I$ is regular, so $\overline{A} \to \overline{A}/I$ is a GL-fibration. Thus by Prop. 3.10 of [5] M.V. holds. Furthermore $K^h_i(k) = K_i(k)$ (Quillen K-theory) so $\overline{K}^h_i(A) =$ (coker α) $\otimes_Z K_{i+1}(k)$ for any X of type II .

(6) There is a surjection $SK_1(A) \to \overline{K}^h_1(A)$ hence a surjection $SK_1(A) \to$ (coker α) $\otimes_Z K_2(k)$. By application (2) this is an isomorphism for a curve of type I . If $K_1(A,I) \to K_1(\overline{A},I)$ is an isomorphism, diagram chasing extends M.V. to the K_2 terms and we would also have $SK_1(A) =$ (coker α) $\otimes_Z K_2(k)$. Krusemeyer has shown in [7] §14 that $K_1(A,I) \to K_1(\overline{A},I)$ is an isomorphism for the node $A = k[t^2-t, t(t^2-t)]$ for k quadratically closed (and unpublished for k additively generated by $\pm$ squares).

(7) A curve of type I is K_1-regular by application (4) above. By Theorem 6.1 of [6] there is a surjection $K_2(A) \to K^h_2(A)$, hence a surjection $K_2(A) \to$ (coker α) $\otimes_Z K_3(k)$. Because of the calculation of Dennis and Krusemeyer [3], this homomorphism is not injective.

We saw earlier that there is an equivalence of functors $\overline{F}_0(A) =$ (coker α) $\otimes_Z F_1(k)$. Thus in each of the above cases we can tensor with $F_1(k)$ in Theorem 3, thus yielding a corresponding statement about $\overline{F}_0(A)$.

6. A further calculation.

The introduction of the cycle group makes it fairly easy to see what happens to the K-theory under a morphism in $\underline{C}$. Let $X = \operatorname{Spec} A$ be the reduced variety over a field k consisting of N planes π_i in affine 3-space A^3_k, no three of which pass through one line, and no four through one point. Let W be the one skeleton of X, and $W_i = W \cap \pi_i$ (W, W_i both taken as reduced varieties). I will assume that X contains three planes whose normals are independent. This implies that W and all the W_i are connected.

The normalization of X is Spec $\overline{A}$, where $\overline{A} = \Pi_{i=1}^{N} k[t_i,u_i]$. Suppose that g_i is the equation (in $k[T_1,T_2,T_3]$) of π_i, and $f_i = \Pi_{j \neq i}\, g_j$. Then, denoting the image of f_i in A by $\overline{f}_i$ the conductor of A in $\overline{A}$ is $I = (\overline{f}_1,\ldots,\overline{f}_N)$.

Consider the Cartesian square

$$\text{(C2)} \qquad \begin{array}{ccc} A & \to & \overline{A} \\ \downarrow & & \downarrow \\ A/I & \to & \overline{A}/I \end{array}$$

We have $\overline{A}/I = \Pi_{i=1}^{N} k[t_i,u_i]/(\overline{f}_i)$ so Spec $\overline{A}/I$ is the disjoint union of the (connected) curves $W_i = W \cap \pi_i$ = Spec $k[t_i,u_i]/\overline{f}_i$. As a point set we have W = Spec (A/I) and there is an inclusion $A/I \to \overline{A}/I$. The latter is reduced since no three planes of X pass through one line. Hence A/I is reduced. Since no four planes of X pass through one point, W and the W_i are connected curves of type I.

Now suppose that we have three product preserving functors F_2, F_1, F_0 from rings to abelian groups which satisfy (H) of §2 and the appropriate Mayer-Vietoris sequences for squares C1 and C2. Suppose also $F_0(k) \to F_0 k[t_1,t_2]$ is an isomorphism (t_1,t_2 indeterminates). Then we can write $F_0(A) = F_0(k) \oplus \overline{F}_0(A)$, the direct sum decomposition induced by the split surjection $A \to k$ given by any closed k-rational point of X. (Again this is independent of the choice.) The Mayer-Vietoris sequence for (C2) then yields an exact sequence

$$\overline{F}_1(A/I) \xrightarrow{\alpha_1} \overline{F}_1(\overline{A}/I) \to \overline{F}_0(A) \to \overline{F}_0(A/I) \xrightarrow{\alpha_0} \overline{F}_0(\overline{A}/I)$$

From §3, $\overline{F}_i(A/I) = (\text{coker } \alpha_W) \otimes_Z F_{i+1}(k)$ and $\overline{F}_i(\overline{A}/I) = [\bigoplus_i (\text{coker } \alpha_{W_i})] \otimes F_{i+1}(k)$ $(i=0,1)$. It is clear from the geometry that the cycle homomorphism $\bigoplus_{i=1}^{N} c(W_i) \to c(W)$ is onto hence a split surjection. (In fact, one of the $c(W_i)$ can be left out). Taking duals and tensoring with $F_2(k)$ or $F_1(k)$ shows that α_0 and α_1 are split inclusions. Let n be the number of lines in W and m = number of vertices in W. Then by §3 we have $\overline{F}_1(A/I) = (2m-n+1)F_2(k)$ and $\overline{F}_1(\overline{A}/I) = (3m-2n+N)F_2(k)$. Thus $\overline{F}_0(A) = [(3m-2n+N) - (2m-n+1)]F_2(k) = (m-n+N-1)F_2(k)$.

The examples I have in mind are

(1) $F_2 = K_2$, $F_1 = K_1$, $F_0 = K_0$. This yields $\widetilde{K}_0(A)$ $(=\overline{K}_0 A) = (m-n+N-1)K_2(k)$.

(2) Similarly (using [1] p674) we get $K_{-1}(A) = (m-n+N-1)k^*$, $K_{-2}(A) = (m-n+N-1)Z$ and $K_{-i}(A) = 0$ for $i>2$.

(3) We have seen that $\overline{A}/I$ is K_1-regular. Therefore (as indicated in [6] Remark 2 p.72) $\overline{A} \to \overline{A}/I$ is a GL-fibration and the groups K_i^h satisfy the Mayer-Vietoris sequence for C2 as well as C1. Therefore $\overline{K}_i^h(A) = (m-n+N-1)K_{i+2}(k)$

(4) There is a surjection $SK_1(A) \to \overline{K}_1^h(A) = (m-n+N-1)K_3(k)$. I do not know if A is K_1-regular.

Note that over the reals $N-n+m-1$ is the number of bounded three cells in the resulting subdivision of the plane. Undoubtedly this is not a coincidence. Probably a result similar to Theorem 3 holds, with fundamental circuits replaced by "fundamental tetrahedra". The dimension shifting process then perhaps could be continued.

Using the Mayer-Vietoris sequence for Pic ([1] p482) it can be seen that Pic $A = 0$. In [13] I wrote Pic A by mistake instead of $\widetilde{K}_0(A)$.

7. Non-rational intersections

Again let $X = \operatorname{Spec} A$ be an affine curve /k as in §1. We no longer assume (1) and (2) of §2. The analogue of [1] ⇔ [2] of Theorem 1 still holds. Assume that $\bar{A}/I$ is reduced and that (3) of §2 holds. We can then think of X as obtained by identification, in a manner similar to [12] §2. Let F_1, F_0 be a Mayer-Vietoris pair as in §3. Then if we try to use the Mayer-Vietoris sequence as in §3 then some of the entries of the matrix α_1 may be homomorphisms $F_1(k) \to F_1(k_1)$ where k_1 is an extension field of k of finite degree. Furthermore $\bar{F}_0$ may not be well defined, in that $\ker(F_0(A) \to F_0(k))$ may depend on the closed k-rational point chosen. However if either (α) X is connected by k-rational identifications or (β) $F_0(k) \to F_0(k_1)$ is an inclusion for any finite extension field k_1 of k, then $\bar{F}_0$ can be defined as before and in M.V. $d\bar{F}_0(A) = 0$ and d maps $F_0(k)$ injectively. Thus as before $\bar{F}_0(A) = \operatorname{coker} \alpha_1$. In general there appears to be no reason why image α_1 should be a direct summand. For simplicity I will assume

(2') Let $P \in \bar{X}$ lie over the non-regular point $Q \in X$. Then P and Q have the same residue class field.

A circuit of X can be defined as before, that is, a sequence $X_1Q_1X_2Q_2 \ldots X_rQ_rX_1$ where the X_i, Q_i are distinct and each X_i is incident with the Q's on either side (not necessarily in a k-rational point). In many cases $\bar{F}_0(A)$ can be calculated by the following rules:

Rule (a): Suppose there is a non-regular point Q (with residue class field k_1) lying on components $X_1, \ldots, X_a$, only one of which (say X_1) contains any non-regular point besides Q. Then $\bar{F}_0(A) = \bar{F}_0(A_1) \oplus (a-1)(F_1(k_1)/\operatorname{im} F_1(k))$, where A_1 is the curve obtained by leaving out the a-1 components $X_2, \ldots, X_a$.

Rule (b): Suppose there is a circuit $X_1Q_1X_2Q_2 \ldots X_rQ_rX_1$ such that k_1 contains all the other k_i. (k_i = residue class field of Q_i). Then $\overline{F}_0(A) = \overline{F}_0(A_1) \oplus F_1(k_1)$ where A_1 is the curve obtained by performing all identifications as in A except that the point $P_2 \in X_2$ of Spec $\overline{A}$ lying over Q_1 is not identified with Q_1. (Instead of P_2 we could use $P_1 \in X_1$, thus sometimes giving two possibilities for A_1).

To prove rule (a) write

$$\alpha_1 = \begin{pmatrix} i & & & & & 1 & \\ & i & & & & 1 & \\ & & i & & & \cdot & \\ & & & \cdot & & \cdot & \\ & & & & \cdot & \cdot & \\ & & & & \cdot & \cdot & \\ & & & & i & 1 & \\ \beta & 0 & 0 & \cdots & 0 & 0 & \alpha' \end{pmatrix} \quad \left.\right\} a \text{ rows}$$

where all absent entries are zero. The first a rows correspond to the points in Spec $\overline{A}$ lying over Q, the first a columns to the components $X_1, X_2, \ldots, X_a$ and the $a+1$ st column to the point Q. The i is induced by the inclusion $k \to k_1$ and 1 is induced by $1 : k_1 \to k_1$. Add columns 2 through a to the first column and then subtract the first row from rows 2 through a. The resulting block iI_{a-1} (I_{a-1} = identity matrix of size $a-1$) contributes the $(a-1)$ $F_1(k_1)/\text{im}F_1(k)$. The rest of the matrix α_1 (after omitting rows and columns 2 through a and interchanging 2 columns) is

$$\tilde{\alpha}_1 = \begin{pmatrix} 1 & i & 0 \\ 0 & \beta & \alpha' \end{pmatrix} .$$

Clearly coker $\tilde{\alpha}_1$ = coker $(\beta\ \alpha')$, and $(\beta\ \alpha')$ is the incidence matrix for A_1, as required.

To prove (b) note that the row of α_1 corresponding to

P_2 (or P_1) can be reduced to zero by elementary row operations. This contributes the copy of $F_1(k_1)$. If other points besides P_1 and P_2 have been identified to form Q removing the row corresponding to P_2 already gives the incidence matrix of A_1. If only P_1 and P_2 have been identified to form Q we can then eliminate the row corresponding to P_1 and the column corresponding to Q as in the proof of (a). This does not change the coker and yields the matrix for A_1.

Examples are $F_1 = K_1$ and $F_0 = K_0$ and $F_1 = K_2$, $F_0 = K_1$ (here we need (3') of §2). As an example formulas (2) and (3) of [12] for $k = R$ (reals) with some complex non-regular points can now be derived easily. Let r be the number of components with only real identification. Let the other notation also be the same as in [12]. We calculate $SK_1(A)$. Applying rule (b) to circuits with only real intersection points we get $(M-m-n+r)K_2(R)$. By connectivity the total number of independent circuits is $M+M'-m-m'-n+1$ leaving $(M+M'-m-m'-n+1) - (M-m-n+r) = M'-m'-r+1$ circuits to yield by rule (b) $(M'-m'-r+1)K_2(C)$. Each time we apply rule (b) to a circuit with a complex intersection point the difference $M' - m'$ is reduced by one. After all circuits have been eliminated we are left with $(M'-m') - (M'-m'-r+1) = r-1$ yielding by rule (a) $(r-1)K_2(\mathbb{C})/\mathrm{im}K_2(R)$. This proves formula (2) of [12], with (3) for Pic following similarly.

Bibliography

[1] H.Bass, Algebraic K-theory, Benjamin, New York, 1968 .

[2] E.D. Davis, On the geometric interpretation of seminormality. To appear.

[3] K. Dennis, M. Krusemeyer, $K_2k[X,Y]/(XY)$ and a problem of Swan. To appear.

[4] K. Dennis, M. Stein, The functor K_2, a survey of computations and problems, pp 243-280 of Lecture Notes in Math. 342, Springer-Verlag, Berlin, 1973.

[5] S. Gersten, Higher K-theory of rings, pp 3-37 of Lecture Notes in Math. 341, Springer-Verlag, Berlin, 1973.

[6] S. Gersten, On Mayer-Vietoris functors and algebraic K-theory, J. of Alg. 18 (1971), 51-88 .

[7] M. Krusemeyer, Fundamental groups, algebraic K-theory and a problem of Abhyankar, Invent. Math. 19 (1973), 15-47.

[8] J. Milnor, Introduction to algebraic K-theory, Princeton University Press, 1971.

[9] C. Pedrini, Incollamenti di ideali primi e gruppi di Picard, Rendiconti del Seminario Matematico della Universita di Padova 48 (1973) 39-66 .

[10] C. Pedrini, On the algebraic K-theory of affine curves, Bollettino Unione Matematica Italiana, Serie IV, Vol. 9 (1974), 856-873 .

[11] M. Platzeck, O. Villamayor, The functors K^n for the ring of a curve, Revista de la Union Matematica Argentina, 25 (1971) 389-394 .

[12] L. Roberts, The K-theory of some reducible affine varieties, J. of Alg. 35 (1975), 516-527.

[13] L. Roberts, The Picard group of planes in 3-space, Queen's Mathematical Preprint No. 1975-6, Queen's University, Kingston, Ontario.

[14] L. Roberts, SK_1 of n lines in the plane, to appear in Transactions of the American Mathematical Society.

[15] J.P. Serre, Groupes algebriques et corps de classes, Hermann, Paris, 1959.

[16] P. Slepian, Mathematical foundations of network analysis, Springer-Verlag, New York Inc. 1968 .

[17] H. Whitney, On the abstract properties of linear dependence, Amer. J. Math., 57 (1935), 509-533 .

[18] R. J. Wilson, Introduction to graph theory, Oliver and Boyd (Edinburgh) and Academic Press (New York) 1972) .

[19] H. Bass, M.P, Murthy, Grothendieck groups and Picard groups of abelian group rings, Annals of Math 86 (1967), 16-73 .

SK_n of orders and G_n of finite rings

by

A. O. Kuku

INTRODUCTION

In §1, we obtain explicit computations for G_n *of finite rings* R *in terms of the characteristic of* R. *§2 is concerned with orders over a Dedekind domain. In §3, we give explicit estimate for* $SK_n(\Gamma)$ *where* Γ *is the maximal order in a p-adic semi-simple algebra and also note that if* F *is global field with integers* R, *then* $SK_n(\Gamma) = \coprod_p SK_n(\Gamma_p)$ *where* Γ_p *is the p-adic completion of* Γ *and* p *ranges over the prime ideals of* R.

NOTATION: The functors K_n refer to Quillen's functors (See [1]) and if A is a Noetherian ring, we write $\underline{\underline{M}}(A)$ for the category of finitely generated A-modules and $G_n(A)$ for $K_n(\underline{\underline{M}}(A))$. For any ring R, we write rad R for the Jacobson's radical of R.

§1

Definition 1.1 An associative ring R with unit is said to be quasi-regular if R has a two-sided nilpotent ideal J such that R/J is left regular. Recall that a ring R is left regular if R is left Noetherian and every finitely generated left R-module has finite homological dimension.

Theorem 1.2 Let R be a quasi-regular ring, J a 2-sided nilpotent ideal of R such that $\overline{R} = R/J$ is regular, then $G_n(R) \approx K_n(\overline{R})$ for all $n \geq 0$.

Proof: Follows from corollary 2 to theorem 4 in [7] and the fact that $K_n(\overline{R}) \approx G_n(\overline{R})$, $\overline{R}$ being regular (also see [7]).

Theorem 1.3: If R is any finite ring, then $G_n(R)$ is a finite group and $G_{2n}(R) = 0$ for all $n \geqslant 0$.

Proof: Let $\overline{R} = R/\text{rad } R$.. Then $\overline{R}$ is semi-simple and hence regular. Also $\overline{R}$ is a finite product of matrix algebras over finite fields. So computing $K_n(\overline{R})$ reduces to computing K_n of finite fields and the result follows by applying Quillen's results on K_n of a finite field (see [8]).

We now indicate how to obtain explicit computations of $G_n(R)$ in terms of the characteristic of R. First we consider some special cases.

Theorem 1.4: Let p be a rational prime and t a positive integer. Let $R = \mathbb{Z}/(p^t\mathbb{Z})$ and $S = M_k(R)$. Then (i) $G_{2n}(R) = 0 = G_{2n}(S)$ (ii) $G_{2n-1}(R) \approx G_{2n-1}(S)$ is a cyclic group of order $p^n - 1$.

Proof: Note that $G_n(R) \approx G_n(S)$ and that $\text{rad } R = p\mathbb{Z}/(p^t\mathbb{Z})$. So, $G_n(R) \approx K_n(\overline{R}) \approx K_n(\mathbb{Z}/p\mathbb{Z}) \approx K_n(F_p)$ where F_p is a field with p elements. Hence by Quillen's results [6], $G_{2n}(R) = 0$ and $G_{2n-1}(R) \approx \mathbb{Z}/(p^n - 1)\mathbb{Z}$ as required.

We now consider the case of an arbitrary finite commutative ring R.

Theorem 1.5: Suppose that R is a finite commutative ring, $\underline{p}_1, \ldots \underline{p}_m$ are the prime ideals of R, and $R/\underline{p}_i$ is a field of order q_i. Then $G_{2n-1}(R)$ is a finite abelian group of order $(q_1^{\,n} - 1)(q_2^{\,n} - 1) \ldots (q_m^{\,n} - 1)$.

Proof: Note that rad $R = \underline{p}_1 \cap \underline{p}_2 \cap \ldots \cap p_m$ and that since the p_i's are coprime, (rad R) $= \underline{p}_1 \underline{p}_2 \ldots \underline{p}_m$. Let j be the smallest positive integer such that (rad R) $= \underline{p}_1^j \ldots \underline{p}_m^j$. Then we have $R \simeq (R/\underline{p}_i^j) \oplus \ldots \oplus (R/\underline{p}_m^j)$. Now $G_n(R) = \oplus\ G_n(R/\underline{p}_i^j)$. Now rad $(R/\underline{p}_i^j) \simeq \underline{p}_i/\underline{p}_i^j$. So, $G_n(R/\underline{p}_i^j) \approx K_n(R/\underline{p}_i)$ for all $i \geq 1$. Hence $K_{2n}(R/\underline{p}_i) = 0$ and $K_{2n-1}(R/\underline{p}_i) = \mathbb{Z}/(q_i^n - 1)\mathbb{Z}$ by Quillen's result see [6]. So we have the required result.

We now consider the non-commutative case. The building blocks for finite non-commutative rings are called "Galois rings". First we define this notion.

Definition 1.6: Let p be a rational prime and put $\mathbb{Z}_{p^t} = \mathbb{Z}/p^t\mathbb{Z}$. A ring S is called a Golois ring of characteristic p^t and rank r if $S = \mathbb{Z}_{p^t}[x]/(f(x))$ where $f(x) \in \mathbb{Z}_{p^t}[x]$ is a monic irreducible polynomial of degree r. S is usually denoted by $GR(p^t, r)$.

Theorem 1.7: Let S be a Galois ring of characteristic p^k and rank r. Then (i) $G_{2n}(S) = 0$ and (ii) $G_{2n-1}(S) =$ finite cyclic group of order $p^n - 1$.

Proof: Note that the radical of $S = pS$ and S/pS is a finite field of order p^r. See [2]. Hence $G_{2n-1}(S) = K_{2n-1}(F_{p^r}) = \mathbb{Z}/(p^{nr} - 1)\mathbb{Z}$.

Remark: We observe that any finite ring R can be expressed as a direct sum of rings of prime power order, each of which has characteristic, of prime power. So, if R is any finite ring, then $R = \oplus_{i=1}^{\ell} R_{p_i}$ where each R_{p_i} is a finite p_i-ring of characteristic $p_i^{\lambda_i}$ say. $G_n(R) = \oplus_{i=1}^{\ell} G_n(R_{p_i})$. $G_n(R_{p_i}/\text{rad } R_{p_i}) \approx G_n(T_i/p_i T_i)$ where each T_i is a subring of R_{p_i} and each T_i is a direct sum of Galois rings $GR(p_i^{u_j}, r_{ij})$, $j = 1, \dots v$ say. So, $G_n(R_{p_i}) \approx K_n(R_{p_i}/\text{rad } R_{p_i}) \approx K_n(T_i/p_i T_i) \approx \oplus_{j=1}^{v} K_n(GR(p_i^{u_j}, r_{ij})/\text{rad } (GR(p_i^{u_j}, r_{ij}))$. But by 1.7, we have explicit form for each $K_n(GR(p_i^{u_j}, r_{ij})/\text{rad } (GR(p_i^{u_j}, r_{ij}))$. Hence we have explicit computation for $G_n(R)$, in terms of the characteristic of R.

§2.

2.1 Now let R be a commutative regular ring, A an R algebra which is finitely generated and projective as an R-module. Suppose that $\underline{P}_R(A)$ is the category of A-modules which are finitely generated and projective as R-modules, $\underline{M}(A)$ the category of finitely generated A-modules, the next result shows that the K_n functors $n \geq 0$ do not distinguish between $M(A)$ and $\underline{P}_R(A)$.

Theorem 2.2; Let A, R, $\underline{P}_R(A)$, $\underline{M}(A)$ be as in 2.1. Then the inclusion $\underline{P}_R(A) \to \underline{M}(A)$ induces an isomorphism $K_n(\underline{P}_R(A)) \approx G_n(A)$ for all $n \geq 0$. In particular if R is a Dedekind ring and Π a finite group, we have $G_n(R\Pi) \approx K_n(\underline{P}_R(R\Pi))$.

Proof: Let $M \in \underline{\underline{M}}(A)$. Since A is Noetherian, there exists a resolution $0 \to D_n \to P_{n-1} \to P_{n-2} \to \quad \to P_o \to M \to 0$ where $P_i \in \underline{P}(A)$, $0 \leq i \leq n-1$, where for any ring B, $\underline{P}(B)$ is the category of finitely generated projective B-modules. Also $M \in \underline{\underline{M}}(A) \Rightarrow M \in \underline{\underline{M}}(R)$, since A is a finitely generated R-module. Since R is regular and $M \in \underline{M}(R)$, $P_i \in \underline{P}(R)$ $0 \leq i \leq n-1$, then $D_n \in \underline{P}(R)$ also. So each P_i and D_n are in $\underline{P}_R(A)$ and by applying the resolution theorem of Quillen, we have that $K_n(\underline{P}_R(A)) \simeq G_n(A)$.

We also have the following result.

Theorem 2.3: Let R be a Dedekind domain with quotient field F, $X = \max(R)$ the set of prime = maximal ideals of R, A a left Noetherian R-algebra, $S = R - 0$, $\Sigma = S^{-1} A$. Then we have the following exact sequence

$$G_{n+1}(\Sigma) \to \bigoplus_{\underline{p} \in X} G_n(A/\underline{p}A) \to G_n(A) \to G_n(\Sigma) \to \bigoplus_{\underline{p} \in X} G_{n-1}(A/\underline{p}A) \to$$

Proof: Let $\underline{M}_S(A)$ be the full sub category of $\underline{M}(A)$ consisting of S-torsion left A-modules. Quillen's localization sequence (see [7]) gives

$$\to G_{n+1}(\Sigma) \to K_n(\underline{M}_S(A)) \to G_n(A) \to G_n(\Sigma) \to K_{n-1}(\underline{M}_S(A)) \to$$

So we only have to show that $K_n(\underline{M}_S(A)) \approx \bigoplus_{\underline{p} \in X} G_n(A/\underline{p}A)$.

Let $M \in \underline{M}_S(A)$, then $\underline{a}M = 0$ for some non-zero ideal $\underline{a}$ in R. Suppose that $\underline{a} = \underline{p}_1^{n_1} \underline{p}_2^{n_2} \ . \ . \ . \ \underline{p}_r^{n_r}$ say, is a prime decomposition of $\underline{a}$, then by Chinese remainder theorem $R/\underline{a} = \Pi(R/\underline{p}_i^{n_i})$. So $M \approx M_i \oplus \ . \ . \ . \oplus M_k$ where each M_i consists of elements of M annihilated by some power of $\underline{p}_i$.

So if for $\underline{p} \in X$, $M_{\underline{p}}(A)$ is the category of $M \in \underline{M}(A)$ annihilated by a power of $\underline{p}$, then $K_n(\underline{M}_S(A)) \simeq \bigoplus_{\underline{p} \in X} K_n(\underline{M}_{\underline{p}}(A))$. Now any $M \in \underline{M}_{\underline{p}}(A)$ has a finite filtration $0 = \underline{p}^{\ell-1} M \subset \ldots \subset \underline{p} M \subset M$ with successive quotients in $\underline{M}(A/\underline{p}A)$. So by Devissage $K_i(\underline{M}_{\underline{p}}(A) \approx G_i(A/\underline{p}A)$.

Remark: The above result (2.3) holds for A any order in a finite dimensional F-algebra Σ and hence for $A = R\Pi$ where Π is any finite group.

Corollary 2.4: If R, A, Σ, are as in 2.3 and A is regular then we have the following exact sequence.

$$K_{n+1}(\Sigma) \to \bigoplus_{\underline{p} \in X} G_n(A/\underline{p}A) \to K_n(A) \to K_n(\Sigma) \to \bigoplus_{\underline{p} \in X} G_{n-1}(A/\underline{p}A)$$

Proof: Follows since $K_n(R) \simeq G_n(R)$ for a regular ring R.

Remark: 2.4 holds for any maximal R-order A in a finite dimensional algebra over F.

§3

Let R be a Dedekind domain with quotient field F, and Γ any R-order in a semi-simple F-algebra Σ. For $n \geq 0$, let $SK_n(\Gamma)$ denote the kernel of the canonical map $K_n(\Gamma) \to K_n(\Sigma)$. In this section we give an explicit estimate for $SK_n(\Gamma)$ $n \geq 0$ where Γ is the maximal order in a semi-simple algebra Σ over a p-adic field F.

Recall that if p is a rational prime, a p-adic field F is any finite extension of $\hat{Q}_p$, and the integers of F is the integral closure of $\hat{Z}_p$ in F. We first consider the case of a central division algebra over F

and prove the following result which is a generalization to higher K_n's of theorem 2.2 of [5], proved by direct computation within K_1.

Theorem 3.1: Let Γ be the maximal order in a central division algebra of dimension t^2 over a p-adic field F, $q = p^{\ell}$ the residue class degree of F, then

(i) $SK_{2n}(\Gamma) = 0 \quad n \geq 1$

(ii) $SK_{2n-1}(\Gamma)$ is a finite cyclic group of order $\leq p^{\ell n t} - 1$ for all $n \geq 1$.

NOTE: It is conjectured that $SK_{2n-1}(\Gamma)$ is a cyclic group of order $\frac{p^{n\ell t} - 1}{p^{\ell n} - 1}$.

Proof of 3.1: That $SK_{2n}(\Gamma) = 0$ for all $n \geq 1$ follows from [5], 1.3(i).

Now suppose L is the inertia field of D i.e. L is the maximal unramified extension of F. Then by [9], theorem 6.9, D is totally ramified over L and has residue class degree 1 over L. Let A be the ring of integers of L. Then D has dimension t over L and Γ has rank t over A. Let $\underline{d}$ = radical of A, $\underline{m}$ = radical of Γ, then L has the property that $A/\underline{d} = \Gamma/\underline{m}$ see [9]. Now $[A/d: R/\text{rad } R] = t$. So, $\Gamma/\underline{m}$ is a finite field of order $q^t = p^{\ell t}$. Now, in the exact sequence

$$\text{(I)}\quad 0 \to K_{2n}(\Gamma) \to K_{2n}(D) \to K_{2n-1}(\Gamma/\underline{m}) \overset{\rho}{\to} K_{2n-1}(\Gamma) \overset{\delta}{\to} K_{2n-1}(D) \to 0$$

of [5] 1.3, $K_{2n-1}(\Gamma/\underline{m}) = K_{2n-1}(F_{q^t})$ and by Quillen's result $K_{2n-1}(\Gamma/\underline{m}) \approx \mathbb{Z}/(q^{nt} - 1)\mathbb{Z} \approx \mathbb{Z}/(p^{n\ell t} - 1)\mathbb{Z}$. Since $SK_{2n-1}(\Gamma) = \ker \delta = \operatorname{Im}\rho$, we have the result.

Remarks 3.2: (i) If F is a global field with integers R, and Γ is the maximal R-order in a central division algebra, then $SK_n(\Gamma) = \coprod_p SK_n(\Gamma_p)$ where Γ_p is the p-adic completion of Γ and p ranges over the prime ideals of R.

(ii) Now let R, F be as in 3.1, Γ the maximal order in a semi-simple F-algebra Σ. Then $\Sigma = \Pi M_{n_i}(D_i)$ say, where each D_i is a central division algebra of dimension t_i^2 over its centre C_i containing F. Let $q_i = p^{\ell_i}$ be the residue class degree of C_i. Then $|\Sigma| \sum_{i=1}^{k} n_i^2 t_i^2$, say. If Γ_i is the maximal order in D_i, $\underline{m}_i$ the radical of Γ_i, Γ the maximal order in Σ and $\underline{m}$ the radical of Γ, then $\Gamma/\underline{m} \approx \bigoplus^{k} M_{n_i}(\Gamma_i/\underline{m}_i)$.

The following result is a direct conseqnence of 3.1.

Corollary 3.3.: Let Γ be the maximal R-order in a p-adic semi-simple Σ satisfying the hypothesis of 3.2.(ii)

Then (i) $SK_{2n}(\Gamma) = 0 \quad n > 0$

(ii) $SK_{2n-1}(\Gamma)$ is a finite Abelian group of

$$\text{order} \leq \prod_{i=1}^{k} (p^{n \ell_i t_i} - 1)$$

REFERENCES

[1] Bass, H. Algebraic K-theory, W.A. Benjamin 1968.

[2] Clark, W.E. and Drake, D.A. Finite chain rings. Abhaudlunger Math. Sem. Univ. Hamburg (3a) 147 - 153, 1973.

[3] Keating, M.E. Whitehead group of some metacyclic groups and orders - Journal of algebra 22 (1972) 332 - 349.

[4] Kuku, A.O. Whitehead group of orders in p-adic semi-simple algebras. Journal of algebra (25) No. 3, 1973 415 - 418.

[5] Kuku, A.O. - Some finiteness theorems in the K-theory of orders in p-adic algebras. Journal of London Maths. Society. (to appear).

[6] Quillen, D. On the cohomology and K-theory of the general linear group over a field. Ann. of Math. 96 1972; 552 - 586.

[7] Quillen, D. Higher algebraic K-theory I. Spruger Lecture notes No. 341, 77 - 139.

[8] Reiner, I. Maximal orders. Academic Press, London.

[9] Roggenkanp, K.W. and Huber-Dyson, Verana Lattices over orders (I) Springer lecture notes (115) 1970.

[10] Swan, R.G. K-theory of finite groups and orders. Springer lecture notes No. 149, 1970.

University of Ibadan,
Ibadan - Nigeria.

K_2 OF A GLOBAL FIELD CONSISTS OF SYMBOLS

H.W. Lenstra, Jr.
Mathematisch Instituut
Universiteit van Amsterdam
Amsterdam, The Netherlands

Introduction. It is well known that K_2 of an arbitrary field is generated by symbols $\{a, b\}$. In this note we prove the curious fact that every element of K_2 of a global field is not just a product of symbols, but actually a symbol. More precisely, we have:

Theorem. Let F be a global field, and let $G \subset K_2(F)$ be a finite subgroup. Then $G \subset \{a, F^*\} = \{\{a, b\} \mid b \in F^*\}$ for some $a \in F^*$.

The proof is given in two sections. In section 1 we prove the analogous assertion for a certain homomorphic image of $K_2(F)$, by a rearrangement of the proof of Moore's theorem given by Chase and Waterhouse [3]. In section 2 we lift the property to $K_2(F)$, using results of Garland and Tate.

1. A sharpening of Moore's theorem.

Let F be a global field, i. e., a finite extension of $\mathbb{Q}$ or a function field in one variable over a finite field. The multiplicative group of F is denoted by F^*, the group of roots of unity in F by μ, and its finite order by m. By a prime v of F we shall always mean a prime divisor of F which is not complex archimedean. If v is non-archimedean, then we also use the symbol v to denote the associated normalized exponential valuation. For a prime v of F, let F_v be the completion of F at v. The group of roots of unity in F_v is called μ_v, and its finite order $m(v)$. The $m(v)$-th power norm residue symbol $F_v^* \times F_v^* \to \mu_v$ is denoted by $(\ ,\ )_v$. For all but finitely many v this map is given by the so-called "tame formula", cf. [1, sec. 1]. This formula implies that, for those v, and for all $a, b \in F_v^*$ with $v(a) = 0$, the symbol $(a, b)_v$ is the unique root of unity in F_v^* which modulo the maximal ideal is congruent to $a^{v(b)}$. It follows that, for any $a, b \in F^*$, we have $(a, b)_v = 1$ for almost all v. Thus a bimultiplicative map

$$\phi: F^* \times F^* \longrightarrow \bigoplus_v \mu_v, \qquad \phi(a, b) = ((a, b)_v)$$

is induced; here v ranges over the primes of F. The image of ϕ is, by the m-th power reciprocity law, contained in the kernel of the homomorphism

$$\psi: \bigoplus_v \mu_v \longrightarrow \mu$$

defined by

$$\psi(\zeta) = \prod_v \zeta_v^{m(v)/m}, \qquad \zeta = (\zeta_v).$$

We need the following converse, which is a sharpening of Moore's theorem [3].

Proposition. Let H be a finite subgroup of the kernel of ψ. Then $H \subset \phi(a, F^*) = \{\phi(a, b) \mid b \in F^*\}$ for some $a \in F^*$.

The proof is a bit technical. The ingredients are taken from [3], but the strengthened conclusion requires a reorganization of the argument which does not add to its transparency. The reader may find the table at the end of this section of some help.

Proof of the proposition. We begin by selecting four finite sets S, T, U, V of primes of F.

For S we take the set of real archimedean primes of F. It can be identified with the set of field orderings of F. If F is a function field it is empty.

For T we take a finite set of non-archimedean primes of F containing those v for which at least one of (1), (2), (3), (4) holds:

(1) $\zeta_v \neq 1$ for some $\zeta = (\zeta_v) \in H$;

(2) $v(h) > 0$, where h is the order of H;

(3) $v(m) > 0$;

(4) $(\ ,\ )_v$ is not tame.

Note that in the function field case (2), (3) and (4) do not occur.

If F is a function field, then choose an arbitrary prime v_∞ of F which is not in T, and put $U = \{v_\infty\}$. In the number field case let $U = \emptyset$.

The selection of V requires some preparation. Let $R \subset F$ be the Dedekind domain $R = \{x \in F \mid v(x) \geq 0$ for all primes $v \notin S \cup U\}$. Every prime $v \notin S \cup U$ corresponds to a prime ideal of R, denoted by P_v. For any rational prime number ℓ dividing the order h of H, consider the abelian extension $F \subset F(\eta_\ell)$, where η_ℓ denotes a primitive ℓm-th root of unity. Clearly, $F \neq F(\eta_\ell)$, and the extension $F \subset F(\eta_\ell)$ is unramified at every $v \notin S \cup T$. So for every $v \notin S \cup T \cup U$ the Artin symbol $(P_v, F(\eta_\ell)/F) \in \mathrm{Gal}(F(\eta_\ell)/F)$ is defined. By Čebotarev's density theorem, cf. [2, p.82], it assumes every value infinitely often. Hence we can choose a finite set V of primes, disjoint from $S \cup T \cup U$, such that

(5) for every rational prime ℓ dividing h there exists $u \in V$ with $(P_u, F(\eta_\ell)/F) \neq 1$.

Next, using the approximation theorem, we choose $a \in F^*$ such that

(6) $a < 0$ for every ordering of F,

(7) $v(a) = 1$ for all $v \in T$,

$v(a) = 0$ for all $v \in U$,

$a \sim 1$ at all $v \in V$

(here "$\sim$" means "close to"). We claim that this element a has the required property Before proving this, we split the remaining primes of F in two parts:

$W = \{v \mid v \notin S \cup T \cup U \cup V,\ v(a) \neq 0\}$

$X = \{v \mid v \notin S \cup T \cup U \cup V,\ v(a) = 0\}$.

Thus, we are in the situation described by the first two columns of the table. Notice that W is finite.

Now let $\zeta = (\zeta_v) \in H$ be an arbitrary element. To prove the proposition, we must find an element $b \in F^*$ such that $\zeta = \phi(a, b)$, i. e., $\zeta_v = (a, b)_v$ for all v.

By (6) and (7) we can find, for each $v \in S \cup T$, an element $c_v \in F_v^*$ with $(a, c_v)_v = \zeta_v$, cf. [4, lemma 15.8]. Choose $c \in F^*$ close to c_v at all $v \in S \cup T$ and close to 1 at all $v \in W \cup U$. Then for $v \in X$ the tame formula tells us that $(a, c)_v$ is the unique root of unity which modulo the maximal ideal is congruent to $a^{v(c)}$. For the value of $(a, c)_v$ if $v \notin X$, see the table.

We fix, temporarily, a rational prime number ℓ dividing h. We make some choices depending on ℓ. First, using (5), choose $u \in V$ such that $(P_u, F(\eta_\ell)/F) \neq 1$. Next, choose $k \in \{0, 1\}$ such that the fractional R-ideal

$$Q = P_u^k \cdot \prod_{v \in X} P_v^{v(c)}$$

satisfies $(Q, F(\eta_\ell)/F) \neq 1$. Finally, using a generalized version of Dirichlet's theorem on primes in arithmetic progressions [2, pp. 83-84], we select a prime $w \in X$ such that

(8) $\quad P_w \cdot Q = (d)$ (as fractional R-ideals)

where d satisfies the following conditions:

(9) $\quad d > 0$ for every ordering of F,

(10) $\quad d \sim 1$ at all $v \in T$,

(11) $\quad v(d) \equiv 0 \bmod N$, where $N = m(v) \cdot [F(\eta_\ell):F]$, for all $v \in U$,

$\quad d \sim 1$ at all $v \in W$.

Then d has the properties indicated in the sixth column of the table, and $(a, d)_v$ is given by the seventh column. Also, (9), (10) and (11) imply that $((d), F(\eta_\ell)/F) = 1$, so (8) and the choice of Q give

$$(P_w, F(\eta_\ell)/F) = (Q, F(\eta_\ell)/F)^{-1} \neq 1.$$

Therefore, P_w does not split completely in the extension $F \subset F(\eta_\ell)$, which is easily seen to be equivalent to

$$m(w)/m \not\equiv 0 \bmod \ell.$$

The table tells us that $(a, c/d)_v = \zeta_v$ for all $v \neq w$, so

$$\phi(a, c/d) = \zeta \cdot \theta$$

where $\theta = (\theta_v)$ is such that $\theta_v = 1$ for all $v \neq w$. Since ζ and $\phi(a, c/d)$ are in the kernel of ψ, the same must hold for θ. That means $\theta^{m(w)/m} = 1$, so

$$\phi(a, (c/d)^{m(w)/m}) = \zeta^{m(w)/m}.$$

We conclude that for every rational prime ℓ dividing h we can find a positive integer $n(\ell) = m(w)/m$ and an element $b(\ell) = (c/d)^{n(\ell)}$ of F^* such that

$$\phi(a, b(\ell)) = \zeta^{n(\ell)}, \qquad n(\ell) \not\equiv 0 \bmod \ell.$$

Clearly, if ℓ ranges over the rational primes dividing h, the numbers $n(\ell)$ have a greatest common divisor which is relatively prime to h. Hence we can choose integers $k(\ell)$ with $\Sigma_\ell\, k(\ell)n(\ell) \equiv 1 \bmod h$, and putting $b = \Pi_\ell\, b(\ell)^{k(\ell)}$ we find

$$\phi(a, b) = \prod_\ell \phi(a, b(\ell))^{k(\ell)} = \zeta^{\Sigma\, k(\ell)n(\ell)} = \zeta.$$

This proves the proposition.

The table:

$v \in$	a	ζ_v	c	$(a,c)_v$	d	$(a,d)_v$	$(a,c/d)_v$
S	<0	$(a,c_v)_v$	$\sim c_v$	$(a,c_v)_v$	>0	1	$(a,c_v)_v$
T	$v(a)=1$	$(a,c_v)_v$	$\sim c_v$	$(a,c_v)_v$	~ 1	1	$(a,c_v)_v$
U	$v(a)=0$	1	~ 1	1	$N \mid v(d)$	1	1
V	~ 1	1	-	1	-	1	1
W	$v(a)\neq 0$	1	~ 1	1	~ 1	1	1
X	$v(a)=0$	1	-	$\equiv a^{v(c)}$	$v(d)=v(c)$ $(v\neq w)$	$\equiv a^{v(d)}$	1 $(v\neq w)$

2. <u>Proof of the theorem</u>. We preserve the notations of section 1. There is a group homomorphism

$$\lambda\colon K_2(F) \longrightarrow \bigoplus_v \mu_v$$

sending $\{a, b\}$ to $\phi(a, b)$, for $a, b \in F^*$. A theorem of Bass, Tate and Garland [1, sections 6 and 7] asserts that

(12) $\quad \mathrm{Ker}(\lambda)$ is finite.

Further, Tate [1, sec. 9, cor. to th. 9] has proved that

(13) $\quad \mathrm{Ker}(\lambda) \subset (K_2(F))^p$ for every prime number p.

From (12) and (13) it is easy to see that there exists a <u>finite</u> subgroup $A \subset K_2(F)$ such that $\mathrm{Ker}(\lambda) \subset A^p$ for each prime number p.

We turn to the proof of the theorem. Let $G \subset K_2(F)$ be a finite subgroup. Replacing G by $G\cdot A$ we may assume that

(14) $\quad \mathrm{Ker}(\lambda) \subset G^p$ for every prime number p.

By the proposition of section 1, applied to $H = \lambda(G)$, there exists $a \in F^*$ such that $\lambda(G) \subset \lambda(\{a, F^*\})$. We claim that $G \subset \{a, F^*\}$.

To prove this, let $N = \{a, F^*\} \cap G$. Then $\lambda(G) = \lambda(N)$ so $G = N\cdot\mathrm{Ker}(\lambda)$, and using (14) we find

$$(G/N) = (N\cdot\mathrm{Ker}(\lambda))/N \subset (N\cdot G^p)/N = (G/N)^p$$

for every prime number p. Thus, the finite group G/N is divisible, and consequently $G/N = \{1\}$. It follows that $G = N$, so $G \subset \{a, F^*\}$.

This concludes the proof of the theorem.

References.

1. H. BASS, K_2 des corps globaux, Sém. Bourbaki 23 (1970/71), exp. 394; Lecture Notes in Math. 244, Berlin 1971.
2. H. BASS, J. MILNOR, J.-P. SERRE, Solution of the congruence subgroup problem for SL_n $(n \geq 3)$ and Sp_{2n} $(n \geq 2)$, Pub. Math. I. H. E. S. 33 (1967), 59-137.
3. S.U. CHASE, W.C. WATERHOUSE, Moore's theorem on uniqueness of reciprocity laws, Invent. Math. 16 (1972), 267-270.
4. J. MILNOR, Introduction to algebraic K-theory, Ann. of Math. Studies 72, Princeton 1971.

GENERATORS AND RELATIONS FOR K_2 OF A DIVISION RING

Sherry M. Green

Let R be a division ring, R^* its group of units. If $u, v \in R^*$ let ${}^u v = uvu^{-1}$ and $[u,v] = uvu^{-1}v^{-1}$, and let $[R^*,R^*]$ denote the commutator subgroup of R^*.

Throughout we will assume n is an integer greater than or equal to four. Let $SL(n,R)$ be the special linear group, $St(n,R)$ the Steinberg group, $\pi : St(n,R) \rightarrow SL(n,R)$ and $K_2(n,R)$ the kernel of π. For $x \in [R^*,R^*]$, let $a_i(x) \in SL(n,R)$ be the matrix $\mathrm{diag}(1, \ldots, 1, x, 1, \ldots, 1)$, with x in the i-th place. We define elements $\bar{b}_i(x) \in St(n,R)$ as follows. Let $b_1(x) \in St(n,R)$ be any cross-section for $a_1(x)$, $\bar{b}_1(1) = 1$. For $i \neq 1$, let

$$\bar{b}_i(x) = w_{1i}(1)\, \bar{b}_1(x)\, w_{1i}(1)^{-1}.$$

Therefore, if $u, v \in R^*$, there exist elements $c_{ij}(u,v) \in K_2(n,R)$ such that

$$c_{ij}(u,v)\bar{b}_i([u,v]) = \bar{h}_{ij}(u)\bar{h}_{ij}(v)\bar{h}_{ij}(vu)^{-1}. \tag{1}$$

One easily shows that $c_{ij}(u,v)$ is independent of i and j, and letting $c(u,v) = c_{ij}(u,v)$, (1) becomes

$$c(u,v)\bar{b}_i([u,v]) = h_{ij}(u)h_{ij}(v)h_{ij}(vu)^{-1}.$$

For $x, y \in [R^*,R^*]$ we define elements $d(x,y) \in K_2(n,R)$ to be

$$d(x,y) = \bar{b}_1(x)\bar{b}_1(y)b_1(xy)^{-1}$$

The main theorem is

THEOREM: *The abelian group* K_2R *has a presentation in terms of generators and relations, as follows. The given generators* $c(u,v)$, $d(x,y)$ *with* $u, v \in R^*$, *and* $x,y \in [R^*,R^*]$ *are subject only to the following relations and their consequences.* (Let $t,u,v \in R^*$ and $x,y,z \in [R^*,R^*]$)

(R1) d *is a normalized 2-cocycle*, i.e.

$$d(x,y)d(xy,z) = d(y,z)d(x,yz)$$

and

$$d(1,z) = 1 = d(z,1)$$

(R2) $c(tu,v) = c(v,t)^{-1}c(u,v)c(t,[u,v])d([t,[u,v]],[u,v])d([tu,v],[v,t])^{-1}$

(R3) $c(u,zv) = c(u,v)c(u,z)d({}^{u}z,[u,v]z^{-1})d([u,v],z^{-1})d({}^{u}z,z^{-1})^{-1}$

(R4) $c(u-v^{-1},v) = c(u,v-u^{-1})$

One easily shows that the $c(u,v), d(x,y)$ generate $K_2(n,R)$, hence K_2R, and satisfy (R1) through (R4). We will only sketch the proof here, the details will appear elsewhere.

PROPOSITION 1: *Let* A *be a multiplicative abelian group with symbols*

$$c : R^* \times R^* \to A$$

$$d : [R^*,R^*] \times [R^*,R^*] \to A$$

satisfying (R1) *through* (R3). *Then there exists a normalized* 2*-cocycle* $f : D \times D \to A$ *such that* $f(d_{ij}(u),d_{ij}(v)) = c(u,v)$ *and* $f(a_1(x),a_1(y)) = d(x,y)$.

The proposition is proved by constructing such an f. We have the following proposition.

PROPOSITION 2: *Let* A *be a multiplicative abelian group with symbols* c,d *as in Proposition 1, satisfying* (R1) *through* (R4). *Then there exists a central extension*

$$1 \to A \to G \to SL(n,R) \to 1$$

The proof uses Proposition 1 and ideas similar to those found in the proof of Matsumoto's Theorem.

Now let A be the group generated by $C(u,v)$, $D(x,y)$ subject only to (R1) through (R4). We then have a map $\xi : A \to K_2R$ such that $\xi(C(u,v)) = c(u,v)$, $\xi(D(x,y)) = d(x,y)$. By Proposition 2 (passing to the direct limit), there exists a central extension

$$1 \to A \to G \to SL(R) \to 1$$

One then shows that G is generated by symbols $x_{ij}(u)$, $u \in R$, subject to the same relations as the Steinberg group. Hence there exists $\lambda : St(R) \to G$ and

$\lambda(K_2R) \subset A$. One then shows $\lambda|_{K_2R} : K_2R \longrightarrow A$ inverts ξ, so that $K_2R \cong A$.

Example: Let R be the quaternion algebra over the field $\mathbb{R}$ of real numbers. Then

$$K_2R \cong {}^{K_2R}/_{<c(r,\,-1)\,|\,r \in \mathbb{R}>} \times \mathcal{D}$$

where $\mathcal{D}$ is the group generated by $d(x,y)$, $x,y \in [R^*,R^*]$ subject to (R1).

REFERENCES

1. S. Green, Generators and Relations for the Special Linear Group Over a Division Ring, Proc. Amer. Math. Soc. (to appear)

2. H. Matsumoto, Sur les sous-groupes arithmetiques des groupes semi-simples déployés, Ann. Sci. Ecole Norm. Sup. (4). 2 (1969), 1-62.

3. J. Milnor, Introduction to Algebraic K-Theory, Annals of Math Studies No. 72, Princeton University Press, Princeton 1971.

4. R. Steinberg, Générateurs, relations, et revêtments de groupes algébriques, Colloq. Theorie des Algébriques (Bruxelles, 1962), Libraire Universitaire, Louvain; Gauthier-Villars, Paris, (1962), 113-127.

5. R. Steinberg, Lectures on Chevalley Groups, Notes taken by J. Faulkner and R. Wilson, Yale University Lectures Notes, (1967).

University of Utah
Salt Lake City, Utah 84111

Injective Stability for K_2

Wilberd van der Kallen

§1. Introduction

1.1 We want to prove the following Theorem and some non-commutative variations on it.

Theorem 1 Let R be a commutative ring with noetherian maximal spectrum of dimension d, $d < \infty$. Let $n \geq d + 2$. Then the natural map $K_2(n,R) \to K_2(R)$ is surjective and the natural map $K_2(n + 1,R) \to K_2(R)$ is an isomorphism.

1.2 The proof of Theorem 1 is given in §§2,3,4,5. (In §5 we deal with a special case). In §6 we extend the Theorem to some non-commutative rings. In §7 we give some examples of non-stability for K_2, based on homotopy theory of real orthogonal groups. In §8 we recall the connection with second homology groups of $E(n,R)$ and with Quillen's non-stable K-groups.

1.3 The statement on surjectivity in Theorem 1 has been proved by Keith Dennis and also by L. N. Vaserstein. (See [5], [22], [28]). We refer to it as "surjective stability for K_2". In particular, this surjective stability implies that $K_2(n + 1,R) \to K_2(R)$ is surjective. (Use $n + 1 \geq d + 3 \geq d + 2$). So what we still have to prove is that it is an injective map. ("injective stability"). We will show that $St(n + 1,R) \to St(n + 2,R)$ is injective. This implies that $K_2(n + 1,R) \to K_2(n + 2,R)$ is injective, and, substituting $n + k$ for n, one sees that $K_2(n + k + 1,R) \to K_2(n + k + 2,R)$ is injective for $k \geq 0$. Taking the limit gives the required result.

We will also need "injective stability for K_1", i.e. the fact that $K_1(n,R) \to K_1(n + 1,R)$ is injective for $n \geq d + 2$. This result is due to Bass and Vaserstein and also follows from the same arguments as surjective stability for K_2. Actually these earlier results are valid in the more general case of a ring satisfying Bass's stable range condition SR_n. (The ring R is the Theorem satisfies SR_n by a well known result of Bass). For our Theorem we will need more than SR_n however. In the case $n = 2$ we will use that we are dealing with semi-local rings ($d = 0$). When $n \geq 3$ we will use that the ring satisfies a very technical variation on the condition SR_n. We will prove our variation on the statement that R satisfies SR_n essentially by repeating the Eisenbud-Evans proof, following Swan. (see [9], [25]).

Once we have shown (in §2) that R satisfies the technical condition we start our construction of a map ρ from $St(n + 2,R)$ into a structure called <u>left</u>. The injective stability will follow from the fact that the composition of ρ with the homomorphism $St(n + 1,R) \to St(n + 2,R)$ is injective. The global features of the construction of ρ and <u>left</u> are based on Matsumoto's proof for his presentation of the K_2 of a field. (see [19], [20]). We modify Matsumoto's approach by the introduction of a chunk, in analogy with the construction of a group scheme from a group chunk, cf [2]. (I have used the same idea before, in presenting the K_2 of a "3-fold stable" ring, cf [16], [15]).

1.4 As I have mentioned before, I rely on earlier stability results. The proofs given by Keith Dennis for these results inspired some of the arguments in the present proof. What is more, I use the definition of the chunk which he suggested to me when we both attended Queen's Conference on Commutative Algebra in July 1975. I would like to thank him for the very instructive discussions we had there. I am also indebted to E. Friedlander, D. Kahn and M. Barratt for telling me the basic facts of homotopy theory of orthogonal groups which I use in §7. And I thank Vicki Davis for typing the manuscript. I enjoyed the hospitality of Northwestern University during the time this research was done.

1.5 Let us now discuss in more detail the construction of ρ and left. First one defines a chunk C which is intended as a model for a piece of $St(n + 2,R)$. The building block for constructing C is $St(n + 1,R)$, which is considered to be "known". (In the case of 3-fold stable rings the building block was R^*, the group of units of R. That made it possible to find a presentation for $K_2(R)$. But in the present situation (i.e. for $d > 0$) the old chunk is too small and we don't get a presentation for $K_2(R)$.) The chunk allows a natural map $\pi: C \to St(n + 2,R)$ which is hoped to be injective. (If π is injective then C can be considered as a good model for $\pi(C)$. The problem of injective stability is actually equivalent to injectivity of π). The purpose of using C is to avoid the "unknown" set $\pi(C)$ which lies inside the "unknown" group $St(n + 2,R)$. Instead we now have the "known" set C constructed from the "known" group $St(n + 1,R)$. In $St(n + 2,R)$ one has for each element x a left multiplication $L_x: y \mapsto xy$. We can restrict its domain and codomain to $\pi(C)$ and obtain a partially defined map $\pi(C) \to \pi(C)$ which has domain $\pi(C) \cap (x^{-1}\pi(C))$. One now looks for its counterpart in the chunk, i.e. one looks for a partially defined map $\mathcal{L}(x)$ from C to C with $\pi \circ \mathcal{L}(x) = L_x \circ \pi$. For some x the choice of $\mathcal{L}(x)$ will be obvious, but not for all x. In any case, it is clear that one wants $\mathcal{L}(x)$ to be defined on the full set $\pi^{-1}(\pi(C) \cap x^{-1}\pi(C))$. Otherwise it gives incomplete information. Suppose one has a formula for $\mathcal{L}(x)$ which gives values on a domain that is too small. One way to enlarge the domain of $\mathcal{L}(x)$ is to use the counterparts $\mathcal{R}(y)$ of right multiplications R_y: $z \mapsto zy$. If the model is going to be correct then $\mathcal{L}(x)$ and $\mathcal{R}(y)$ will commute, because L_x and R_y do. That gives conditions for the values of $\mathcal{L}(x)$ at points where one doesn't yet have a formula. In order to define the extension of $\mathcal{L}(x)$ by means of these conditions, one has to find out whether the conditions are consistent with each other. That leads to the problem: Does $\mathcal{L}(x)$ commute with $\mathcal{R}(y)$ as far as the maps are defined? (That problem arises each time one introduces a new $\mathcal{L}(x)$ or $\mathcal{R}(y)$). We define <u>left</u> as the set of maps $\mathcal{L}$ which have domains of the proper size, satisfy $\pi \circ \mathcal{L} = L_x \circ \pi$ for some x, and commute with a selection from the maps $\mathcal{R}(y)$. Another way to enlarge the domain of a map $\mathcal{L}(x)$ is to use the fact that one wants $\mathcal{L}(p)\mathcal{L}(q)$ to coincide with $\mathcal{L}(pq)$ at points where the composite map $\mathcal{L}(p)\mathcal{L}(q)$ is defined. This leads to the problem: Do the $\mathcal{L}(x)$ combine in the expected way? Some of the answers will also be needed in the construction of ρ.

If $\mathcal{L}$, $\tilde{\mathcal{L}}$ are elements of <u>left</u> then $\mathcal{L} \circ \tilde{\mathcal{L}}$ denotes their composition as partially defined maps. We can show that there exists exactly one element $\mathcal{L} * \tilde{\mathcal{L}}$ of <u>left</u> which extends $\mathcal{L} \circ \tilde{\mathcal{L}}$. So <u>left</u> is now a set with composition $*$. This composition is associative. One expects <u>left</u> to be a group, isomorphic to $St(n + 2,R)$. Anyway, the units of <u>left</u> form a group <u>Uleft</u> with $*$ as composition. We look at those elements of <u>Uleft</u> which correspond to generators of $St(n + 2,R)$. They satisfy a set of defining relations for $St(n + 2,R)$. This yields a homomorphism ρ: $St(n + 2,R) \to$ <u>Uleft</u>. Because $St(n + 1,R)$ has been built into the chunk it is easy to check that the composition of ρ with $St(n + 1,R) \to St(n + 2,R)$ is injective. End of sketch.

1.6 Professor A. Suslin recently informed me that he obtained, in collaboration with M. Tulenbayev, a result similar to the main results of this paper. I quote from his letter:

"Let Λ be an associative ring. Then under $n \geq$ s.r. $\Lambda + 2$ the canonical map $St(n,\Lambda) \to St(n+1,\Lambda)$ is injective and consequently $K_{2,n}(\Lambda) \to K_2(\Lambda)$ is an isomorphism."

I presume that s.r. is the same as s.rk. in [27], but at this time no further information is available.

§2. Multiple Stable Range Conditions.

2.1 Rings are associative and have a unit. Let R be a ring. Recall that $(b_1,\dots,b_m)\in R^m$ is called <u>unimodular</u> if $\sum_{i=1}^{m} Rb_i = R$. If R is commutative then we may also say that $(b_1,\dots,b_m)$ is unimodular if $\sum_{i=1}^{m} b_iR = R$. We say that R satisfies SR_n if the following holds: Given a unimodular sequence (or column) $(b_1,\dots,b_n)$ there are $r_1,\dots,r_{n-1}\in R$ such that $(b_1 + r_1b_n,\dots,b_{n-1} + r_{n-1}b_n)$ is unimodular. One reason to recall this definition is that the literature is not unanimous: One also finds the notation SR_{n-1} for what we call SR_n.

2.2 <u>Definition</u> Let c,u,n,p be natural numbers with $c \geq u \geq n - 1$, $p \geq 1$. We say that R satisfies $SR_n^p(c,u)$ if the following holds: Let $A_1,\dots,A_p$ be matrices of size $(n-1) \times c$. For each i, let U_i be the submatrix of A_i consisting of the last u columns. Assume that for each i the matrix U_i can be completed, by adding rows, to a product of $u \times u$ elementary matrices. Then there is a column $\lambda\in R^{c-1}$ such that $A_i\binom{1}{\lambda}$ is a unimodular column for each i. So the property $SR_n^p(c,u)$ gives, for each set of matrices $A_1,\dots,A_p$, which satisfy the condition on the U_i, a column λ which behaves well with respect to $A_1,\dots,A_p$ simultaneously.

<u>Comment</u> If $c > u \geq n - 1$ then $SR_n^1(c,u)$ is automatic. One can show by an argument of Vaserstein that SR_n implies $SR_n^1(c,u)$ for any c,u with $c = u \geq n - 1$. This explains why we use the subscript n, given the convention in 2.1. (See also 3.37).

2.3 <u>Notation</u> We say that R satisfies $\widetilde{SR}_n$ if it satisfies SR_n, $SR_n^3(n + 2, n + 1)$, $SR_{n+1}^4(n + 2, n + 1)$, $SR_{n+2}^3(n + 2, n + 2)$. So $\widetilde{SR}_n$ is just shorthand for a list of conditions which we happen to need. It is not clear what the hierarchy is for the conditions in the list. It may be that $SR_n^3(n+2, n+1)$ actually implies $\tilde{\tilde{SR}}_n$.

2.4 THEOREM 2 Let R be a commutative ring with noetherian maximal spectrum of dimension d, $d < \infty$. Then R satisfies $\tilde{\tilde{SR}}_n$ for $n \geq \max(3, d+2)$.

Comments Theorem 2 is certainly not the strongest result one can obtain along these lines. See for instance Theorem 3 below (in 2.11) and remark 2.12. One should also prove a non-commutative version of Theorem 2. This is done in Section 6.

2.5 The proof of Theorem 2 is given in the remainder of Section 2. (The idea is to copy §3 of [25], with minor adaptations). Instead of working with the maximal spectrum it is more convenient to work with the so-called j-spec. Its points are the prime ideals which are intersections of maximal ideals, and the topology on j-spec is (induced from) the Zariski topology. As the points of j-spec correspond to the irreducible closed subsets of the maximal spectrum, it is clear that j-spec has the same dimension as the maximal spectrum. Fix R as in Theorem 2. It is well known that R satisfies SR_n so we need not prove that. As an illustration we will prove $SR_n^3(n+2, n+1)$. Then we will indicate how to get $SR_{n+2}^3(n+2, n+2)$, $SR_{n+1}^4(n+2, n+1)$ and, more generally, how to prove Theorem 3 below.

2.6 Let $m \geq 1$, $s \geq 1$, $y \in$ j-spec. Let $(a_1, \ldots, a_m)$, $(b_1, \ldots, b_m)$, $(c_1, \ldots, c_m)$ be sequences of elements of R^s. (So $a_i \in R^s$ etc.) The letters a,b,c represent A_1, A_2, A_3 respectively, where A_i is as in 2.2.

Definitions Let V(y) be the irreducible subset of j-spec corresponding to y. So V(y) = closure of $\{y\}$, and y is the generic point of $V(y)$. We put $d(y)$, the "depth" of y, equal to the dimension of $V(y)$. Let $k(y)$ be the quotient field of R/y. There is a natural map $R^s \to k(y)^s$ which we denote by $f \mapsto \overline{f}$. We say that the system $(a_1,\ldots,a_m)$, $(b_1,\ldots,b_m)$, $(c_1,\ldots,c_m)$ is <u>y-basic</u> if (A) or (B) holds, where

(A): The field $k(y)$ has two or three elements and there are $\mu_i \in k(y)$ such that the three vectors $\overline{a}_1 + \mu_2\overline{a}_2 + \cdots + \mu_m\overline{a}_m$, $\overline{b}_1 + \mu_2\overline{b}_2 + \cdots + \mu_m\overline{b}_m$, $\overline{c}_1 + \mu_2\overline{c}_2 + \cdots + \mu_m\overline{c}_m$ are non-zero.

(B): The field $k(y)$ contains at least four elements; the vectors $\overline{a}_1,\ldots,\overline{a}_m \in k(y)^s$ form a system of rank $\geq \min(m, 1 + d(y))$ and the same holds for $\overline{b}_1,\ldots,\overline{b}_m$ and for $\overline{c}_1,\ldots,\overline{c}_m$.

We say that $(a_1,\ldots,a_m),(b_1,\ldots,b_m),(c_1,\ldots,c_m)$ is <u>basic</u> if it is y-basic for all $y \in$ j-spec. We use this definition for any pair of integers m,s with $m \geq 1$, $s \geq 1$. We call m the <u>length</u>.

2.7 <u>Lemma</u> Let $(a_1,\ldots,a_m)$, $(b_1,\ldots,b_m)$, $(c_1,\ldots,c_m)$ be basic, $m > 1$. Then there are $t_1,\ldots,t_{m-1} \in R$ such that $(a_1 + t_1a_m,\ldots,a_{m-1} + t_{m-1}a_m)$, $(b_1 + t_1b_m,\ldots,b_{m-1} + t_{m-1}b_m)$, $(c_1 + t_1c_m,\ldots,c_{m-1} + t_{m-1}c_m)$ is also basic, with length $m - 1$.

<u>Proof</u> We will first show that, at all but finitely many primes, the new system is automatically y-basic, regardless of the choice of the t_i. So suppose $(a_1 + t_1a_m,\ldots,a_{m-1} + t_{m-1}a_m)$, $(b_1 + t_1b_m,\ldots)$, $(c_1 + t_1c_m,\ldots)$ is not y-basic. If $k(y)$ has less than four elements, then $V(y)$ is an irreducible component of the closed set $\{z \in \text{j-spec} \mid f^3 - f \in z \text{ for all } f \in R\}$, because this set only contains maximal ideals. So there are only finitely many y such that $k(y)$ has less than four elements. We may therefore assume that $k(y)$ has at least four elements. Without loss of generality (3 times finite is finite) we may assume that the rank of $\overline{a}_1 + \overline{t_1a_m},\ldots,\overline{a}_{m-1} + \overline{t_{m-1}a_m}$

is strictly smaller than $\min(m-1, 1+d(y))$. (Note that $m-1$ is the new length). As we also know that the rank of $\bar{a}_1,\ldots,\bar{a}_m$ is at least $\min(m, 1+d(y))$, the rank must have dropped when passing from $(a_1,\ldots,a_m)$ to $(a_1+t_1a_m,\ldots,a_{m-1}+t_{m-1}a_m)$. It cannot have dropped by more than one, so $m > 1+d(y)$ and the rank of $\bar{a}_1,\ldots,\bar{a}_m$ is $1+d(y)$. We therefore want to show that it occurs only for finitely many y's that at the same time $m > 1+d(y)$ and $1+d(y) = \operatorname{rank}(\bar{a}_1,\ldots,\bar{a}_m)$. As $0 \leq d(y) \leq d$ it is sufficient to show this for a fixed value of $d(y)$, say $d(y) = r-1$, $r\in\mathbb{N}$. We claim that y is a generic point of a component of the closed set $X = \{x \in j\text{-spec} \mid \text{the images of } a_1,\ldots,a_m \text{ in } k(x)^s \text{ form a system of rank} \leq r\}$. (From this claim it follows that there are only finitely many possibilities for y). So suppose y is not such a generic point. Then there is $x < y$ with $x\in X$. One gets $d(x) > d(y)$, so $\min(m, 1+d(x)) > r$. But $(a_1,\ldots,a_m)$, $(b_1,\ldots,b_m)$, $(c_1,\ldots,c_m)$ is x-basic, so this is impossible. (Note that $k(x)$ is infinite).

We have proved now that it only can go wrong at finitely many primes, say $y_1,\ldots,y_g$ We may assume that $y_i < y_j$ implies $j < i$. (otherwise renumber). Then there exist $\tau_i\in R$ with $\tau_i \notin y_i$ but $\tau_i \in y_j$ for $j < i$. (Well known). Writing $t_i = \sum_j \rho_{ij}\tau_j$ we discuss the primes $y_1,\ldots,y_g$ one by one, starting with y_1, and choosing ρ_{ij} to fit the needs of y_j. In other words, we suppose ρ_{ij} to be given for $j < q$ and we look for ρ_{iq} such that the result will be y_q-basic. (this doesn't depend on the ρ_{ij} with $j > q$). So fix $y = y_q$. If $k(y_q)$ has less than four elements, we may as well assume that $\bar{\tau}_q = 1$, because $\bar{\tau}_q^2 = 1$. But then it is obvious from the definition of y-basic that one can choose the ρ_{ij} appropriately. If $k(y_q)$ has at least four elements then we have $m > 1+d(y_q)$ and we have $\operatorname{rank}(\bar{a}_1,\ldots,\bar{a}_m) = 1+d(y)$ or $\operatorname{rank}(\bar{b}_1,\ldots,\bar{b}_m) = 1+d(y)$ or $\operatorname{rank}(\bar{c}_1,\ldots,\bar{c}_m) = 1+d(y)$. The worst case is that all three of the ranks equal $1+d(y)$. (If $\operatorname{rank}(\bar{b}_1,\ldots,\bar{b}_m) \neq 1+d(y)$ then

$\operatorname{rank}(\bar{b}_1,\ldots,\bar{b}_m) \geq 2 + d(y)$ and we don't have to look at the $\bar{b}_i$). We have to make sure that neither of the three ranks drops below $1 + d(y)$, when passing to the new system. This is achieved as follows. First one checks that we can choose the ρ_{iq}, with induction on i, such that, for any choice of those ρ's that are still to be considered, $\operatorname{rank}(\bar{a}_1 + \bar{t}_1\bar{a}_m,\ldots,\bar{a}_i + \bar{t}_i\bar{a}_m) \geq 2 + d(y) + i - m$. (This is an exercise in rank counting). One observes that at each step at most one value of $\bar{\rho}_{iq}$ fails to give the inequality. Now $k(y_q)$ has at least four elements, so R/y_q has at least four elements. Therefore we can avoid the failing values of $\bar{\rho}_{iq}$ for the a's, the b's, the c's simultaneously.

2.8 Corollary Let $(a_1,\ldots,a_m)$, $(b_1,\ldots,b_m)$, $(c_1,\ldots,c_m)$ be basic, $m > 1$. Then there are $t_2,\ldots,t_m$ such that $a_1 + t_2a_2 + \cdots + t_ma_m$, $b_1 + t_2b_2 + \cdots + t_mb_m$, $c_1 + t_2c_2 + \cdots + t_mb_m$ is basic with length 1.

Proof Apply Lemma 2.7 repeatedly.

2.9 We want to apply the Corollary to the columns of the matrices A_1,A_2,A_3 occurring in the conditions $SR^3_n(n + 2,n + 1)$ and $SR^3_{n+2}(n + 2,n + 2)$. Let us do $SR^3_n(n + 2,n + 1)$ first. So we have matrices A_1,A_2,A_3 of size $(n - 1) \times (n + 2)$ and the last $n + 1$ columns of A_1 form a system of rank $n - 1$ for all $y \in$ j-spec. Note that $n - 1 \geq 1 + d \geq 1 + d(y)$ for all $y \in$ j-spec. Let $a_1,\ldots,a_{n+2}$ be the columns of A_1, let $b_1,\ldots,b_{n+2}$ be the columns of A_2 and let $c_1,\ldots,c_{n+2}$ be the columns of A_3. We want to show that this is a basic system. If $k(y)$ has at least four elements then the system is y-basic. So consider y which has a smaller $k(y)$. The question is whether there are $\mu_i \in k(y)$ such that the vectors

$\bar{a}_1 + \mu_2\bar{a}_2 + \cdots + \mu_{n+2}\bar{a}_{n+2}$, $\bar{b}_1 + \cdots + \mu_{n+2}\bar{b}_{n+2}$, $\bar{c}_1 + \cdots + \mu_{n+2}\bar{c}_{n+2}$

are non-zero. What choices of the μ_i are wrong for the first vector? They form a plane in $k(y)^{n+1}$, because $\text{rank}(\bar{a}_2,\ldots,\bar{a}_{n+2}) = n - 1$. So to see whether we can get all three vectors non-zero, we just look whether $k(y)^{n+1}$ can be filled by three planes. It can't if $n \geq 3$, even if the field has only two elements. So that is the reason we have $n \geq 3$ in Theorem 2. We now apply the Corollary. It gives us $\lambda = (t_2,\ldots,t_m)$ such that $a_1 + t_2a_2 + \cdots + t_{n+2}a_{n+2}$, $b_1 + t_2b_2 + \cdots + t_{n+2}b_{n+2}$, $c_1 + \cdots + t_{n+2}c_{n+2}$ is a basic system with length 1. This means that we get three vectors which have non-zero images in $k(y)^{n-1}$ for all $y \in$ j-spec. In other words, we get three unimodular vectors. So that proves $SR_n^3(n + 2,n + 1)$. The proof of $SR_{n+2}^3(n + 2,n + 2)$ is similar: This time it boils down to checking the property $SR_{n+2}^3(n + 2,n + 2)$ for small fields $k(y)$. The wrong points in $k(y)^{n+1}$ fill at most three lines that don't pass through the origin or two lines and the origin. So not all points of $k(y)^{n+1}$ are wrong, even if $n = 2$. That proves $SR_{n+2}^3(n + 2,n + 2)$ for R. And, as we didn't need the restriction $n \geq 3$ here, we see that $SR_4^3(4,4)$ holds for a commutative semi-local ring. This is easy to prove anyway, but let us record it:

2.10 <u>Proposition</u> A commutative semi-local ring satisfies $SR_4^3(4,4)$.

2.11 What the method of proof actually shows is the following:

<u>THEOREM 3</u> Let R be a commutative ring with noetherian maximal spectrum of dimension $d < \infty$. Let $c \geq u \geq n - 1 \geq d + 1$ and $p \geq 1$. If $SR_n^p(c,u)$ holds for all residue fields of R then it holds for R.

<u>Proof</u> If $k(y)$ has at most p elements then $\bar{f}^{(p-1)!}$ is zero or one for $\bar{f} \in k(y)$. So y will be a generic point of a component of $\{ z \in \text{j-spec} \mid f(f^{(p-1)!} - 1) \in z \text{ for all } f \in R\}$. There are only finitely

many such primes y. One treats them as the y's whose residue fields have less than four elements in 2.6 through 2.9. The remaining y's are treated as the y's whose residue fields have at least four elements in 2.6 through 2.9. And instead of using three sequences $(a_1,\ldots,a_m)$, $(b_1,\ldots,b_m)$, $(c_1,\ldots,c_m)$ one now uses p sequences of length m.

2.12 Remark One can refine the result, cf. Bass, as follows: Say one has finitely many subspaces of the maximal spectrum (not j-spec) with the full maximal spectrum as the union. Then d can be replaced by the maximum dimension of these subspaces. This is not always the same as the original d. (See page 173, §2, Ch. IV in [4]). One adapts the proof by defining for each of the subspaces the analogues of j-spec and the depth function $d(y)$.

2.13 For proving Theorem 2 we still have to show that $SR^4_{n+1}(n + 2, n + 1)$ holds for fields, when $n \geq 3$. This follows from the fact that one cannot fill $k(y)^{n+1}$ with four lines.

2.14 Remark. Note that $\widetilde{\widetilde{SR}}_2$ holds for a semi-local commutative ring which doesn't have any residue field with 2 or 3 elements.

§3. The Chunk

3.1 In Sections 3 and 4 we will prove

THEOREM 4 Let R satisfy $\tilde{\tilde{SR}}_n$, $n \geq 2$. Then the natural map $K_2(n+1,R) \to K_2(n+2,R)$ is an isomorphism.

Comments We don't require R to be commutative. As surjective stability is known even under SR_n (or SR_{n+1}) we only have to prove that the map is injective. In most of the proof we only use SR_n, $SR^3_{n+2}(n+2,\ n+2)$. So most of the proof also works for commutative semi-local rings. In Section 5 we take a closer look at the case of commutative semi-local rings. There we will repair the proofs which involve $SR^4_{n+1}(n+2,\ n+1)$ or $SR^3_n(n+2,\ n+1)$, using properties of commutative semi-local rings instead. We only need to repair proofs for n = 2 because this is the case of Theorem 1 which is not covered by Theorems 2 and 4. It turns out that our proofs in Section 5 are at least as complicated as the proofs they are replacing. So in that sense the higher dimensional case is easier! (Of course multiple stable range conditions, if true, are much easier to prove in the semi-local case).

3.2 In the proof of Theorem 4 we never use SR_n directly, but only some of its known consequences. If we take that into account we get the following version of Theorem 4:

THEOREM 4' Let R be a ring, n an integer, $n \geq 2$. Assume that (i),(ii),(iii) are satisfied, where

(i) $E(n,R)$ acts transitively on the set of unimodular columns of length n and $E(n+1,R)$ acts transitively on the set of unimodular columns of length n + 1.

(ii) The natural map $K_1(n,R) \to K_1(n+2,R)$ is injective.

(iii) $SR_n^3(n+2,n+1)$, $SR_{n+1}^4(n+2,n+1)$, $SR_{n+2}^3(n+2,n+2)$ hold for R.

Then the natural map $K_2(n+1,R) \to K_2(n+2,R)$ is injective.

Comment It is not clear whether Theorem 4' is actually sharper than Theorem 4. We will not mention Theorem 4' after this, but just prove Theorem 4. Note that (i),(ii) imply that the natural map $K_1(n,R) \to K_1(n+1,R)$ is an isomorphism, so that the map $K_1(n+1,R) \to K_1(n+2,R)$ is also injective. (See [4],Ch.V,(3.3) (iii)).

3.3 So let us assume that R satisfies $\tilde{\tilde{SR}}_n$, $n \geq 2$. (We will indicate which arguments use more than SR_n, $SR_{n+2}^3(n+2, n+2)$).

3.4 Notations Let I and J be sets. Then $St(I \times J,R)$, or just $St(I \times J)$, is the group with generators $x_{ij}(r)$, where $i \in I$, $j \in J$, $i \neq j$, $r \in R$, and defining relations

(1) $x_{ij}(r)x_{ij}(s) = x_{ij}(r+s)$. (Here one assumes, of course, that $i \in I$, $j \in J$, $i \neq j$, $r \in R$, $s \in R$).

(2) $[x_{ij}(r),x_{jk}(s)] = x_{ik}(rs)$, if this makes sense, where $[p,q]$ stands for $pq\,p^{-1}q^{-1}$. (We need $j \in I \cap J$ and $i \neq k$ among other things).

(3) $[x_{ij}(r),x_{k\ell}(s)] = 1$ if i,j,k,ℓ are distinct and the expression makes sense.

(4) $[x_{ij}(r),x_{ik}(s)] = 1$ if this makes sense.

(5) $[x_{ij}(r),x_{kj}(s)] = 1$ if this makes sense.

In the case that $I = J = \{1,\ldots,m\}$ we just write $St(m,R)$ for

$St(I \times J,R)$, as usual. We also write $St(m)$ for it. If $I \subset I'$, $J \subset J'$, there is an obvious map from $St(I \times J)$ into $St(I' \times J')$. We will abuse notations and denote both the generators of $St(I \times J)$ and the generators of $St(I' \times J')$ by $x_{ij}(r)$. This is a major abuse because the natural map need not be injective. In fact, that is what this paper is about. Instead of using different notations for an element x of $St(I \times J)$ and its image in $St(I' \times J')$, we will indicate in what group the notation is to be interpreted. So if $x,y \in St(I \times J)$, the statement "$x = y$ in $St(I' \times J')$" will mean that the images of x and y in $St(I' \times J')$ are equal. We use this convention in order to avoid complicated notations. Let us give one more example to show how the convention works: Consider $x = x_{12}(r)$ in $St(2)$, $y = x_{23}(s)$ in $St(\{2,3\} \times \{2,3\})$. Then $[x,y] = x_{13}(rs)$ in $St(3)$. Here x stands for the image of $x_{12}(r)$ in $St(3)$, y stands for the image of $x_{23}(s)$, under a different map!, and $x_{13}(rs)$ is just a generator of $St(3)$.

It will be convenient to have notations for certain subsets of $\{1,2,\ldots,n+2\}$. (The convention which we just introduced forces us to mention groups of type $St(I \times J)$ all the time). We use $[k]$ for the set $\{1,\ldots,k\}$ and stars for complements: $\{1\}^* = \{2,3,\ldots,n+2\}$, $\{n+2\}^* = [n+1]$ etcetera. Notice that the groups $St(\{1\} \times [n+2])$ and $St(\{1\} \times \{1\}^*)$ are identical. We will use both notations.

If $I,J \subseteq [n+2]$ then there is a natural map <u>mat</u> from $St(I \times J)$ into the elementary group $E(n+2,R) = E_{n+2}(R)$. (cf.[20]). We call its image $E(I \times J)$. The image of $St(m)$ is called $E(m)$, for $m \leq n+2$. (We never go beyond $n+2$). We will say that $[x,y] = x_{13}(rs)$ in $E(3)$, where $x,y,x_{13}(rs)$ are as in the example above. So we could as well write $\underline{mat}[x,y] = \underline{mat}(x_{13}(rs))$ or $[\underline{mat}(x),\underline{mat}(y)] = \underline{mat}(x_{13}(rs))$. (The map <u>mat</u> is a homomorphism and $E(3)$ is an honest subset of $E(n+2)$).

3.5 Consider $St(I \times J)$ when $I \cap J = \emptyset$. One easily sees that $(a_{ij})_{i\in I, j\in J} \mapsto \prod_{i\in I, j\in J} x_{ij}(a_{ij})$ provides an isomorphism from $R^{I\times J}$ onto $St(I \times J)$. The homomorphism mat: $St(I \times J) \to E(I \times J)$ is an isomorphism in this case, because one can still read the a_{ij} off from the image in $E(I \times J)$. More generally, say K, L, M are disjoint subsets of $[n + 2]$ and $I = K \cup L$, $J = L \cup M$. Then the map $St(K \times J) \to St(I \times J)$ is injective, because mat: $St(K \times J) \to E(n+2)$ is injective. So we may denote the image of $St(K \times J)$ in $St(I \times J)$ by $St(K \times J)$ again. It is a normal subgroup. Similarly $St(I \times M)$ can be identified with a normal subgroup of $St(I \times J)$. The action by conjugation of $St(I \times J)$ on $St(K \times J)$ can be studied inside $E(n+2)$, using the isomorphism mat: $St(K \times J) \to E(K \times J)$. (Same for action on $St(I \times M)$). Sending $x_{ij}(r)$ to $x_{ij}(r)$ for $r\in R$, $i\in L$, $j\in J$ and $St(K \times J)$ to 1 gives a homomorphism $\pi_{L\times J}$: $St(I \times J) \to St(L \times J)$ with the natural map $St(L \times J) \to St(I \times J)$ as a cross section. One sees that $St(I \times J)$ is the semi-direct product of $St(L \times J)$ and $E(K \times J)$, with the action coming from conjugation in $E(n+2)$. Recall that a semi-direct product $H \rtimes G$ is given by three data: A group G, a group H and an action of G on H. Say ${}^g h$ denotes the value resulting from the action of $g\in G$ on $h\in H$. Then $H \rtimes G$ consists of pairs (h,g), $h\in H$, $g\in G$, with multiplication $(h,g)(h_1,g_1) = (h\, {}^g h_1, g\, g_1)$. We can summarize the discussion as follows:

$$St(I \times J) = St(K \times J) \rtimes St(L \times J) \cong E(K \times J) \rtimes St(L \times J)$$

$$St(I \times J) = St(I \times M) \rtimes St(I \times L) \cong E(I \times M) \rtimes St(I \times L).$$

3.6 Definitions <u>Low</u> = $St([n+2] \times \{n+2\}^*)$, <u>Up</u> = $St([n+2] \times \{1\}^*)$ and the mediator is <u>Med</u> = $St([n+2] \times \{1,n+2\}^*)$. The <u>chunk</u> C consists of the orbits of <u>Med</u> in the set <u>Low</u> $\times$ <u>Up</u> under the action <u>shift</u> which is defined as follows: <u>shift</u> $(g)(X,Y) = (Xg^{-1}, gY)$ for $g\in$ <u>Med</u>, $X\in$ <u>Low</u>, $Y\in$ <u>Up</u>, where we abuse notation, as promised. (From the context it follows that (Xg^{-1}, gY) must be an element of <u>Low</u> x <u>Up</u>, so g^{-1} must stand for the inverse of the image of g in <u>Low</u> and the other g must stand for the image in <u>Up</u>). We denote the orbit of (X,Y) by $\langle X,Y\rangle$. One can also say that $\langle X,Y\rangle$ is the equivalence class for the relation: $(X,Y) \sim (X',Y')$ if there is $g \in$ <u>Med</u> such that $X' = Xg^{-1}$ in <u>Low</u> and $Y' = gY$ in <u>Up</u>.

<u>Digression</u> (This piece will not be used).

The proofs have been written without pictures, but of course they were not found that way. In order to understand what is going on, one may want to picture the elements of the $St(I \times J)$ like matrices: Say $n = 3$, so $n + 2 = 5$. Then one would picture an arbitrary element x of $St([5] \times \{5\}^*)$ as

$$\begin{pmatrix} * & * & * & * & 0 \\ * & * & * & * & 0 \\ * & * & * & * & 0 \\ * & * & * & * & 0 \\ \hline * & * & * & * & 1 \end{pmatrix} \text{ or } \left(\begin{array}{c|c} A & \begin{matrix}0\\0\\0\\0\end{matrix} \\ \hline b & 1 \end{array}\right)$$

with $A \in St(4)$, $b \in R^4 \simeq St(\{5\} \times [5])$.

Here we use the semi-direct product: $St([5] \times \{5\}^*) \simeq St(\{5\} \times [5]) \rtimes St(4)$. We can also write

$$x = \left(\begin{array}{c|c} A & \begin{matrix}0\\0\\0\\0\end{matrix} \\ \hline 0\;0\;0\;0 & 1 \end{array}\right) \begin{pmatrix} 1 & & & & 0 \\ & 1 & & & \\ & & 1 & & \\ 0 & & & 1 & \\ & & b & & 1 \end{pmatrix}$$

In

$$\left(\begin{array}{c|c} A & \begin{matrix}0\\0\\0\\0\end{matrix} \\ \hline 0\;0\;0\;0 & 1 \end{array}\right)$$

we do not mean to say that A is a matrix. We mean

to say that A lives on the indices on which it is pictured. (So this is more than saying that the entries of mat(A) fit the picture). Some of the rules for matrix multiplication are still valid. For instance,

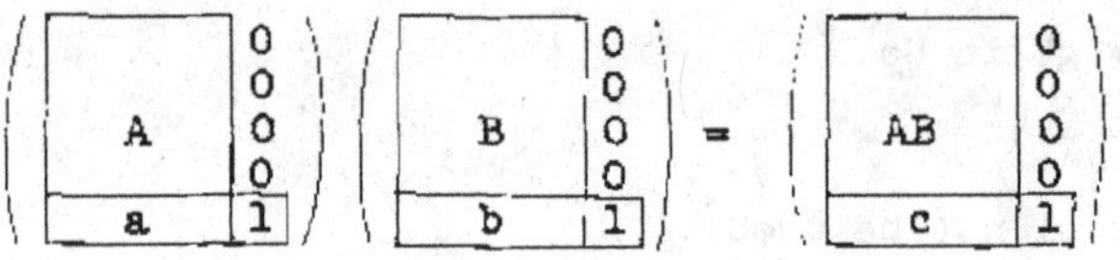

$$\left(\begin{array}{c|c} A & \begin{matrix}0\\0\\0\\0\end{matrix} \\ \hline a & 1 \end{array}\right) \left(\begin{array}{c|c} B & \begin{matrix}0\\0\\0\\0\end{matrix} \\ \hline b & 1 \end{array}\right) = \left(\begin{array}{c|c} AB & \begin{matrix}0\\0\\0\\0\end{matrix} \\ \hline c & 1 \end{array}\right) \quad \text{with } c = b + a \; \underline{\text{mat}}(B).$$

(Notice that one multiplies block-wise).

The following division into blocks will play an important role

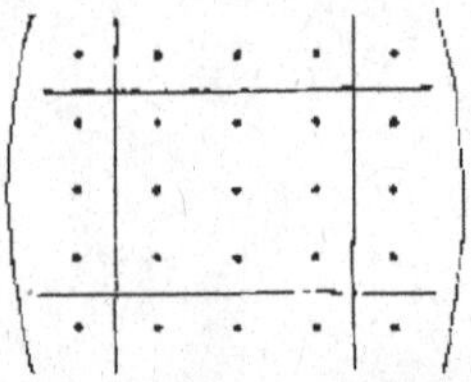

$$\left(\begin{array}{ccc|c} \cdot & \cdot & \cdot & \cdot \\ \cdot & \cdot & \cdot & \cdot \\ \cdot & \cdot & \cdot & \cdot \\ \hline \cdot & \cdot & \cdot & \cdot \end{array}\right) .$$ The big block in the middle corresponds to St([1,5]* × [1,5]*) which is isomorphic to St(3).

We have corresponding to St(4), and

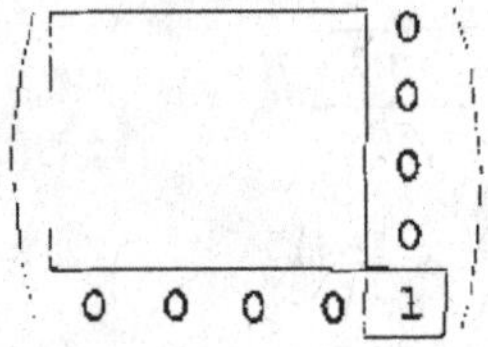

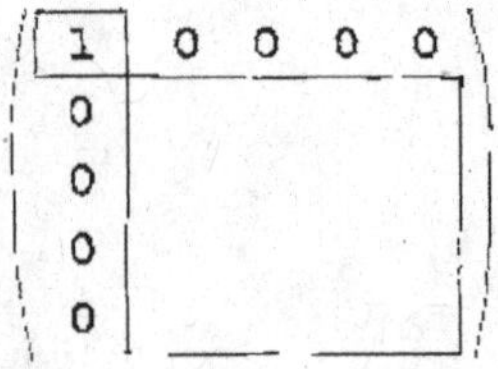

corresponding to St([1]* × [1]*), which is iso-morphic to St(4).

We will use elements of type $\left(\begin{array}{c|c} & \begin{matrix}0\\0\\0\\0\end{matrix} \\ \hline & 1 \end{array}\right)$. They form Low.

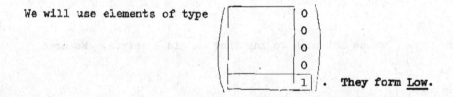

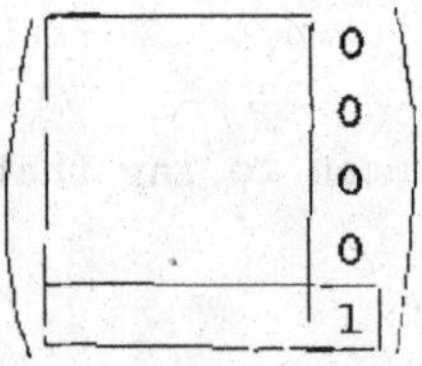

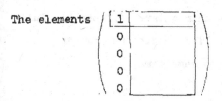

The elements $\begin{pmatrix} 1 & & & & \\ 0 & & & & \\ 0 & & & & \\ 0 & & & & \\ 0 & & & & \end{pmatrix}$ form <u>Up</u>, and the

$$\begin{pmatrix} 1 & & & & 0 \\ 0 & & & & 0 \\ 0 & & & & 0 \\ 0 & & & & 0 \\ 0 & & & & 1 \end{pmatrix} = \begin{pmatrix} 1 & 0 & 0 & 0 & 0 \\ 0 & & & & 0 \\ 0 & & & & 0 \\ 0 & & & & 0 \\ 0 & 0 & 0 & 0 & 1 \end{pmatrix} \begin{pmatrix} 1 & & & & 0 \\ 0 & 1 & 0 & 0 & 0 \\ 0 & 0 & 1 & 0 & 0 \\ 0 & 0 & 0 & 1 & 0 \\ 0 & & & & 1 \end{pmatrix} \text{ form } \underline{\text{Med}}.$$

"Complementary" to <u>Low</u> one has the

$$x_5(v) = \begin{pmatrix} 1 & & 0 & & v_1 \\ & 1 & & & v_2 \\ & & 1 & & v_3 \\ & & & 1 & v_4 \\ & & 0 & & 1 \end{pmatrix}$$

and complementary to <u>Up</u> one has the $x_1(w) = \begin{pmatrix} 1 & & & & \\ & 1 & & 0 & \\ w & & 1 & & \\ & & & 1 & \\ & 0 & & & 1 \end{pmatrix}$,

$w = (w_2, w_3, w_4, w_5)$.

(We also write columns in the form (*,...,*), which is of course more suitable for rows).

The $x_5(v)$, $x_1(w)$ will be very important. We will apply the multiple stable range conditions to v, w. For instance, say one has some freedom of choice for v in

$$\begin{pmatrix} 1 & & & & \\ & 1 & 0 & & \\ & & 1 & & v \\ & & & 1 & \\ & 0 & & & 1 \end{pmatrix} \begin{pmatrix} & & & & 0 \\ & & & & 0 \\ & & T & & 0 \\ & & & & 0 \\ 0 & 0 & 0 & 0 & 1 \end{pmatrix} = \begin{pmatrix} & & & & 0 \\ & & & & 0 \\ & & T & & 0 \\ & & & & 0 \\ 0 & 0 & 0 & 0 & 1 \end{pmatrix} \begin{pmatrix} 1 & & & & \\ & 1 & 0 & & \\ & & 1 & & z \\ & & & 1 & \\ & 0 & & & 1 \end{pmatrix}$$

Then one may arrange that (z_2, z_3, z_4) is unimodular. (If one has full freedom of choice then this is obvious. The conditions come in if one wants to achieve more with v at the same time).

The basic pattern thus is

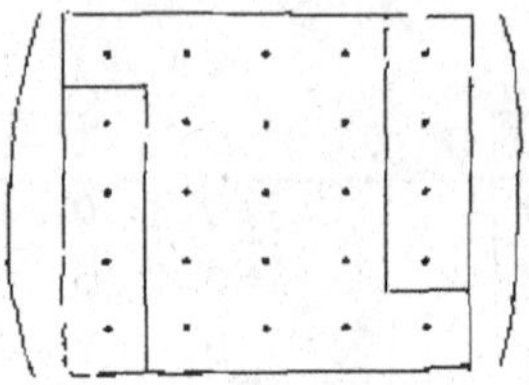

. One sees in it the outlines of Up, Low, Med, and therefore also the places where the co-ordinates of the "complementary" groups are situated, where $\{x_5(v) \mid v \in R^4\}$ is the group complementary to Low, for instance. End of digression.

3.7 Lemma Med → Low and Med → Up have the same kernel N.

Comment So we may identify Med/N with a subgroup of Low and also with a subgroup of Up. Then we can say that $(X,Y) \sim (X',Y')$ is equivalent to: There is $g \in$ Med/N with $X' = Xg^{-1}$, $Y' = gY$. Now there is less abuse of notation. In particular, $(X,Y) \sim (X',Y)$ implies $X = X'$.

Proof of Lemma Let g be in one of the two kernels. Then mat$(g) = 1$ in $E([n + 2] \times \{1,n + 2\}^*)$. The group Med is a semi-direct product of $St(\{1,n + 2\} \times \{1,n + 2\}^*)$ and $St(\{1,n + 2\}^* \times \{1,n + 2\}^*)$, and a similar decomposition holds for $E([n + 2] \times \{1,n + 2\}^*)$. The restriction of mat to the normal subgroup is an isomorphism, so g must be in the subgroup $St(\{1,n + 2\}^* \times \{1,n + 2\}^*)$. (We will use this argument often). Say the image of g in Low is trivial. Then its image in $St(n+1)$ must be trivial, again because of the semi-direct product structure, this time of Low. We claim that its image in St({1}* x {1}*) must also be trivial. One passes from St({1,n + 2}* x {1,n + 2}*) to St(n + 1) by adding an index 1, and one passes from St({1,n + 2}* x {1,n + 2}*) to St({1}* x {1}*) by adding an index n + 2. But it can make no difference for g whether the new index is called 1 or n + 2, whence the claim. As Up contains St({1}* x {1}*), the image of g in Up is trivial. The other part of the proof is similar.

3.8 Notations We have a natural map $\pi: C \to St(n + 2)$ given by $\pi\langle X,Y\rangle = XY$. It is clear that π is well-defined. We denote the composition of π and mat simply by mat again. So now we also have mat: $C \to E(n + 2)$ with mat $\langle X,Y\rangle$ = mat(X)mat(Y).

3.9 Definitions For $g \in$ Low we define L(g): Low x Up → Low x Up by L(g)(X,Y) = (gX,Y). And we define $\mathcal{L}(g): C \to C$ by $\mathcal{L}(g)\langle X,Y\rangle = \langle gX,Y\rangle$. So $\mathcal{L}(g)\langle X,Y\rangle$ is the class of L(g)(X,Y). It is easy to see that $\mathcal{L}(g)$ is well-defined, that it is a permutation of C, that $\mathcal{L}(gh) = \mathcal{L}(g)\mathcal{L}(h)$ for $g,h \in$ Low. Similarly, for $f \in$ Up, we put R(f)(X,Y) = (X,Yf) and $\mathcal{R}(f)\langle X,Y\rangle = \langle X,Yf\rangle$. Taking $\mathcal{L}$ and $\mathcal{R}$ together one gets, for $g \in$ Med the permutation $\mathfrak{Int}(g)$ of C given by $\mathfrak{Int}(g)\langle X,Y\rangle = \langle gX,Yg^{-1}\rangle$.

3.10 Proposition (The squeezing principle).
Let $1 < i < n + 2$, $X \in St(\{i\}^* \times \{n + 2\}^*)$, $Y \in St(\{i\}^* \times \{1\}^*)$. Suppose that XY = 1 in $St(\{i\}^* \times [n + 2])$. Then $\langle X,Y\rangle = \langle 1,1\rangle$ in C.

Comment We call it the squeezing principle because it shows how one can prove an equality in the chunk by squeezing the problem into some $St(\{i\}^* \times [n + 2])$.

Proof of Proposition Write X as X_1X_2 with $X_1 \in St(\{i\}^* \times \{i,n+2\}^*)$, $X_2 \in St(\{i\}^* \times \{i\})$. (This is a new form of the abuse introduced in 3.4. We really mean $X_1 \in St(\{i\}^* \times \{i,n+2\}^*)$, and we refer to an element of $St(\{i\}^* \times \{n+2\}^*)$ when writing X_1X_2). Write Y as Y_1Y_2 with $Y_1 \in St(\{i\}^* \times \{i\})$, $Y_2 \in St(\{i\}^* \times \{1,i\}^*)$. It follows from the decomposition of $E(\{i\}^* \times [n+2])$ as a semi-direct product that $X_2Y_1 = 1$ in $E(\{i\}^* \times \{i\})$, hence in $St(\{i\}^* \times \{i\})$, which is isomorphic to it. So $\langle X,Y\rangle = \langle X_1,Y_2\rangle$ and we may as well assume $X = X_1$, $Y = Y_2$. As $\underline{mat}(X) = \underline{mat}(Y^{-1})$, the matrix $\underline{mat}(X)$ has trivial columns at positions $1,i,n+2$. (A column or row is called trivial if it is the same as in the identity matrix). It easily follows that there is $m \in St(\{1,n+2\} \times \{1,i,n+2\}^*)$ such that $\underline{mat}(X)\underline{mat}(m)$ also has trivial rows at positions 1 and $n+2$. (and at position i of course). Replacing (X,Y) by $(Xm,m^{-1}Y)$ we may now assume that $\underline{mat}(X)$ has trivial rows and columns at positions $1,i,n+2$. The same will hold for its inverse $\underline{mat}(Y)$. So now we have $X \in St(\{i\}^* \times \{i,n+2\}^*)$, $Y \in St(\{i\}^* \times \{1,i\}^*)$ with $XY = 1$ in $St(\{i\}^* \times [n+2])$ and the matrices have these trivial rows and columns. Because $St(\{i\}^* \times [n+2])$ is a semi-direct product it is easy to see that actually $XY = 1$ in $St(\{i\}^* \times \{i\}^*)$. Write X as X_3X_4 with $X_4 \in St(\{n+2\} \times \{i,n+2\}^*)$, $X_3 \in St(\{i,n+2\}^* \times \{i,n+2\}^*)$. As $\underline{mat}(X)$ has a trivial row at position $n+2$ the factor X_4 has to be 1. So now we have $X = X_3$ and we may say $X \in St(\{i,n+2\}^* \times \{i,n+2\}^*)$. Similarly we get $Y \in St(\{1,i\}^* \times \{1,i\}^*)$.

Consider $[X,x_{i,1}(t)]$ in $St(\{n+2\}^* \times \{i,n+2\}^*)$. It lies in the normal subgroup $St(\{i\} \times \{i,n+2\}^*)$, which is mapped isomorphically into $E(n+2)$. But in $E(n+2)$ we know that $\underline{mat}(X)$ has trivial rows and columns at positions 1 and i, so the commutator is trivial. Similarly $[X,x_{1,i}(t)] = 1$ in $St(\{i,n+2\}^* \times \{n+2\}^*)$. So X commutes with $w_{i,1}(1) = x_{i,1}(1)x_{1,i}(-1)x_{i,1}(1)$ in

$St(\{n+2\}^* \times \{n+2\}^*)$. On the other hand
$w_{i,1}(1)x_{1,k}(a)w_{i,1}(1)^{-1} = x_{i,k}(a)$,
$w_{i,1}(1)x_{k,1}(a)w_{i,1}(1)^{-1} = x_{k,i}(a)$, $w_{i,1}(1)x_{k,\ell}(a)w_{i,1}(1)^{-1} = x_{k,\ell}(a)$
if $a \in R$, $k \neq 1$, $k \neq \ell$, $\ell \neq 1$, $k \neq i$, $\ell \neq i$, $k,\ell \in [n+1]$. It follows that conjugation by $w_{i,1}(1)$ corresponds to the automorphism <u>switch</u>$(1,i)$ of $St(n+1)$ which is induced by the permutation of $[n+1]$ which switches 1 and i, and leaves the other indices fixed. We see that $X = w_{i,1}(1)Xw_{i,1}(1)^{-1} =$ <u>switch</u>$(1,i)X$ in $St(n+1)$. Let X' be the counterpart of X in $St(\{1,n+2\}^* \times \{1,n+2\}^*)$ which one gets from $X \in St(\{i,n+2\}^* \times \{i,n+2\}^*)$ by replacing the indices 1 by indices i. We have $X = X'$ in <u>Low</u>, so $\langle X,Y\rangle = \langle X',Y\rangle = \langle 1,X'Y\rangle$. To prove the Proposition it suffices to prove that $X'Y = 1$ in <u>Up</u>. (This is also necessary, by 3.7). The relation $X' =$ <u>switch</u>$(1,i)X'$ in $St(n+1)$ has a counterpart $X' =$ <u>switch</u>$(n+2,i)X'$ in $St(\{1\}^* \times \{1\}^*)$, because for X' there is no difference between 1 and $n+2$. (Here <u>switch</u>$(n+2,i)$ is an automorphism of $St(\{1\}^* \times \{1\}^*)$). Replacing the indices i by indices $n+2$ one gets from X' to an element X'' in $St(\{1,i\}^* \times \{1,i\}^*)$ with $X' = X''$ in $St(\{1\}^* \times \{1\}^*)$. So we have to show that $X''Y = 1$ in $St(\{1\}^* \times \{1\}^*)$. This amounts to the same as proving $X''Y = 1$ in $St(\{i\}^* \times \{i\}^*)$, because for X'' and Y there is no difference between 1 and i. But in $St(\{i\}^* \times \{i\}^*)$ we know that $XY = 1$. And we also know that $X =$ <u>switch</u>$(1,i)X$ in $St(n+1)$. Again, as for X there is no difference between i and $n+2$ we also have $X =$ <u>switch</u>$(1,n+2)X$ in $St(\{i\}^* \times \{i\}^*)$, i.e. $X = X''$ in $St(\{i\}^* \times \{i\}^*)$. So $X''Y = 1$ in $St(\{i\}^* \times \{i\}^*)$ indeed.

3.11 <u>Corollary</u> (Squeezing principle reformulated).
Let $X,X' \in St(\{i\}^* \times [n+1])$, $Y,Y' \in St(\{i\}^* \times \{1\}^*)$ for some $1 < i < n+2$. Suppose that $XY = X'Y'$ in $St(\{i\}^* \times [n+2])$. Then $\langle X,Y\rangle = \langle X',Y'\rangle$ in C.

<u>Proof</u> One has $\langle X^{-1}X', Y'Y^{-1}\rangle = \langle 1,1\rangle$ by the Proposition. Now apply $\mathcal{L}(X)$ and $\mathcal{R}(Y)$.

3.12 <u>Notation</u> Let $v \in R^{n+1}$, say $v = (v_1,\ldots,v_{n+1})$. (One should really write v as a column). Then we write $x_{n+2}(v)$ for the product of the $x_{i,n+2}(v_i)$. Similarly, if $w = (w_2,\ldots,w_{n+2})$ then $x_1(w)$ stands for the product $x_{21}(w_2) \cdots x_{n+2,1}(w_{n+2})$. We also write $x_{n+2}(v_1,\ldots,v_{n+1})$ for $x_{n+2}(v)$ and we write $x_1(w_2,\ldots,w_{n+2})$ for $x_1(w)$.

3.13 We want to define maps $\mathcal{L}(x_{n+2}(v))$ for $v \in R^{n+1}$. The general case is too difficult to do right now. But let us look at the case $v_1 = 0$, $X = x_{n+2,1}(q)$, $Y \in \underline{Up}$. In $St(n+2)$ one has $[x_{n+2}(v), x_{n+2,1}(q)] = x_1(vq)$. So it is reasonable to put $L(x_{n+2}(v))(x_{n+2,1}(q),Y) = (x_1(v_2q,\ldots,v_{n+1}q,q), x_{n+2}(v)Y)$, and $\mathcal{L}(x_{n+2}(v))\langle x_{n+2,1}(q),Y\rangle$ = class of this element $L(x_{n+2}(v))(x_{n+2,1}(q),Y)$.

We have to show that the resulting class only depends on the class of $(x_{n+2,1}(q),Y)$. So suppose $\langle x_{n+2,1}(q),Y\rangle = \langle x_{n+2,1}(r),Y'\rangle$. Then $x_{n+2,1}(q-r) \in \underline{mat}(\underline{Med})$, so $q = r$. But then also $Y = Y'$ by 3.7. So there is no other element of the same class which assumes this simple form.

We have now defined $\mathcal{L}(x_{n+2}(v))\langle X,Y\rangle$ in the case that $v_1 = 0$ and that $\langle X,Y\rangle$ contains a representative of a specific form.

3.14 Now let $T \in St(n+1)$, $Y \in \underline{Up}$, $q \in R$, $v \in R^{n+1}$. Suppose that $x_{n+2}(v)T = Tx_{n+2}(0,w_2,\ldots,w_{n+1})$ in $St(n+2)$. We put $w = (0,w_2,\ldots,w_{n+1})$. So $w_1 = 0$ and $w = \underline{mat}(T^{-1})v$.
One is tempted to define $\mathcal{L}(x_{n+2}(v))\langle Tx_{n+2,1}(q),Y\rangle$ as being $\mathcal{L}(T)\mathcal{L}(x_{n+2}(w))\langle x_{n+2,1}(q),Y\rangle$, where the latter is defined by 3.9 and 3.13. (Its image in $St(n+2)$ is like we want it). However, it is not easy to check that this is a consistent definition: What

happens if $\langle Tx_{n+2,1}(q),Y\rangle = \langle T'x_{n+2,1}(q'),Y'\rangle$ with $x_{n+2}(v)T' = T'x_{n+2}(0,*,\ldots,*)$? (Stars stand for things which don't need names. Two stars need not stand for the same thing).

3.15 __Notation__ In the situation of 3.14 we put $L(x_{n+2}(v))(Tx_{n+2,1}(q),Y) = L(T)(x_1(w_2q,\ldots,w_{n+1}q,q),x_{n+2}(w)Y)$. So we do with the representatives what we wanted to do with the classes. And we extended the definition in 3.13.

3.16 __Lemma__ Let $A\in St(\{1,n+2\}^* \times \{1,n+2\}^*)$, $T\in St(n+1)$, $Y\in \underline{Up}$, $q\in R$, $v\in R^{n+1}$, such that both $L(x_{n+2}(v))(Tx_{n+2,1}(q),Y)$ and $L(x_{n+2}(v))(Tx_{n+2,1}(q)A,A^{-1}Y)$ are defined (as in 3.15). Then they are in the same class.

__Proof__ Using $L(T)$ and $R(Y)$ one reduces to $L(x_{n+2}(w))(x_{n+2,1}(q),1)$ versus $L(x_{n+2}(w))(x_{n+2,1}(q)A,A^{-1}) = L(x_{n+2}(w))(Ax_{n+2,1}(q),A^{-1})$. One shows that they determine the same element of the chunk by executing $L(x_{n+2}(w))$ in both cases and then using A to transform one representative into the other. The semi-direct product structures of $St(\{1\}^* \times \{n+2\}^*)$ and $St(\{n+2\}^* \times \{1\}^*)$ make this easy.

__Remark__ We only needed to require that one of the two is defined as in 3.15. Then the other one is also defined.

3.17 __Lemma__ Suppose in 3.14 that $w_1 = w_2 = 0$. Let $A\in St([n+1] \times \{1,n+2\}^*)$ such that $x_{n+2}(w)A = Ax_{n+2}(0,*,\ldots,*)$. Let $B\in St(\{n+2\} \times \{1\}^*)$ such that $Tx_{n+2,1}(q)AB = T'x_{n+2,1}(*)$ in __Low__ for some $T'\in St(n+1)$. (See 3.14 for notations). Then $L(x_{n+2}(v))(Tx_{n+2,1}(q),Y) \sim L(x_{n+2}(v))(Tx_{n+2,1}(q)AB,B^{-1}A^{-1}Y)$.

__Proof__ First note that, given A, the element B is unique, because

B can be computed in $E([n+2] \times \{n+2\}^*)$. Write $A = A_1A_2$ with $A_1 \in St(\{1\} \times \{n+2\}^*)$, $A_2 \in St(\{1,n+2\}^* \times \{1,n+2\}^*)$. In Med we can write AB as $A_1B_1A_2$ with $B_1 \in St(\{n+2\} \times \{1\}^*)$. By the previous Lemma we can assume that $A_2 = 1$. And using $R(Y)$ we can get rid of Y. Now "execute" $L(x_{n+2}(v))$ in both cases and use $L(T)$ to get rid of T. Then we have to deal with the case $T = 1$, $Y = 1$, $A_2 = 1$, $v = w$. But there we can apply the squeezing principle, with $i = 2$. (see 3.11).

3.18 Definition Let $T \in St(n+1)$, $v \in R^{n+1}$, $q \in R$, $Y \in$ Up be such that $x_{n+2}(v)T = Tx_{n+2}(0,0,*\ldots,*)$. Then we put $\mathcal{L}(x_{n+2}(v))\langle Tx_{n+2,1}(q),Y\rangle$ = class of $L(x_{n+2}(v))(Tx_{n+2,1}(q),Y)$. It is easy to see from Lemma 3.17 that this is a consistent definition. So now we have defined $\mathcal{L}(x_{n+2}(v))\langle X,Y\rangle$ for some more v and X. One checks that our new definition is compatible with the one in 3.13. We say that $\mathcal{L}(x_{n+2}(v))\langle X,Y\rangle$ is <u>defined at the bottom</u> if there is $T \in St(n+1)$ with $\langle X,Y\rangle = \langle Tx_{n+2,1}(*),*\rangle$, $x_{n+2}(v)T = Tx_{n+1,n+2}(*)$. In particular $\mathcal{L}(x_{n+2}(v))\langle X,Y\rangle$ is then defined by the definition above. (We say that it is defined at the bottom because the relevant entries of mat$(x_{n+1,n+2}(*))$ and mat$(x_{n+2,1}(*))$ are in the bottom two rows). We will prefer to talk about the case that $\mathcal{L}(x_{n+2}(v))\langle X,Y\rangle$ is defined at the bottom rather than the more general case covered by the definition. The reason is that the notion "defined at the bottom" has a constant meaning, while "defined" will have a different meaning when we will introduce $\mathcal{L}(x_{n+2}(v))\langle X,Y\rangle$ in cases not covered by the present definition.

3.19 Given $\langle X,Y\rangle \in C$, what are the $T \in St(n+1)$ such that we can write $\langle X,Y\rangle = \langle Tx_{n+2,1}(*),*\rangle$? Write X as X_1X_2 with $X_1 \in St(n+1)$, $X_2 \in St(\{n+2\} \times [n+1])$. Using the semi-direct product structure of Low one sees that a necessary and sufficient condition is that there is $g \in \underline{Med}$ with $X_1 g = T$. So it only depends on X_1. In particular, if $v \in R^{n+1}$, $T \in St(n+1)$, $B \in St(\{n+2\} \times [n+1])$ then $\mathcal{L}(x_{n+2}(v))\langle TB,*\rangle$ is defined at the bottom if and only if $\mathcal{L}(x_{n+2}(v))\langle Tx_{n+2,1}(*),*\rangle$ is.

3.20 Lemma Let $v \in R^{n+1}$, $T \in St(n+1)$, $r \in R$, such that $x_{n+2}(v)T = Tx_{n+2}(w_1, w_2, \ldots, w_{n+1})$ with $(w_2, \ldots, w_{n+1})$ unimodular. Then $\mathcal{L}(x_{n+2}(vr))\langle Tx_{n+2,1}(*),*\rangle$ is defined at the bottom.

Proof By "linearity" it is sufficient to do the case $r = 1$. As $St(n)$ acts transitively on unimodular columns of length n (see [4], Ch. V, Thm. (3.3)), there is $T' \in St(\{1,n+2\}^* \times \{1,n+2\}^*)$ with $x_{n+2}(v)TT' = TT'x_{1,n+2}(w_1)x_{n+1,n+2}(1)$. Choose $T'' = x_{1,n+1}(+w_1)$. Then $x_{n+2}(v)TT'T'' = TT'T''x_{n+1,n+2}(1)$, as required.

3.21 Remark Note that this is the first time that we use a stable range condition.

3.22 Lemma Let $\mathcal{L}(x_{n+2}(v))\langle X,Y\rangle$ be defined at the bottom and let $\langle P,Q\rangle \in C$ with $\underline{mat}\langle X,Y\rangle = \underline{mat}\langle P,Q\rangle$. Then $\mathcal{L}(x_{n+2}(v))\langle P,Q\rangle$ is defined at the bottom.

Proof Put $M = \underline{mat}(X^{-1}P) = \underline{mat}(YQ^{-1})$. Then $M \in E(n+2)$ and M has trivial columns at positions 1 and $n+2$. Choose $B \in \underline{Med}$ such that $M\,\underline{mat}(B)$ has trivial rows at those positions too. Using injective stability for K_1 we see that $M\,\underline{mat}(B) \in \underline{mat}(St(\{1,n+2\}^* \times \{1,n+2\}^*))$, so in particular $M \in \underline{mat}(\underline{Med})$. Therefore $\langle X,Y\rangle = \langle X',Y'\rangle$ with $\underline{mat}(X') = \underline{mat}(P)$, $\underline{mat}(Y') = \underline{mat}(Q)$. But it is not hard to see (and may have been noted by the reader) that being defined at the bottom only depends on the matrices, not on the finer structure of Low.

3.23 Lemma Let $v, w \in R^{n+1}$, $T \in St(n+1)$, $\langle X,Y\rangle \in C$ such that $x_{n+2}(v)T = Tx_{n+2}(w)$ and such that $\mathcal{L}(x_{n+2}(w))\langle X,Y\rangle$ is defined at the bottom. Then $\mathcal{L}(T)\mathcal{L}(x_{n+2}(w))\langle X,Y\rangle = \mathcal{L}(x_{n+2}(v))\langle TX,Y\rangle$ and the latter is also defined at the bottom.

Proof Actually it is easy to see that they are equal if $\mathcal{L}(x_{n+2}(w))\langle X,Y\rangle$ is defined (as in 3.18): Write $\langle X,Y\rangle = \langle T'x_{n+2,1}(*),*\rangle$ with $T' \in St(n+1)$ such that one can execute $L(x_{n+2}(w))(T'x_{n+2,1}(*),*)$. Then compare.

3.24 Lemma If $L(x_{n+1}(v))(X,Y)$ is defined (see 3.15) and if $\mathcal{L}(x_{n+1}(v))\langle X,Y\rangle$ is defined at the bottom, then the former is a representative of the latter.

Proof By the previous Lemma we may assume $X = x_{n+2,1}(q)$, $q \in R$. As $\mathcal{L}(x_{n+2}(v))\langle X,Y\rangle$ is defined at the bottom, there must be $A \in St([1,n+2]^* \times [1,n+2]^*)$ with $x_{n+2}(v)A = Ax_{n+1,n+2}(*)$. Then we may replace (X,Y) by $(XA, A^{-1}Y)$, because of Lemma 3.16. But then it is obvious.

3.25 Lemma (Additivity, first case).
Let $v = (0, v_2, \ldots, v_{n+1})$, $w = (0, w_2, \ldots, w_{n+1})$. Let $\langle X,Y\rangle = \mathcal{L}(x_{n+2}(w))\langle x_{n+2,1}(q),Z\rangle$ and let $\mathcal{L}(x_{n+2}(v))\langle X,Y\rangle$ be defined at the bottom. Then $\mathcal{L}(x_{n+2}(v))\mathcal{L}(x_{n+2}(w))\langle x_{n+2,1}(q),Z\rangle =$ $\mathcal{L}(x_{n+2}(v+w))\langle x_{n+2,1}(q),Z\rangle$.

Proof $L(x_{n+2}(v))L(x_{n+2}(w))(x_{n+2,1}(q),Z) =$ $L(x_{n+2}(v+w))(x_{n+2,1}(q),Z)$ and the left hand side is relevant by the previous Lemma.

3.26 <u>Definition</u> We say that $\mathcal{L}(x_{n+2}(v))$, $\mathcal{L}(x_{n+2}(w))$ <u>slide past each other at $\langle X,Y\rangle$</u> if there is $T\in St(n+1)$ such that $\langle X,Y\rangle = \langle Tx_{n+2,1}(*),*\rangle$ and such that
$x_{n+2}(v)\,T = Tx_{n+2}(0,z_2,\ldots,z_k,0,\ldots,0)$,
$x_{n+2}(w)T = Tx_{n+2}(0,\ldots\ldots,0,z_{k+1},\ldots,z_{n+1})$ for some $2 \leq k \leq n+1$ and some $z_2,\ldots,z_{n+1}$ in R.

The relation is symmetric because there is an element T'' of $St(\{1,n+2\}^* \times \{1,n+2\}^*)$ with $x_{i,n+2}(*)T'' = T''x_{n+3-i,n+2}(*)$ for $2 \leq i \leq n+1$. (Use a product of the elements $w_{p,q}(1)$). From the same observation it follows that both $\mathcal{L}(x_{n+2}(v))\langle X,Y\rangle$, $\mathcal{L}(x_{n+2}(w))\langle X,Y\rangle$ are defined in the fashion described in 3.18. (For the second one this is obvious). Executing the maps one actually sees that both steps in $\mathcal{L}(x_{n+2}(w))\,\mathcal{L}(x_{n+2}(v))\langle X,Y\rangle$ are defined in the way described by 3.18. And again we can use T'' to show from this that both steps in $\mathcal{L}(x_{n+2}(v))\,\mathcal{L}(x_{n+2}(w))\langle X,Y\rangle$ are defined that way.

3.27 <u>Lemma</u> (Additivity, second case).
Let $\mathcal{L}(x_{n+2}(v))$, $\mathcal{L}(x_{n+2}(w))$ slide past each other at $\langle X,Y\rangle$, and let both $\mathcal{L}(x_{n+2}(v))\langle X,Y\rangle$ and $\mathcal{L}(x_{n+2}(v+w))$ be defined at the bottom. Then $\mathcal{L}(x_{n+2}(v))\,\mathcal{L}(x_{n+2}(w))\langle X,Y\rangle = \mathcal{L}(x_{n+2}(v+w))\langle X,Y\rangle$.

<u>Proof</u> We choose a representative $(Tx_{n+2,1}(*),*)$ of $\langle X,Y\rangle$, as in 3.26. One checks that the first execution of the expression $L(x_{n+2}(v))L(x_{n+2}(w))(Tx_{n+2,1}(*),*)$ leaves a result of the form $L(x_{n+2}(v))(P,Q)$ with $P = TAx_{n+2,1}(*)$, $A\in St([n+1] \times \{1\})$. It easily follows that $\mathcal{L}(x_{n+2}(v))\langle P,Q\rangle$ is defined at the bottom and therefore, by Lemma 3.24, we can use
$L(x_{n+2}(v))L(x_{n+2}(w))(Tx_{n+2,1}(*),*)$ for representing

$\mathcal{L}(x_{n+2}(v))\mathcal{L}(x_{n+2}(w))\langle X,Y\rangle$. Similarly one can use $L(x_{n+2}(v + w))(Tx_{n+2,1}(*),*)$ for $\mathcal{L}(x_{n+2}(v + w))\langle X,Y\rangle$. But then it is easy. (Compare Lemma 3.25).

3.28 <u>Lemma</u> Let $v,w,z \in R^{n+1}$, $T \in St(n + 1)$ such that $x_{n+2}(w)T = Tx_{n+1,n+2}(*)$ and $x_{n+2}(v)T = Tx_{n+2}(z)$ with $(z_2,\ldots,z_n)$ unimodular. Then Lemma 3.27 applies for $\langle X,Y\rangle = \langle Tx_{n+2,1}(*),*\rangle$.

<u>Proof</u> By Lemma 3.20 we only have to show that $\mathcal{L}(x_{n+2}(v))$ and $\mathcal{L}(x_{n+2}(w))$ slide past each other at $\langle X,Y\rangle$. There are $a_2,\ldots,a_n$ such that $a_2z_2 + \cdots + a_nz_n = z_1$. We have a representative of $\langle X,Y\rangle$ which takes the form $(Tx_{12}(a_2) \cdots x_{1n}(a_n)x_{n+2,1}(*),*)$. Replacing $(Tx_{n+2,1}(*),*)$ by this representative we reduce to the case $z_1 = 0$. Similarly we can reduce to the case $z_{n+1} = 0$ by using a representative of the form $(Tx_{n+1,2}(b_2) \cdots x_{n+1,n}(b_n)x_{n+2,1}(*),*)$. But if $z_1 = z_{n+1} = 0$ then the situation is the one described in definition 3.26.

3.29 What Lemma 3.28 tells us is that, if $\mathcal{L}(x_{n+2}(w))\langle X,Y\rangle$ is defined at the bottom, one has additivity for those v which make a certain piece of a column unimodular. This is the sort of situation the conditions $SR_n^p(c,u)$ refer to. So let us apply them. (Until now we only used SR_n). The case of commutative semi-local rings diverges from the "general" case at this point (for the first time).

3.30 <u>Proposition</u> (Additivity for maps defined at the bottom). Let $\mathcal{L}(x_{n+2}(v))\langle X,Y\rangle$ be defined at the bottom, with value $\langle P,Q\rangle$, let $\mathcal{L}(x_{n+2}(w))\langle P,Q\rangle$ be defined at the bottom and let $\mathcal{L}(x_{n+2}(v + w))\langle X,Y\rangle$ be defined at the bottom. Then $\mathcal{L}(x_{n+2}(v + w))\langle X,Y\rangle = \mathcal{L}(x_{n+2}(w))\mathcal{L}(x_{n+2}(v))\langle X,Y\rangle$.

<u>Proof</u> We assume $\tilde{\tilde{SR}}_n$ and therefore $SR_n^3(n+2,n+1)$. If $v + w = 0$ then it is easy: apply Lemma 3.23 and Lemma 3.25 for instance. (Or just apply the definitions.) In the general case the idea is to chose z such that repeated application of Lemma 3.28 yields:
$\mathcal{L}(x_{n+2}(z))\mathcal{L}(x_{n+2}(w))\mathcal{L}(x_{n+2}(v))\langle X,Y\rangle =$
$\mathcal{L}(x_{n+2}(z+w))\mathcal{L}(x_{n+2}(v))\langle X,Y\rangle = \mathcal{L}(x_{n+2}(z+v+w))\langle X,Y\rangle =$
$\mathcal{L}(x_{n+2}(z))\mathcal{L}(x_{n+2}(v+w))\langle X,Y\rangle$. Then one applies $\mathcal{L}(x_{n+2}(-z))$ to both ends of the string. So say $\langle P,Q\rangle = \langle Tx_{n+2,1}(*),*\rangle$ with $x_{n+2}(w)T = Tx_{n+1,n+2}(*)$. Then we want that
$x_{n+2}(z)T = Tx_{n+2}(*,a_2,\ldots,a_n,*)$ with $(a_2,\ldots,a_n)$ unimodular. (That will guarantee that the first two members of the string exist and that they are equal). There is an elementary matrix $M_1 = \underline{\text{mat}}(T^{-1})$ in $E(n+1)$ such that $(*,a_2,\ldots,a_n,*)$ is just M_1z. Choose A_1 to be the matrix ($n-1$ by $n+2$) with first column zero and the remaining part taken from rows 2 through n of M_1. Then A_1 is a matrix of the type which one considers in $SR_n^3(n+2,n+1)$. (It doesn't matter that the obvious way to add rows to the "U_1-part" of A_1 is not the way the reader expected when reading 2.2: If a matrix is in $E(n+1)$ up to a permutation of rows, then the matrix is in $E(n+1)$ up to multiplication of one row by a sign. (Use the $w_{ij}(1)$)). We will have unimodular $(a_2,\ldots,a_n)$ if and only if $A_1\binom{1}{z}$ is unimodular. Next we want the third member of the string to be defined and we want it to be equal to the second member. Now there is a matrix $M_2 \in E(n+1)$ such that it works if
$M_2(z+w) = (*,b_2,\ldots,b_n,*)$ with $(b_2,\ldots,b_n)$ unimodular. Choose A_2 to be the matrix ($n-1$ by $n+2$) with the first column equal to the middle part of M_2w and the remainder taken from row 2 through n of M_2. Again we have translated what we want into the form "$A_i\binom{1}{z}$ is unimodular", with A_i of the proper type. Finally there is A_3 which stands for the wish to have the fourth member exist and to have it equal to the third. By $SR_n^3(n+2,n+1)$

we can choose z such that it works. It is easy to see (from the fact that the executions of $\mathcal{L}(x_{n+2}(z))$ are defined at the bottom) that one can apply $\mathcal{L}(x_{n+2}(-z))$ to both ends. This is an instance of the case $v + w = 0$ and we see that the $\mathcal{L}(x_{n+2}(-z))\mathcal{L}(x_{n+2}(z))$ cancel out.

3.31 Definition Let $v \in R^{n+1}$, $\langle X,Y\rangle \in C$ such that $\underline{mat}(x_{n+2}(v))XY) \in \underline{mat}(C)$. We claim there is $z \in R^{n+1}$ such that both steps in $\mathcal{L}(x_{n+2}(-z))\mathcal{L}(x_{n+2}(v + z))\langle X,Y\rangle$ are defined at the bottom. We define $\mathcal{L}(x_{n+2}(v))\langle X,Y\rangle$ to be equal to the result. We need to do some checks to see that this definition is consistent. First of all it is consistent with the earlier definitions in case $\mathcal{L}(x_{n+2}(v))\langle X,Y\rangle$ is defined at the bottom, by Proposition 3.30. One can also check that it is consistent with the other earlier definitions, using Lemma 3.24, but those definitions have served their purpose anyway. (So let's overrule them).

3.32 So let us prove that the definition makes sense. First let us show that z exists. As in 3.30 we see that there is an n by $n + 2$ matrix A_1 with the property that $\mathcal{L}(x_{n+2}(v + z))\langle X,Y\rangle$ is defined at the bottom (in the way described by Lemma 3.20) if $A_1\binom{1}{z}$ is unimodular. Moreover, this matrix is of the type one considers in $SR^4_{n+1}(n + 2, n + 1)$. (This proof doesn't work for arbitrary commutative semi-local rings). Choose $\langle P',Q'\rangle \in C$ such that $\underline{mat}(P'Q') = \underline{mat}(x_{n+2}(v)XY)$. There is an n by $n + 2$ matrix A_2, of the proper type, such that $\mathcal{L}(x_{n+2}(z))\langle P',Q'\rangle$ is defined at the bottom if $A_2\binom{1}{z}$ is unimodular. But if $\mathcal{L}(x_{n+2}(z))\langle P',Q'\rangle$ is defined at the bottom then the second step in $\mathcal{L}(x_{n+2}(-z))\mathcal{L}(x_{n+2}(z))\langle P',Q'\rangle$ is also defined at the bottom. Therefore, if both $A_1\binom{1}{z}$ and $A_2\binom{1}{z}$ are unimodular, both steps in $\mathcal{L}(x_{n+2}(-z))\mathcal{L}(x_{n+2}(v + z))\langle X,Y\rangle$ are defined at the bottom. (Use

Lemma 3.22). So z exists. We have only used A_1 and A_2, so we might still choose A_3, A_4 because we have $SR^4_{n+1}(n+2, n+1)$ available. (So far it was just $SR^2_{n+1}(n+2, n+1)$ which actually holds for commutative semi-local rings too). We need to show that another z gives the same answer. So suppose that z' also fits, i.e. that both steps in $\mathcal{L}(x_{n+2}(-z'))\mathcal{L}(x_{n+2}(v+z'))\langle X,Y\rangle$ are defined at the bottom. The idea is to choose $t \in R^{n+1}$ such that all steps in
$\mathcal{L}(x_{n+2}(t-z))\mathcal{L}(x_{n+2}(v+z))\langle X,Y\rangle$, $\mathcal{L}(x_{n+2}(t+v))\langle X,Y\rangle$,
$\mathcal{L}(x_{n+2}(t-z'))\mathcal{L}(x_{n+2}(v+z'))\langle X,Y\rangle$,
$\mathcal{L}(x_{n+2}(t))\mathcal{L}(x_{n+2}(-z))\mathcal{L}(x_{n+2}(v+z))\langle X,Y\rangle$ are defined at the bottom. That amounts to four conditions on t and they can simultaneously be satisfied because of $SR^4_{n+1}(n+2, n+1)$. (so we don't use the old A_1, A_2 but a new set of four matrices, chosen after z,z'). Using Lemma 3.22 we see that all steps in the following computation are defined at the bottom, which makes that Proposition 3.30 applies: $\mathcal{L}(x_{n+2}(t))\mathcal{L}(x_{n+2}(-z))\mathcal{L}(x_{n+2}(v+z))\langle X,Y\rangle =$
$\mathcal{L}(x_{n+2}(t-z))\mathcal{L}(x_{n+2}(v+z))\langle X,Y\rangle = \mathcal{L}(x_{n+2}(t+v))\langle X,Y\rangle =$
$\mathcal{L}(x_{n+2}(t-z'))\mathcal{L}(x_{n+2}(v+z'))\langle X,Y\rangle =$
$\mathcal{L}(x_{n+2}(t))\mathcal{L}(x_{n+2}(-z'))\mathcal{L}(x_{n+2}(v+z'))\langle X,Y\rangle$. Now apply $\mathcal{L}(x_{n+2}(-t))$ to both ends.

3.33 Proposition (Additivity).
$\mathcal{L}(x_{n+2}(v))\mathcal{L}(x_{n+2}(w))\langle X,Y\rangle = \mathcal{L}(x_{n+2}(v+w))\langle X,Y\rangle$ whenever the left hand side is defined.

Proof Obviously the right hand side is defined if the left hand side is. (Read 3.31). So assume this is the case. Say $\langle P,Q\rangle = \mathcal{L}(x_{n+2}(w))\langle X,Y\rangle$. If both $\mathcal{L}(x_{n+2}(v))\langle P,Q\rangle$ and $\mathcal{L}(x_{n+2}(v+w))\langle X,Y\rangle$ are defined at the bottom then it is an easy consequence of the definition of $\mathcal{L}(x_{n+2}(w))\langle X,Y\rangle$. Choose z such that $\mathcal{L}(x_{n+2}(v))\langle P,Q\rangle = \mathcal{L}(x_{n+2}(-z))\mathcal{L}(x_{n+2}(v+z))\langle P,Q\rangle$, with both steps at the right hand side defined at the bottom, and such that $\mathcal{L}(x_{n+2}(z + v + w))\langle X,Y\rangle$ is defined at the bottom. (This is an $SR^3_{n+1}(n + 2,n + 1)$ type problem so it can be solved by virtue of $SR^4_{n+1}(n + 2,n + 1)$. We still are doing things that don't work for some commutative semi-local rings). We have to show that $\mathcal{L}(x_{n+2}(v + z))\mathcal{L}(x_{n+2}(w))\langle X,Y\rangle = \mathcal{L}(x_{n+2}(z))\mathcal{L}(x_{n+2}(v + w))\langle X,Y\rangle$. But the right hand side is $\mathcal{L}(x_{n+2}(v + w + z))\langle X,Y\rangle$ by the case discussed above. But then we are back at just this same case.

3.34 Lemma Let $v,w \in R^{n+1}$, $T \in St(n + 1)$, $X \in \underline{Low}$, $Y \in \underline{Up}$ such that $x_{n+2}(v)T = Tx_{n+2}(w)$ and such that $\mathcal{L}(x_{n+2}(v))\langle TX,Y\rangle$ is defined. Then $\mathcal{L}(x_{n+2}(v))\langle TX,Y\rangle = \mathcal{L}(T)\mathcal{L}(x_{n+2}(w))\langle X,Y\rangle$.

Proof By definition $\mathcal{L}(x_{n+2}(w))\langle X,Y\rangle = \mathcal{L}(x_{n+2}(-z))\mathcal{L}(x_{n+2}(w + z))$ $\langle X,Y\rangle$ with both steps at the right hand side defined at the bottom. Now apply Lemma 3.23.

3.35 In Section 5 we will have to find an alternative for 3.30, 3.31, 3.32 such that 3.33, 3.34 still hold and such that $\mathcal{L}(x_{n+2}(v))\langle X,Y\rangle$ is defined if $\underline{mat}(x_{n+2}(v)XY) \in \underline{mat}(C)$. The next proofs and definitions will then go through for commutative semi-local rings too.

3.36 Lemma The set $\underline{mat}(C)$ (see 3.8) consists of all matrices in $E(n + 2)$ whose first column is of the form $(a_1,\ldots,a_{n+2})$ with $(a_1,\ldots,a_{n+1})$ unimodular.

Proof Obviously every element of $\underline{mat}(C)$ looks like that. Conversely, let M be such a matrix. Multiplying M from the left by a matrix in $E(n + 1)$ one reduces to the case $(a_1,\ldots,a_{n+2}) = (1,0,\ldots,0,a_{n+2})$, because $E(n + 1)$ acts transitively on unimodular columns of length n + 1 (see [4], Ch. V, Thm. (3.3)). So multiplying M from the left by an element of mat(Low) we reduce to the case that the first column of M is trivial. Multiplying from the right by an element of mat(Up) we can get the first row trivial too. By injective stability for K_1 the matrix M then is in mat $St(\{1\}^* \times \{1\}^*)$, so certainly in mat(Up).

3.37 For the sake of completeness we include the following Lemma. (Compare with 2.2).

Lemma Let $n \geq m \geq 2$. Let A be an $(m-1) \times n$ matrix over a ring S, where S satisfies SR_m. Suppose that A can be completed to an invertible $n \times n$ matrix. Then there is $\lambda \in S^{n-1}$ such that $A\binom{1}{\lambda}$ is unimodular.

Proof We use the notations of 3.4, except that the base ring is now S, not R. Let $M \in GL(n,S)$ be such that A consists of the top m-1 rows of M. It is enough to show the following. There is $L \in GL(n,S)$, with trivial first row, such that in the first column $(a_1,\ldots,a_n)$ of ML the piece $(a_1,\ldots,a_{m-1})$ is unimodular. Using SR_n, which follows from SR_m as $n \geq m$, we choose $L_1 \in E([n] \times \{1\})$, $U \in E(\{1\} \times [n])$ so that the (1,n)-entry of UL_1M^{-1} is zero. As $(UL_1M^{-1})(ML_1^{-1}) = U$, one sees that in the first column $(b_1,\ldots,b_n)$ of ML_1^{-1} the piece $(b_1,\ldots,b_{n-1})$ is unimodular. So if $m = n$ we are done. Otherwise $n - 1 \geq m$ and $E(n - 1)$ acts transitively on unimodular columns of length $n - 1$. So choose $T \in E([n] \times [n-1])$ so that the first column of TML_1^{-1} is trivial. Then choose $B \in E(\{1\} \times [n])$ so that the first row of $L_2 = BTML_1^{-1}$ is also trivial. We may replace M by $ML_1^{-1}L_2^{-1}$. But $ML_1^{-1}L_2^{-1} = T^{-1}B^{-1}$ is an $n \times n$ matrix with an invertible $(n-1) \times (n-1)$ submatrix in the upper left hand corner. Thus the problem has been reduced in size and we can apply induction.

§4. The Group Uleft.

4.1 There is a left-right symmetry in the chunk: Consider the anti-homomorphism inv which sends $x_{ij}(r)$ to $x_{n+3-i,n+3-j}(-r)$. (It is the composite of the map $z \mapsto z^{-1}$ with a homomorphism). If $I,J \subseteq [n+2]$ then we get inv: $St(I \times J) \to St(I' \times J')$ where I' is the set $\{a \in [n+2] \mid n+3-a \in I\}$ or $\{n+3-a \mid a \in I\}$ and J' is the set $\{n+3-b \mid b \in J\}$. One has inv•inv = id, inv(Low) = Up, inv(Med) = Med, inv(Up) = Low. (We should write $\text{inv}_{I \times J}$: $St(I \times J) \to St(I' \times J')$). It is easy to see that $\text{inv}\langle X,Y\rangle = \langle \text{inv}(Y), \text{inv}(X)\rangle$ defines an involution of C, i.e. a map $C \to C$ which is its own inverse. We have defined in Section 3 what $\mathcal{L}(X)\langle P,Q\rangle$ means if $X,P \in$ Low, $Q \in$ Up. We also defined $\mathcal{R}(Y)\langle P,Q\rangle$ if $P \in$ Low, $Y,Q \in$ Up. The connection between the two notions is as follows: $\mathcal{R}(Y) = \text{inv} \cdot \mathcal{L}(\text{inv}(Y)) \cdot \text{inv}$. This suggests to define $\mathcal{R}(x_1(v)) = \text{inv} \cdot \mathcal{L}(\text{inv}\ x_1(v)) \cdot \text{inv}$, i.e. $\mathcal{R}(x_1(v_2,\ldots,v_{n+2}))\langle X,Y\rangle = \text{inv}(\mathcal{L}(x_{n+2}(v_{n+2},\ldots,v_2))\langle \text{inv}(Y), \text{inv}(X)\rangle)$ whenever the right hand side is defined. So let us do that. Then $\mathcal{R}(x_1(v))\langle X,Y\rangle$ is defined if and only if $\text{mat}(XYx_1(v)) \in \text{mat}(C)$. This is just one of the properties we get by translating earlier results by means of inv. Other ones are: $\mathcal{R}(T)\mathcal{R}(x_1(v))\langle X,Y\rangle = \mathcal{R}(x_1(w))\mathcal{R}(T)\langle X,Y\rangle$ if both sides are defined and $T \in St(\{1\}^* \times \{1\}^*)$, $v = (v_2,\ldots,v_{n+2})$, $w = (w_2,\ldots,w_{n+2})$ with $Tx_1(w) = x_1(v)T$. And additivity: $\mathcal{R}(x_1(v))\mathcal{R}(x_1(w))\langle X,Y\rangle = \mathcal{R}(x_1(v+w))\langle X,Y\rangle$ if both sides are defined. It may seem more convenient to write $\langle X,Y\rangle\mathcal{R}(x_1(v))$ instead of $\mathcal{R}(x_1(v))\langle X,Y\rangle$. We don't do that because we want to emphasize the order of execution in expressions like $\mathcal{R}(x_1(w))\mathcal{L}(x_{n+2}(v))\langle X,Y\rangle$. In the alternative notation it would read $(\mathcal{L}(x_{n+2}(v))\langle X,Y\rangle)\mathcal{R}(x_1(w))$. The reader may find however that certain arguments are better understood when one writes $\mathcal{R}$'s at the right.

4.2 We want to show that $\mathcal{L}(X)$, $\mathcal{L}(x_{n+2}(v))$ commute with $\mathcal{R}(T)$, $\mathcal{R}(x_1(w))$ if $X \in \underline{\mathrm{Low}}$, $v \in R^{n+1}$, $w = (w_2,\ldots,w_{n+2})$, $T \in St(\{1\}^* \times \{1\}^*)$. (So T is less arbitrary than X). As some of these maps are only defined on part of C it only makes sense to prove, for instance, that $\mathcal{L}(x_{n+2}(v))\mathcal{R}(x_1(w))$ equals $\mathcal{R}(x_1(w))\mathcal{L}(x_{n+2}(v))$ where both compositions are defined.

4.3 Notation If f is a map defined on part of C and with values in C and if g is also such a map then $f \circ g$ is the map which sends $\langle X,Y\rangle \in C$ to $f(g\langle X,Y\rangle)$ whenever the latter is defined. We say that $f \approx g$ if $f\langle X,Y\rangle = g\langle X,Y\rangle$ whenever both sides are defined. This is not an equivalence relation.

4.4 So we want to prove that $\mathcal{L}(X)\circ\mathcal{R}(T) \approx \mathcal{R}(T)\circ\mathcal{L}(X)$ etc. (See 4.2). In fact $\mathcal{L}(X)\circ\mathcal{R}(T) \approx \mathcal{R}(T)\circ\mathcal{L}(X)$ is a triviality (T and X as in 4.2). The non-trivial case to consider is the one of $\mathcal{L}(x_{n+2}(v))\circ\mathcal{R}(x_1(w))$ versus $\mathcal{R}(x_1(w))\circ\mathcal{L}(x_{n+2}(v))$.

4.5 Definition We say that $\mathcal{L}(x_{n+2}(v))$, $\mathcal{R}(x_1(w))$ slide past each other at $\langle \underline{X},\underline{Y}\rangle$ if there are $T \in St(n+1)$, $U \in St(\{1\}^* \times \{1\}^*)$, $A \in St(\{n+2\} \times [n+2])$, $B \in St(\{1\} \times [n+2])$, $2 \leq k \leq n$, such that $\langle X,Y\rangle = \langle TA,BU\rangle$ and such that

$x_{n+2}(v)T = Tx_{n+2}(0,0,\ldots,0,f_{k+1},\ldots,f_{n+1})$,

$Ux_1(w) = x_1(g_2,\ldots,g_k,0,\ldots,0)U$. We could also require that actually $A \in St(\{n+2\} \times [k])$ because one can take the part of A which comes from $St(\{n+2\} \times [k]^*)$ to the right without spoiling anything. Having done that one can do the same sort of thing to B and reduce to the case that $B \in St(\{1\} \times [k]^*)$. So if we want to prove that $\mathcal{L}(x_{n+2}(v))$, $\mathcal{R}(x_1(w))$ slide past each other at $\langle X,Y\rangle$

we only need to have $A \in St(\{n+2\} \times [n+2])$, $B \in St(\{1\} \times [n+2])$. But if we apply that they slide past each other we usually take $A \in St(\{n+2\} \times [k])$, $B \in St(\{1\} \times [k]^*)$. We refer to this particular choice by saying that (TA, BU) is <u>separated with respect to</u> <u>$x_{n+2}(v)$, $x_1(w)$</u>.

4.6 <u>Proposition</u> Let $\mathcal{L}(x_{n+2}(v))$, $\mathcal{R}(x_1(w))$ slide past each other at $\langle X,Y \rangle$. Then $\mathcal{L}(x_{n+2}(v))\mathcal{R}(x_1(w))\langle X,Y\rangle = \mathcal{R}(x_1(w))\mathcal{L}(x_{n+2}(v))\langle X,Y\rangle$ (Both sides are defined).

<u>Proof</u> So take $(TA,BU) \in \langle X,Y\rangle$ separated with respect to $x_{n+2}(v)$, $x_1(w)$. One easily sees that $\mathcal{R}(U)\cdot\mathcal{L}(x_{n+2}(v)) \approx \mathcal{L}(x_{n+2}(v))\cdot\mathcal{R}(U)$. (Compare 4.4). Its counterpart at the other side states $\mathcal{L}(T)\cdot\mathcal{R}(x_1(w)) \approx \mathcal{R}(x_1(w))\cdot\mathcal{L}(T)$. From this and Lemma 3.34 one sees that we may assume $T = 1$. Similarly we may assume $U = 1$. Say $A = x_{n+2,1}(a_1) \cdots x_{n+2,k}(a_k)$ and $B = x_{1,k+1}(b_{k+1}) \cdots x_{1,n+2}(b_{n+2})$. We just compute both $\mathcal{L}(x_{n+2}(v))\mathcal{R}(x_1(w))\langle A,B\rangle$ and $\mathcal{R}(x_1(w))\mathcal{L}(x_{n+2}(v))\langle A,B\rangle$ and compare: $\mathcal{L}(x_{n+2}(v))\mathcal{R}(x_1(w))\langle A,B\rangle = \mathcal{L}(x_{n+2}(v))\mathcal{R}(x_1(w))\langle Ax_{1,k+1}(b_{k+1}) \cdots x_{1,n+1}(b_{n+1}), x_{1,n+2}(b_{n+2})\rangle = \mathcal{L}(x_{n+2}(v))\langle Ax_{1,k+1}(b_{k+1}) \cdots x_{1,n+1}(b_{n+1})x_1(w), x_{n+2}(b_{n+2}, -w_2b_{n+2}, \ldots, -w_kb_{n+2}, 0, \ldots, 0)\rangle =$ $\mathcal{L}(x_{n+2}(v))\langle Ax_1(w), PBx_{n+2}(0, -w_2b_{n+2}, \ldots, w_kb_{n+2}, 0, \ldots, 0)\rangle$ with $P =$ product of the $x_{ij}(-w_ib_j)$ with $2 \leq i \leq k$, $k+1 \leq j \leq n+1$. So we get $\mathcal{L}(x_{n+2}(v))\langle x_1(w)x_{n+2,1}(a_1 + a_2w_2 + \cdots + a_kw_k)$, $x_{n+2,2}(a_2) \cdots x_{n+2,k}(a_k)PBx_{n+2}(0, -w_2b_{n+2}, \ldots, -w_kb_{n+2}, 0, \ldots, 0)\rangle$. Say $q = a_1 + a_2w_2 + \cdots + a_kw_k$. Then it equals $\langle x_1(w_2, \ldots, w_k, v_{k+1}q, \ldots, v_{n+1}q, q), x_{n+2}(v)x_{n+2,2}(a_2) \cdots$ $x_{n+2,k}(a_k)PBx_{n+2}(0, -w_2b_{n+2}, \ldots, -w_kb_{n+2}, 0, \ldots, 0)\rangle$, or $\langle x_1(w_2, \ldots, w_k, v_{k+1}q, \ldots, v_{n+1}q, q)x_{n+2,2}(a_2) \cdots x_{n+2,k}(a_k)Q$, $x_{n+2}(v)PBx_{n+2}(0, -w_2b_{n+2}, \ldots, -w_kb_{n+2}, 0, \ldots, 0)\rangle$ where $Q =$ product of the $x_{ij}(v_ia_j)$ with $k+1 \leq i \leq n+1$, $2 \leq j \leq k$. That again is the

same as $\langle x_1(w_2,\ldots,w_k,v_{k+1}q,\ldots,v_{n+1}q,q)x_{n+2,2}(a_2) \cdots x_{n+2,k}(a_k)Q,$ $Px_{1,k+1}(b_{k+1}) \cdots x_{1,n+1}(b_{n+1})x_{n+2}(p,-w_2p,\ldots,-w_kp,v_{k+1},\ldots,v_{n+1})\rangle$ where $p = b_{n+2} - b_{k+1}v_{k+1} - \cdots - b_{n+1}v_{n+1}$. So $\mathcal{L}(x_{n+2}(v))\,\mathcal{R}(x_1(w))\langle A,B\rangle$ is equal to this very symmetric expression. Using inv or doing the same sort of computation for $\mathcal{R}(x_1(w))\,\mathcal{L}(x_{n+2}(v))\langle A,B\rangle$ one sees that the results are the same. Other proof: Use that $\mathcal{L}(x_{n+2}(v))\,\mathcal{R}(x_1(w))\langle A,B\rangle$ apparently can be written in the form $\langle$(product of $x_{ij}(r)$'s with $i > j$), (product of $x_{ij}(r)$'s with $i < j$)$\rangle$. Then, applying inv, derive the same result for $\mathcal{R}(x_1(w))\,\mathcal{L}(x_{n+2}(v))\langle A,B\rangle$. And show that two elements of this particular form are equal as soon as their images under mat are equal. (Reduce for instance to the case that one of the two elements is trivial). For $n > 2$ there still is another proof, based on writing $v = z + (v-z)$ where z and $v - z$ have more zeroes than we assumed for v. The squeezing principle will then do the job. At any rate, the computation may look horrendous but there really is no problem.

4.7 Lemma Let $A \in St(\{n+2\} \times [n+2])$, $B \in (St\{1\} \times [n+2])$, $T \in St(n+1)$, $U \in St(\{1\}^* \times \{1\}^*)$, $v = (v_1,\ldots,v_{n+1})$, $w = (w_2,\ldots,w_{n+2})$, $z = (z_2,\ldots,z_{n+2})$ such that $x_{n+2}(v)T = Tx_{n+1,n+2}(*)$, $Ux_1(w) = x_1(z)U$, $(z_2,\ldots,z_n)$ is unimodular. Then $\mathcal{L}(x_{n+2}(v))$, $\mathcal{R}(x_1(w))$ slide past each other at $\langle TA,BU\rangle$.

Proof We may assume $T = 1$ and $U = 1$. We want to get rid of w_{n+1} and w_{n+2}. As $(w_2,\ldots,w_n)$ is unimodular, there is $P \in St(\{n+1,n+2\} \times \{1,n+2\}^*)$ with $Px_1(w) = x_1(w_2,\ldots,w_n,0,0)P$. We have $\langle A,B\rangle = \langle AP^{-1},PB\rangle$ and inspection shows that $\mathcal{L}(x_{n+2}(v))$, $\mathcal{R}(x_1(w))$ slide past each other at $\langle AP^{-1},PB\rangle$.

4.8 So now we are in a situation comparable with 3.29. We can now take up the problem mentioned in 4.4 but we will handle it in a way which does not apply to some commutative semi-local rings.

4.9 Proposition Let $v = (v_1,\ldots,v_{n+1})$, $w = (w_2,\ldots,w_{n+2})$. Then $\mathcal{L}(x_{n+2}(v))\cdot\mathcal{R}(x_1(w)) = \mathcal{R}(x_1(w))\cdot\mathcal{L}(x_{n+2}(v))$.

Proof So say $\mathcal{L}(x_{n+2}(v))\mathcal{R}(x_1(w))\langle X,Y\rangle$ and $\mathcal{R}(x_1(w))\mathcal{L}(x_{n+2}(v))\langle X,Y\rangle$ are both defined. We have to show they are equal. Using $SR^4_{n+1}(n+2,n+1)$ as in 3.30, 3.32 one can choose z such that in $\mathcal{L}(x_{n+2}(-z))\mathcal{L}(x_{n+2}(v+z))\mathcal{R}(x_1(w))\langle X,Y\rangle$ and $\mathcal{R}(x_1(w))\mathcal{L}(x_{n+2}(-z))\mathcal{L}(x_{n+2}(v+z))\langle X,Y\rangle$ both executions of $\mathcal{L}(x_{n+2}(-z))$ and both executions of $\mathcal{L}(x_{n+2}(v+z))$ are defined at the bottom. Recall that $\mathcal{R}(x_1(w))\langle P,Q\rangle$ is defined when $\underline{mat}(PQx_1(w))\in\underline{mat}(C)$. So $\mathcal{R}(x_1(w))\mathcal{L}(x_{n+2}(v+z))\langle X,Y\rangle$ is defined. We want to write $\mathcal{R}(x_1(w))\mathcal{L}(x_{n+2}(v))\langle X,Y\rangle =$
$\mathcal{R}(x_1(w))\mathcal{L}(x_{n+2}(-z))\mathcal{L}(x_{n+2}(v+z))\langle X,Y\rangle =$
$\mathcal{L}(x_{n+2}(-z))\mathcal{R}(x_1(w))\mathcal{L}(x_{n+2}(v+z))\langle X,Y\rangle =$
$\mathcal{L}(x_{n+2}(-z))\mathcal{L}(x_{n+2}(v+z))\mathcal{R}(x_1(w))\langle X,Y\rangle =$
$\mathcal{L}(x_{n+2}(v))\mathcal{R}(x_1(w))\langle X,Y\rangle$. So what we need is the statement: "$\mathcal{R}(x_1(w))\mathcal{L}(x_{n+2}(v))\langle X,Y\rangle = \mathcal{L}(x_{n+2}(v))\mathcal{R}(x_1(w))\langle X,Y\rangle$ holds when both executions of $\mathcal{L}(x_{n+2}(v))\langle X,Y\rangle$ are defined at the bottom." So we may assume both executions of $\mathcal{L}(x_{n+2}(v))$ are such. Now we can use $SR^2_n(n+2,n+1)$, which certainly holds, because even $SR^3_n(n+2,n+1)$ holds. We see that there is $u = (u_2,\ldots,u_{n+2})$ such that $\mathcal{L}(x_{n+2}(v))$, $\mathcal{R}(x_1(u))$ slide past each other at $\langle X,Y\rangle$ (in the way described by Lemma 4.7) and such that $\mathcal{L}(x_{n+2}(v))$, $\mathcal{R}(x_1(u-w))$ slide past each other at $\mathcal{R}(x_1(w))\langle X,Y\rangle$ (in same way). Then $\mathcal{R}(x_1(w))\mathcal{L}(x_{n+2}(v))\langle X,Y\rangle =$
$\mathcal{R}(x_1(w-u))\mathcal{R}(x_1(u))\mathcal{L}(x_{n+2}(v))\langle X,Y\rangle =$
$\mathcal{R}(x_1(w-u))\mathcal{L}(x_{n+2}(v))\mathcal{R}(x_1(u))\langle X,Y\rangle =$
$\mathcal{R}(x_1(w-u))\mathcal{L}(x_{n+2}(v))\mathcal{R}(x_1(u-w))\mathcal{R}(x_1(w))\langle X,Y\rangle =$

$\mathcal{R}(x_1(w-u))\mathcal{R}(x_1(u-w))\mathcal{L}(x_{n+2}(v))\mathcal{R}(x_1(w))\langle X,Y\rangle =$
$\mathcal{L}(x_{n+2}(v))\mathcal{R}(x_1(w))\langle X,Y\rangle$.

4.10 Definition Recall that π: $C \to St(n+2)$ is given by $\pi\langle X,Y\rangle = XY$. We define the set left to be the set of maps $\mathcal{L}$ which satisfy

(1) There is an element $g \in St(n+2)$ such that

 (1a) The domain of $\mathcal{L}$ consists of the $\langle X,Y\rangle \in C$ with $\underline{mat}(gXY) \in \underline{mat}(C)$.

 (1b) The image of $\mathcal{L}$ is contained in C and for $\langle X,Y\rangle$ in the domain one has $\pi\mathcal{L}\langle X,Y\rangle = g\pi\langle X,Y\rangle$.

(2) For $U \in St(\{1\}^* \times \{1\}^*)$ one has $\mathcal{R}(U)\cdot\mathcal{L} \approx \mathcal{L}\cdot\mathcal{R}(U)$.

(3) Let $w = (w_2,\ldots,w_{n+2})$. Then $\mathcal{R}(x_1(w))\cdot\mathcal{L} \approx \mathcal{L}\cdot\mathcal{R}(x_1(w))$.

Remark This also makes sense for commutative semi-local rings We will give a different proof of 4.9 in that case. The remainder of the proofs goes through for commutative semi-local rings too.

4.11 From what we have proved until now it should be clear that left contains at least the following elements: $\mathcal{L}(X)$ for $X \in \underline{Low}$, $\mathcal{L}(x_{n+2}(v))$ for $v \in R^{n+1}$.

4.12 Lemma Let $\mathcal{L}_1, \mathcal{L}_2 \in \underline{left}$. There is a unique $\mathcal{L}_3 \in \underline{left}$ with $\mathcal{L}_1\cdot\mathcal{L}_2 \approx \mathcal{L}_3$.

Proof We first prove uniqueness. Suppose $\mathcal{L}_1\cdot\mathcal{L}_2 \approx \mathcal{L}_3$ and $\mathcal{L}_1\cdot\mathcal{L}_2 \approx \mathcal{L}_4$ with $\mathcal{L}_i \in \underline{left}$ for $i = 1,2,3,4$. We have $SR^3_{n+2}(n+2,n+2)$, even for commutative semi-local rings, so with Lemma 3.36 we see there is $w = (w_2,\ldots,w_{n+2})$ such that $\mathcal{L}_1\cdot\mathcal{L}_2\langle x_1(w),1\rangle$, $\mathcal{L}_3\langle x_1(w),1\rangle$ are defined. So if $g_i \in St(n+2)$ correspond to $\mathcal{L}_i$ then $g_1g_2x_1(w) = g_3x_1(w)$, or $g_1g_2 = g_3$.

Similarly $g_1g_2 = g_4$, so $g_3 = g_4$ and $\mathcal{L}_3, \mathcal{L}_4$ have the same domains. Let $\mathcal{L}_3\langle X,Y\rangle$ be defined. Using $SR^3_{n+2}(n+2,n+2)$ again we can choose $z = (z_2,\ldots,z_{n+2})$ such that $\mathcal{L}_1\mathcal{L}_2\mathcal{R}(x_1(z))\langle X,Y\rangle$ is defined. Then $\mathcal{R}(x_1(z))\mathcal{L}_3\langle X,Y\rangle = \mathcal{L}_3\mathcal{R}(x_1(z))\langle X,Y\rangle = \mathcal{L}_1\mathcal{L}_2\mathcal{R}(x_1(z))\langle X,Y\rangle =$ $\mathcal{L}_4\mathcal{R}(x_1(z))\langle X,Y\rangle = \mathcal{R}(x_1(z))\mathcal{L}_4\langle X,Y\rangle$. Now apply $\mathcal{R}(x_1(-z))$ to both ends. So far for uniqueness. Next existence. What we just did indicates how to define $\mathcal{L}_3$: it is clear what the domain should be and if $\langle X,Y\rangle$ is in that domain then $\mathcal{R}(x_1(-w))\mathcal{L}_1\mathcal{L}_2\mathcal{R}(x_1(w))\langle X,Y\rangle$ is defined for some w, again by $SR^3_{n+2}(n+2,n+2)$. (One step, the last one, is automatically defined because $\langle X,Y\rangle$ is "in the domain"). So put $\mathcal{L}_3\langle X,Y\rangle$ equal to this expression. First note that the result does not depend on the particular choice of w because

$\mathcal{R}(x_1(-w-v))\mathcal{L}_1\mathcal{L}_2\mathcal{R}(x_1(v+w))\langle X,Y\rangle =$
$\mathcal{R}(x_1(-w))\mathcal{R}(x_1(-v))\mathcal{L}_1\mathcal{L}_2\mathcal{R}(x_1(v+w))\langle X,Y\rangle =$
$\mathcal{R}(x_1(-w))\mathcal{L}_1\mathcal{R}(x_1(-v))\mathcal{L}_2\mathcal{R}(x_1(v+w))\langle X,Y\rangle =$
$\mathcal{R}(x_1(-w))\mathcal{L}_1\mathcal{L}_2\mathcal{R}(x_1(-v))\mathcal{R}(x_1(v+w))\langle X,Y\rangle =$
$\mathcal{R}(x_1(-w))\mathcal{L}_1\mathcal{L}_2\mathcal{R}(x_1(w))\langle X,Y\rangle$ whenever the two ends are defined.

So $\mathcal{L}_3$ is well-defined and obviously $\mathcal{L}_1\cdot\mathcal{L}_2 \approx \mathcal{L}_3$. We have to show that $\mathcal{L}_3\in$<u>left</u>. The first condition is satisfied. Next let $U\in St(\{1\}^* \times \{1\}^*)$ and let w, $\langle X,Y\rangle$ be as above. Choose y such that $x_1(w)U = Ux_1(y)$. Then

$\mathcal{L}_3\mathcal{R}(U)\langle X,Y\rangle = \mathcal{R}(x_1(-y))\mathcal{L}_1\mathcal{L}_2\mathcal{R}(x_1(y))\mathcal{R}(U)\langle X,Y\rangle =$
$\mathcal{R}(x_1(-y))\mathcal{L}_1\mathcal{L}_2\mathcal{R}(U)\mathcal{R}(x_1(w))\langle X,Y\rangle =$
$\mathcal{R}(x_1(-y))\mathcal{R}(U)\mathcal{L}_1\mathcal{L}_2\mathcal{R}(x_1(w))\langle X,Y\rangle =$
$\mathcal{R}(U)\mathcal{R}(x_1(-w))\mathcal{L}_1\mathcal{L}_2\mathcal{R}(x_1(w))\langle X,Y\rangle = \mathcal{R}(U)\mathcal{L}_3\langle X,Y\rangle$. Finally
$\mathcal{R}(x_1(v))\cdot\mathcal{L}_3 \approx \mathcal{L}_3\cdot\mathcal{R}(x_1(v))$ follows from
$\mathcal{R}(x_1(v))\mathcal{R}(x_1(-w))\mathcal{L}_1\mathcal{L}_2\mathcal{R}(x_1(w))\langle X,Y\rangle =$
$\mathcal{R}(x_1(v-w))\mathcal{L}_1\mathcal{L}_2\mathcal{R}(x_1(w-v))\mathcal{R}(x_1(v))\langle X,Y\rangle$, where w is chosen such that the first member is defined.

4.13 <u>Definition</u> Let $\mathcal{L}_1, \mathcal{L}_2, \mathcal{L}_3$ be as in Lemma 4.12. We write

$\mathcal{L}_3 = \mathcal{L}_1 * \mathcal{L}_2$. So that makes <u>left</u> into a set with composition $*$. There is a neutral element: $\mathcal{L} * \mathrm{id} = \mathrm{id} * \mathcal{L} = \mathcal{L}$ for $\mathcal{L} \in$ <u>left</u>.

4.14 <u>Lemma</u> Let $\mathcal{L}_1, \mathcal{L}_2, \mathcal{L}_3 \in$ <u>left</u>. Then $(\mathcal{L}_1 * \mathcal{L}_2) * \mathcal{L}_3 = \mathcal{L}_1 * (\mathcal{L}_2 * \mathcal{L}_3)$. (So we will write $\mathcal{L}_1 * \mathcal{L}_2 * \mathcal{L}_3$).

<u>Proof</u> It is clear that $(\mathcal{L}_1 * \mathcal{L}_2) * \mathcal{L}_3$ and $\mathcal{L}_1 * (\mathcal{L}_2 * \mathcal{L}_3)$ have the same domains. (See proof of 4.12). Let $(\mathcal{L}_1 * \mathcal{L}_2) * \mathcal{L}_3 \langle X,Y \rangle$ be defined. Choose u such that $(\mathcal{L}_1 * \mathcal{L}_2)\mathcal{L}_3 \mathcal{R}(x_1(u))\langle X,Y \rangle$ is defined and choose v such that $\mathcal{L}_1 \mathcal{L}_2 \mathcal{R}(x_1(v)) \mathcal{L}_3 \mathcal{R}(x_1(u))\langle X,Y \rangle$ is defined, both times applying $SR^3_{n+2}(n+2, n+2)$. Then
$(\mathcal{L}_1 * \mathcal{L}_2) * \mathcal{L}_3 \langle X,Y \rangle = \mathcal{R}(x_1(-u))(\mathcal{L}_1 * \mathcal{L}_2)\mathcal{L}_3 \mathcal{R}(x_1(u))\langle X,Y \rangle =$
$\mathcal{R}(x_1(-u-v))\mathcal{L}_1 \mathcal{L}_2 \mathcal{R}(x_1(v)) \mathcal{L}_3 \mathcal{R}(x_1(u))\langle X,Y \rangle$. Next choose w, z such that $\mathcal{L}_1(\mathcal{L}_2 * \mathcal{L}_3)\mathcal{R}(x_1(w))\langle X,Y \rangle$ and $\mathcal{L}_2 \mathcal{L}_3 \mathcal{R}(x_1(z))\langle X,Y \rangle$ are defined. (Again by $SR^3_{n+2}(n+2, n+2)$). Then

$$\begin{aligned}
&\mathcal{L}_1 * (\mathcal{L}_2 * \mathcal{L}_3)\langle X,Y \rangle = \mathcal{R}(x_1(-w))\mathcal{L}_1(\mathcal{L}_2 * \mathcal{L}_3)\mathcal{R}(x_1(w))\langle X,Y \rangle = \\
&\mathcal{R}(x_1(-w))\mathcal{L}_1 \mathcal{R}(x_1(w-z))\mathcal{L}_2 \mathcal{L}_3 \mathcal{R}(x_1(z))\langle X,Y \rangle = \\
&\mathcal{R}(x_1(-w))\mathcal{L}_1 \mathcal{R}(x_1(w-z))\mathcal{L}_2 \mathcal{R}(x_1(z-u))\mathcal{L}_3 \mathcal{R}(x_1(u))\langle X,Y \rangle = \\
&\mathcal{R}(x_1(-w))\mathcal{L}_1 \mathcal{R}(x_1(w-u-v))\mathcal{L}_2 \mathcal{R}(x_1(v))\mathcal{L}_3 \mathcal{R}(x_1(u))\langle X,Y \rangle = \\
&\mathcal{R}(x_1(-u-v))\mathcal{L}_1 \mathcal{L}_2 \mathcal{R}(x_1(v))\mathcal{L}_3 \mathcal{R}(x_1(u))\langle X,Y \rangle = (\mathcal{L}_1 * \mathcal{L}_2) * \mathcal{L}_3 \langle X,Y \rangle.
\end{aligned}$$

4.15 <u>Definition</u> $\mathcal{L}_1 \in$ <u>left</u> is called a unit if there is $\mathcal{L}_2 \in$ <u>left</u> with $\mathcal{L}_1 * \mathcal{L}_2 = \mathcal{L}_2 * \mathcal{L}_1 = \mathrm{id}$. In fact one can show that every element of <u>left</u> is a unit, but we will not need that. (The easiest proof is when Theorem 4 is established). Define <u>Uleft</u> to be the group of units in <u>left</u>, with $*$ as composition. This group contains the maps $\mathcal{L}(x_{n+2}(v))$ and $\mathcal{L}(X)$ for $v \in R^{n+1}$, $X \in$ <u>Low</u> resp. In particular all the $\mathcal{L}(x_{ij}(r))$ are in <u>Uleft</u> for $i,j \in [n+2]$, $i \neq j$. We want to show that $x_{ij}(r) \mapsto \mathcal{L}(x_{ij}(r))$ defines a homomorphism ρ: $St(n+2) \to$ <u>Uleft</u>. Therefore we want to prove relations between the $\mathcal{L}(x_{ij}(r))$.

4.16 Put $y_{ij}(r) = \mathcal{L}(x_{ij}(r))$ for $r \in R$, $i,j \in [n+2]$, $i \neq j$. We will see that the $y_{ij}(r)$ satisfy, in <u>Uleft</u>, the Steinberg relations (i.e. the relations (1) through (5) in 3.4 with x_{ij} replaced by y_{ij}). That will give us a homomorphism ρ: $St(n+2) \to$ <u>Uleft</u>, with $x_{ij}(r)$ going to $y_{ij}(r)$ for $r \in R$, $i,j \in [n+2]$, $i \neq j$. One easily sees that $\rho(T) = \mathcal{L}(T)$ for $T \in St(n+1)$. (At this moment we are not interested in what ρ does to other elements). The homomorphism ρ thus induces a homomorphism ρ': $St(n+1) \to$ <u>Uleft</u> which is injective by 3.7. That is what we want: Now $St(n+1) \to St(n+2)$ is seen to be injective and by restriction $K_2(n+1,R) \to K_2(n+2,R)$ is injective. (cf. introduction.)

4.17 So let us try and prove the necessary relations between the $y_{ij}(r)$ in <u>Uleft</u>. As $\mathcal{L}(X) * \mathcal{L}(X') = \mathcal{L}(XX')$ for $X,X' \in$ <u>Low</u>, it is easy to see that the relations are satisfied which don't involve any $y_{i,n+2}(r)$.

4.18 <u>Lemma</u> $\mathcal{L}(x_{n+2}(v)) * \mathcal{L}(x_{n+2}(w)) = \mathcal{L}(x_{n+2}(v+w))$ for $v,w \in R^{n+1}$.

<u>Proof</u> We have to show that $\mathcal{L}(x_{n+2}(v)) \cdot \mathcal{L}(x_{n+2}(w)) \approx \mathcal{L}(x_{n+2}(v+w))$. But that is known as "additivity."

4.19 Let $v,w \in R^{n+1}$, $T \in St(n+1)$ such that $Tx_{n+2}(v) = x_{n+2}(w)T$. Then $\mathcal{L}(T) * \mathcal{L}(x_{n+2}(v)) = \mathcal{L}(x_{n+2}(w)) * \mathcal{L}(T)$, by Lemma 3.3 4.

4.20 Only the following type of relation is not covered by 4.17, 4.18, 4.19: $[y_{j,n+2}(r), y_{n+2,i}(s)] = y_{ji}(rs)$, where $r,s \in R$, $i,j \in [n+1]$, $i \neq j$. So we will be done if we prove this type of relation.

4.21 <u>Proposition</u> $[y_{j,n+2}(r), y_{n+2,i}(s)] = y_{ji}(rs)$ with i,j,r,s as above.

Proof Put $\mathcal{L} = y_{j,n+2}(r)*y_{n+2,i}(s)*y_{j,n+2}(-r)*y_{n+2,i}(-s)*y_{ji}(-rs)$ and $f = y_{j,n+2}(r)\cdot y_{n+2,i}(s)\cdot y_{j,n+2}(-r)\cdot y_{n+2,i}(-s)\cdot y_{ji}(-rs)$, so that $\mathcal{L}$ extends f. We have to show that $\mathcal{L}$ = id. So let $\langle X,Y\rangle \in C$. Using $SR^2_{n+2}(n+2,n+2)$ we can choose $w = (w_2,\dots,w_{n+2})$ such that $\underline{mat}(XYx_1(w))$ and $\underline{mat}(x_{j,n+2}(-r)x_{n+2,i}(-s)x_{ji}(-rs)XYx_1(w))$ are both in $\underline{mat}(C)$. It is clear that the element corresponding to $\mathcal{L}$ in $St(n+2)$ is 1, so $\mathcal{L}$ is defined on all of C. We have $\mathcal{L}\,\mathcal{R}(x_1(w))\langle X,Y\rangle = \mathcal{R}(x_1(w))\mathcal{L}\langle X,Y\rangle$, so to prove that $\mathcal{L}\langle X,Y\rangle$ is equal to $\langle X,Y\rangle$, we may replace $\langle X,Y\rangle$ by $\mathcal{R}(x_1(w))\langle X,Y\rangle$. In other words, we may assume that $f\langle X,Y\rangle$ is defined. (Use Lemma 3.36). So now we have to prove that $f\langle X,Y\rangle = \langle X,Y\rangle$. If $U \in \underline{Up}$ then $f\cdot\mathcal{R}(U) = \mathcal{R}(U)\cdot f$ because $y_{pq}(b)\cdot\mathcal{R}(U) = \mathcal{R}(U)\cdot y_{pq}(b)$ for all $b\in R$, $p,q\in[n+2]$, $p \neq q$. So we can further reduce to the case $Y = 1$, or more precisely, the case $Y = 1$, $X = Tx_{n+2,1}(a)$, where $T\in St(n+1)$, $a\in R$. Using $\underline{inv}$ (see 4.1), additivity, and Lemma 3.24, one checks that $\mathcal{R}(x_{n+2,1}(-a))\langle X,1\rangle =$ $\mathcal{R}(x_{n+2,1}(-a)x_{n+1,1}(-1))\mathcal{R}(x_{n+1,1}(1))\langle X,1\rangle = \langle Xx_{n+2,1}(-a),1\rangle$. Thus it reduces further to proving that $\mathcal{L}\langle T,1\rangle = \langle T,1\rangle$ for $T\in St(n+1)$. But $f\langle T,1\rangle$ is defined for such T, as one sees from inspecting the intermediate result on the matrix level (use Lemma 3.36 again). So we still have to prove $f\langle T,1\rangle = \langle T,1\rangle$ for such T. Now observe that $f\cdot\mathcal{R}(x_{pq}(b)) = \mathcal{R}(x_{pq}(b))\cdot f$ both for $q = 1$ and for $q \neq 1$, if $p \neq n+2$ ($p \neq q$, $b\in R$). In particular, $\mathcal{R}(x_{pq}(b))f\langle T,1\rangle =$ $f\,\mathcal{R}(x_{pq}(b))\langle T,1\rangle = f\langle Tx_{pq}(b),1\rangle$ for $p,q\in[n+1]$, $p \neq q$, $b\in R$. Therefore it reduces to the case $T = 1$ where it is easy. (Apply squeezing principle or compute).

§5. The case d = 0, n = 2.

5.1 We still have to prove Theorem 1 in the case d = 0, n = 2, because $\tilde{\tilde{SR}}_2$ doesn't always hold for d = 0. (d = 0 corresponds to semi-local commutative rings). From now on let n = 2 (in this section). We will use a few properties of commutative semi-local rings instead of the properties $SR_n^3(n+2,n+1)$, $SR_{n+1}^4(n+2,n+1)$ which we used in the previous sections. For future applications let us list which properties we need.

5.2 <u>Definition</u> Let R be a ring. We say that R satisfies SR_2^* if the following holds:

(1) R satisfies SR_2 and $SR_4^3(4,4)$.

(2) If $\binom{a}{b} \in R^2$, there is $T \in E(2,R)$ such that $T\binom{a}{b} = \binom{p}{q}$ with p contained in the Jacobson radical of R.

5.3 Examples: Commutative semi-local rings satisfy SR_2^* and division rings satisfy SR_2^*. For further examples see Section 6.

5.4 Theorem 1 will follow from Theorems 2,4, and 5 where Theorem 5 is the Theorem we prove in this section. It reads:

<u>THEOREM 5</u> Let R satisfy SR_2^* (see 5.2). Then the natural map $K_2(3,R) \to K_2(4,R)$ is an isomorphism.

We only need to show that the map is injective, as surjectivity is known. (For division rings the full Theorem is known. See [5]).

5.5 <u>Lemma</u> (Kaplansky, Lenstra). Let R satisfy SR_2 and let $xy = 1$ in R. Then $yx = 1$.

<u>Proof</u> The column (x,1-yx) is unimodular, so there are $a,b \in R$ with $a(x + b(1-yx)) = 1$. Put $z = x + b(1-yz)$. One checks that $zy = 1$. So now $xy = zy = az = 1$. Then z is a two-sided unit, y is its

inverse and x is the inverse of y. Or, in formula's:
$yx = (az)yx(a(zy)z) = a(zy)(x(az)y)z = 1$.

5.6 Let R satisfy SR_2^*. We can use 3.4 through 3.29, so let us start from 3.29. Our immediate goal is to prove "additivity" and to define $\mathcal{L}(x_4(v))\langle X,Y\rangle$ when $\underline{mat}(x_4(v)XY)\in\underline{mat}(C)$. (As $n = 2$ we now write 4 instead of $n + 2$).

<u>Notations</u> R^* denotes the group of units (all units are two-sided by Lemma 5.5) and Rad(R) denotes the Jacobson radical of R.

5.7 <u>Definition</u> Let $a,q\in R$ with $1 + aq\in R^*$. Let $Y\in\underline{Up}$. We define $\mathcal{L}(x_{14}(a))\langle x_{41}(q),Y\rangle$ to be the element $\langle P,QY\rangle$ of C with $P\in St(\{3\}^* \times \{4\}^*)$, $Q\in St(\{3\}^* \times \{1\}^*)$ chosen such that $PQ = x_{14}(a)x_{41}(q)$ in $St(\{3\}^* \times [4])$. So we have to show that such P,Q exist. Then the definition will be consistent by the squeezing principle with $i = 3$. (Although P,Q are not unique, $\langle P,Q\rangle$ is.) And just as in 3.13 one sees that $(x_{41}(q),Y)$ is the only representative of its class which looks like $(x_{41}(*),*)$. We also should compare with earlier definitions, i.e. with those preceding 3.29. But these earlier definitions just don't cover the present case, so there is no conflict.

5.8 So let us show that P,Q exist. We will actually show more. Recall that $w_{ij}(t) = x_{ij}(t)x_{ji}(-t^{-1})x_{ij}(t)$, $h_{ij}(t) = w_{ij}(t)w_{ij}(1)^{-1}$ if t is a unit. Take $t = 1 + aq$. We have $h_{12}(t)\in St(2)$ and its image in $E(4)$ is a diagonal matrix. Choose $P = h_{12}(t)x_{41}(q)$ in $St(\{1,2,4\} \times \{1,2\})$. (Note that in 5.7 we only needed $P\in St(\{3\}^* \times \{4\}^*)$. Such a P is the image of the P we have chosen.) We have to show that there is $Q\in St(\{3\}^* \times \{1\}^*)$ with $PQ = x_{14}(a)x_{41}(q)$ in $St(\{3\}^* \times [4])$. We will actually show that there is $Q \in St(\{1,2,4\} \times \{2,3\})$ with this property. (That's a stronger result). So look at $\underline{mat}(P^{-1}x_{14}(a)x_{41}(q))$. It has a second column of the form $(0,t,0,0)$ and a fourth column of the form $(t^{-1}a,0,0,1-qt^{-1}a)$. The other two columns are trivial. Choose $D' = P^{-1}x_{14}(a)x_{41}(q)x_{14}(-t^{-1}a) \in St(\{1,2,4\} \times \{1,2,4\})$. Then $\underline{mat}(D')$ is a diagonal matrix with diagonal $(1,t,1,1-qt^{-1}a)$. By injective stability for K_1 we have $\underline{mat}(D') \in E(\{2,4\} \times \{2,4\})$ and by surjective stability for K_2 we actually have some $D \in St(\{2,4\} \times \{2,4\})$ with $D = D'$ in $St(\{1,2,4\} \times \{1,2,4\})$. (One can avoid the stability results by giving an explicit description of such an element D). The element $Q = Dx_{14}(t^{-1}a)$ in $St(\{1,2,4\} \times \{2,4\})$ satisfies the requirements.

<u>Remark</u> In the commutative case one would use Steinberg symbols and/or the symbols $\langle a,q \rangle$ from [7], [16] to give an expression for Q. In the context of the notations in this note such a description would be very confusing however.

5.9 <u>Notation</u> Let $a,q \in R$ with $1 + aq \in R^*$. Let $Y \in \underline{Up}$. We put $L(x_{14}(a))(x_{41}(q),Y) = (P,QY) = (P,Dx_{14}(t^{-1}a)Y)$ where P,D,Q are chosen as in 5.8. So $P \in St(\{1,2,4\} \times \{1,2\})$, $\underline{mat}(P)$ is a lower triangular matrix with the only non-zero off-diagonal element at position $(4,1)$. And $Q \in St(\{1,2,4\} \times \{2,4\})$, $D \in St(\{2,4\} \times \{2,4\})$, $\underline{mat}(D)$ is a diagonal matrix. As usual the notation is such that $L(x_{14}(a))(x_{41}(q),Y)$ is a representative of $\mathcal{L}(x_{14}(a))\langle x_{41}(q),Y \rangle$.

5.10 <u>Lemma</u> Let a,q,Y be as above and let $T \in St(\{2,3\} \times \{2,3\})$. Then $\mathcal{L}(T)\mathcal{L}(x_{14}(a))\langle x_{41}(q),Y \rangle = \mathcal{L}(x_{14}(a))\mathcal{L}(T)\langle x_{41}(q),Y \rangle$.

<u>Proof</u> We may assume that T is one of the generators of $St(\{2,3\} \times \{2,3\})$. If $T = x_{2,3}(*)$ we can use the squeezing prin-

ciple or just the definition of $\mathcal{L}(x_{14}(a))$. So let $T = x_{32}(*)$. Say $L(x_{14}(a))(x_{41}(q),Y) = (P,QY)$. Then
$\langle TP,QY\rangle = \langle x_{32}(*)h_{12}(t)x_{41}(q),Dx_{14}(*)Y\rangle =$
$\langle h_{12}(t)x_{32}(*)x_{41}(q),Dx_{14}(*)Y\rangle = \langle P,x_{32}(*)Dx_{14}(*)Y\rangle =$
$\langle P,DTx_{14}(*)Y\rangle = \langle P,QTY\rangle$ where we use semi-direct products to move the $x_{32}(*)$ through. (Although the coefficient of x_{32} changes along the way it returns to its old value because mat(T) commutes with mat(PQ).)

5.11 As a special case look at $T = w_{3,2}(1)$. We get from the Lemma that $\mathcal{L}(x_{14}(a))\langle x_{41}(q),Y\rangle = \langle TPT^{-1},TQT^{-1}Y\rangle$, in the usual notations. But conjugation by T replaces the 2's in the subscripts by 3's. So we see that we can also define $\mathcal{L}(x_{14}(a))\langle x_{41}(q),Y\rangle$ to be $\langle P',Q'Y\rangle$ where $P' \in St(\{2\}^* \times \{4\}^*)$, $Q' \in St(\{2\}^* \times \{1\}^*)$ are such that $P'Q' = x_{14}(a)x_{41}(q)$ in $St(\{2\}^* \times [4])$. So our preference for row 2 in definition 5.7 has no influence on the result; we could have used row 3 instead of row 2. This leads to:

5.12 <u>Notation</u> In 5.9 replace the indices 2 by indices 3 in order to get elements $P' \in St(\{1,3,4\} \times \{1,3\})$, $Q' \in St(\{1,3,4\} \times \{3,4\})$ from P,Q resp. Then put $L'(x_{14}(a))(x_{41}(q),Y) = (P',Q'Y)$. So $L'(x_{14}(a))$ represents another way to evaluate $\mathcal{L}(x_{14}(a))$, this time with preference for the third row instead of the second one.

5.13 <u>Lemma</u> Let a,q,Y be as in 5.7 and let $b,c \in R$. Then
$\mathcal{L}(x_{14}(a))\mathcal{L}(x_{12}(c)x_{13}(b))\langle x_{41}(q),Y\rangle =$
$\mathcal{L}(x_{12}(c)x_{13}(b))\mathcal{L}(x_{14}(a))\langle x_{41}(q),Y\rangle$.

<u>Proof</u> The squeezing principle applies.

5.14 <u>Definition</u> Let a,q,Y be as above. Let $v \in R^3$, $T \in St(3)$ such that $x_4(v)T = Tx_{14}(a)$. (Note that $n + 2 = 4$). Then we put $\mathcal{L}(x_4(v))\langle Tx_{41}(q),Y\rangle = \mathcal{L}(T)\mathcal{L}(x_{14}(a))\langle x_{41}(q),Y\rangle$. It is easy to see from Lemma 5.13 and Lemma 5.10 that this is a consistent definition. It is also easy to see that this particular $\mathcal{L}(x_1(v))\langle Tx_{41}(q),Y\rangle$ has not been defined before 3.29. And the present definition is clearly compatible with the one in 5.7.

5.15 Lemma Let a,q,Y be as in 5.13 and let $\mathcal{L}(x_{14}(a))\langle x_{41}(q),Y\rangle = \langle P,QY\rangle$. Then both steps in $\mathcal{L}(x_{24}(*))\mathcal{L}(x_{34}(*))\langle P,QY\rangle$ are defined at the bottom.

Proof We may take P,Q as in 5.9. Then apply Lemma 3.20.

5.16 Lemma Let a,q,Y be as above, $b,c\in R$. Let $T\in St(\{2,3\} \times \{2,3\})$, $v\in R^3$ such that $Tx_4(a,b,c) = x_4(v)T$. Then $\mathcal{L}(T)\mathcal{L}(x_{24}(b))\mathcal{L}(x_{34}(c))\mathcal{L}(x_{14}(a))\langle x_{41}(q),Y\rangle = \mathcal{L}(x_{24}(v_2))\mathcal{L}(x_{34}(v_3))\mathcal{L}(x_{14}(v_1))\langle Tx_{41}(q),Y\rangle$.

Proof Note that $v_1 = a$. Take P,Q as usual, such that $(P,QY) = L(x_{14}(a))(x_{41}(q),Y)$. Then, by Lemma 5.10, we have $\mathcal{L}(x_{14}(v_1))\langle Tx_{41}(q),Y\rangle = \langle TP,QY\rangle$. Therefore it remains to show that $L(T)L(x_{24}(b))L(x_{34}(c))(P,QY)$ is equivalent to $L(x_{24}(v_2))L(x_{34}(v_3))(TP,QY)$. They are actually equal. (Compare proof of 3.23).

5.17 Proposition Let $q\in R, Y\in Up$, $U\in St(\{1\} \times \{2,3\})$. Let $v,w\in R^3$ be such that $Ux_4(v) = x_4(w)U$. Suppose that both $\mathcal{L}(U)\mathcal{L}(x_{24}(v_2))\mathcal{L}(x_{34}(v_3))\mathcal{L}(x_{14}(v_1))\langle x_{41}(q),Y\rangle$ and $\mathcal{L}(x_{24}(w_2))\mathcal{L}(x_{34}(w_3))\mathcal{L}(x_{14}(w_1))\mathcal{L}(U)\langle x_{41}(q),Y\rangle$ are defined. Then they are equal.

Proof Say $U = x_{12}(-u_2)x_{13}(-u_3)$. Choose $B \in St(\{1,4\} \times \{2,3\})$ such that $x_{41}(q)B = Ux_{41}(q)$. Using 5.2(2), choose $T \in St(\{2,3\} \times \{2,3\})$ such that $Tx_4(v) = x_4(v')T$ with $v'_2 \in Rad(R)$. Applying the previous Lemma we can reduce to the case $v_2 \in Rad(R)$. (Bring $\mathcal{L}(T)$ in from the left and push it over to the right, finally replacing Y by TY). We have $\mathcal{L}(U)\mathcal{L}(x_{24}(v_2))\mathcal{L}(x_{34}(v_3))\mathcal{L}(x_{14}(v_1))\langle x_{41}(q),Y\rangle =$ $\mathcal{L}(x_4(-u_2v_2,v_2,0))\mathcal{L}(x_4(-u_3v_3,v_3,0))\mathcal{L}(x_{14}(v_1))\langle Ux_{41}(q),Y\rangle$ by Lemma 5.15, 3.23, 5.13. So the point is to show that $\mathcal{L}(x_4(-u_2v_2,v_2,0))\mathcal{L}(x_4(-u_3v_3,v_3,0))\mathcal{L}(x_{14}(v_1))\langle x_{41}(q),BY\rangle$ equals $\mathcal{L}(x_{24}(v_2))\mathcal{L}(x_{34}(v_3))\mathcal{L}(x_{14}(v_1-u_2v_2-u_3v_3))\langle x_{41}(q),BY\rangle$. (Note that $v_2 = w_2$, $v_3 = w_3$). As $\mathcal{R}(BY)$ commutes with all the $\mathcal{L}(*)$ we can assume $BY = 1$. Now look at $\mathcal{L}(x_4(-u_3v_3,0,v_3))\mathcal{L}(x_{14}(v_1))\langle x_{41}(q),1\rangle$. We can execute the first step using $L'(x_{14}(v_1))$. (See 5.12). That will leave us with (P,Q) where $P \in St(\{2\}^* \times \{4\}^*)$, $Q \in St(\{2\}^* \times \{1\}^*)$. As <u>mat</u>(P) has a specific form it is easy to see that there is $S \in St(\{1,4\} \times \{3\})$ such that in E(4) $x_4(-u_3v_3,0,v_3)PS = S'x_{34}(*)x_{41}(q)$ with $S' \in St(\{1,3\} \times \{1,3\})$, i.e. such that $(PS,S^{-1}Q)$ is a suitable representative to compute $\mathcal{L}(x_4(-u_3v_3,0,v_3))\langle P,Q\rangle$ with. One sees that there are $P' \in St(\{2\}^* \times \{4\}^*)$, $Q' \in St(\{2\}^* \times \{4\}^*)$ with $\langle P',Q'\rangle = \mathcal{L}(x_4(-u_3v_3,0,v_3))\langle P,Q\rangle = \mathcal{L}(x_4(-u_3v_3,0,v_3))\mathcal{L}(x_{14}(v_1))$ $\langle x_{41}(q),1\rangle$. Similarly $\mathcal{L}(x_{34}(v_3))\mathcal{L}(x_{14}(v_1-u_3v_3))\langle x_{41}(q),1\rangle$ can be computed "inside $St(\{2\}^* \times [4])$", provided that it exists at all. But it does, because the execution of $\mathcal{L}(x_{14}(v_1-u_2v_2-u_3v_3))\langle x_{41}(q),1\rangle$ is defined, with $u_2v_2 \in Rad(R)$. (So deleting u_2v_2 cannot change it into something that isn't defined). We get $\mathcal{L}(x_4(-u_3v_3,0,v_3))\mathcal{L}(x_{14}(v_1))\langle x_{41}(q),1\rangle =$ $\mathcal{L}(x_{34}(v_3))\mathcal{L}(x_{14}(v_1-u_3v_3))\langle x_{41}(q),1\rangle$, by the squeezing principle with $i = 2$. Thus we still have to prove that $\mathcal{L}(x_4(-u_2v_2,v_2,0))\mathcal{L}(x_{34}(v_3))\mathcal{L}(x_{14}(v_1-u_3v_3))\langle x_{41}(q),1\rangle =$ $\mathcal{L}(x_{24}(v_2))\mathcal{L}(x_{34}(v_3))\mathcal{L}(x_{14}(v_1-u_2v_2-u_3v_3))\langle x_{41}(q),1\rangle$. We claim

that the right hand side is equal to
$\mathcal{L}(x_{34}(v_3))\mathcal{L}(x_{24}(v_2))\mathcal{L}(x_{14}(v_1-u_2v_2-u_3v_3))\langle x_{41}(q),1\rangle$. The reason is that $\mathcal{L}(x_{34}(v_3))$, $\mathcal{L}(x_{24}(v_2))$ slide past each other at $\mathcal{L}(x_{14}(v_1-u_2v_2-u_3v_3))\langle x_{41}(q),1\rangle$. This makes that one can execute both steps in the way of Lemma 3.24, both in the case that one does $\mathcal{L}(x_{34}(v_3))$ first and in the case that one does $\mathcal{L}(x_{24}(v_2))$ first. Comparing the two results confirms the claim. (Compare with proof of 3.27). But $\mathcal{L}(x_4(-u_2v_2,v_2,0))$, $\mathcal{L}(x_{34}(v_3))$ also slide past each other at $\mathcal{L}(x_{14}(v_1-u_3v_3))\langle x_{41}(q),1\rangle$, as it is easy to clear the $-u_2v_2$ by means of an element of <u>Med</u>. (Compare proof of 3.28). So the same argument applies to the left hand side of our problem and we are left with proving that
$\mathcal{L}(x_{24}(v_2))\mathcal{L}(x_{14}(v_1-u_2-u_3v_3))\langle x_{41}(q),1\rangle$ is the same as
$\mathcal{L}(x_4(-u_2v_2,v_2,0))\mathcal{L}(x_{14}(v_1-u_3v_3))\langle x_{41}(q),1\rangle$. This is the same sort of problem as we met above, now with 3 replaced by 2 in the indices. So the squeezing principle applies again, with $i = 3$ this time.

5.18 <u>Definition</u> Let $Y \in$ <u>Up</u> and $v,w \in R^3$, $T \in St(3)$ such that $x_4(v)T = Tx_4(w)$. We put
$\mathcal{L}(x_4(v))\langle Tx_{41}(q),Y\rangle = \mathcal{L}(T)\mathcal{L}(x_{24}(w_2))\mathcal{L}(x_{34}(w_3))\mathcal{L}(x_{14}(w_1))$ $\langle x_{41}(q),Y\rangle$ when the right hand side is defined. It is easy to see from 5.16 and 5.17 that this gives a consistent definition. It generalizes the earlier definitions in 5.14, 3.18, 3.13.

Now consider $v \in R^3$, $q \in R$, $Y \in$ <u>Up</u> such that <u>mat</u>$(x_4(v)x_{41}(q)Y) \in$ <u>mat</u>(C). It follows from SR_2 that there are $u_2,u_3 \in R$ such that $1+(v_1-u_2v_2-u_3v_3)q \in R^*$. (See 3.36). Then
$\mathcal{L}(x_4(v))\langle x_{41}(q),*\rangle = \mathcal{L}(x_4(v))\langle x_{12}(u_2)x_{13}(u_3)x_{41}(q),*\rangle$ is defined. More generally one sees that $\mathcal{L}(x_4(v))\langle X,Y\rangle$ is defined if and only if <u>mat</u>$(x_4(v)XY) \in$ <u>mat</u>(C). That is nice. (cf. 5.6). As we want to apply the squeezing principle in the sequel, let us look at the

special case that $\mathcal{L}(x_4(v))\langle X,Y\rangle$ is defined with $x_4(v), X, Y \in St(\{2\}^* \times [4])$. (i.e. the row index 2 is not needed in the expression for these elements in the groups in which they live naturally). In the argument above we don't need u_2 and, at any rate if we use 5.12 for evaluating, we get an answer of the form $\langle P',Q'\rangle$ with $P' \in St(\{2\}^* \times \{4\}^*)$, $Q' \in St(\{2\}^* \times \{1\}^*)$. So this situation leads to a result which is susceptible to the squeezing principle with $i = 2$. (i.e. one can test against a similar element).

The situation is the same if 2 is replaced by 3: If $v = (v_1, v_2, 0)$, $X \in St(\{3\}^* \times \{4\}^*)$, $Y \in St(\{3\}^* \times \{1\}^*)$, then $\mathcal{L}(x_4(v))\langle X,Y\rangle = \langle P',Q'\rangle$ with $P' \in St(\{3\}^* \times \{4\}^*)$, $Q' \in St(\{3\}^* \times \{1\}^*)$ and $P'Q' = x_4(v))XY$ in $St(\{3\}^* \times [4])$.

5.19 __Proposition__ (Additivity) (cf 3.33).
$\mathcal{L}(x_4(v))\mathcal{L}(x_4(w))\langle X,Y\rangle = \mathcal{L}(x_4(v + w))\langle X,Y\rangle$ if both sides are defined.

__Proof__ We may assume that $X = x_{41}(q)$.

Case 1: $w_2 = v_1 = v_3 = 0$. We can use SR_2 and Proposition 5.17 to reduce to the case $1 + w_1 q \in R^*$, and still $w_2 = v_1 = v_3 = 0$. (Also compare 5.18 where we describe how to evaluate $\mathcal{L}(x_4(w))$.) Then it is obvious.

Case 2: $w_2 = 0$. We can now reduce to the case that $\mathcal{L}(x_4(v_1 + w_1, 0, v_3 + w_3))\langle X,Y\rangle$ is defined, by the same method. Then the squeezing principle applies to $\mathcal{L}(x_4(v_1,0,v_3))\mathcal{L}(x_4(w_1,0,w_3))$ $\langle X,Y\rangle$, and $\mathcal{L}(x_{24}(v_2))\mathcal{L}(x_4(v_1,0,v_3))\mathcal{L}(x_4(w_1,0,w_3))\langle X,Y\rangle =$ $\mathcal{L}(x_{24}(v_2))\mathcal{L}(x_4(v_1 + w_1, 0, v_3 + w_3))\langle X,Y\rangle$. Essentially case 1 applies to both sides. (cf. 5.21 case 2).

Case 3: $w_3 = 0$. Reduce to case 2 by means of Lemma 5.16.

Case 4: General case. We may assume $1 + w_1 q \in R^*$. Then $\mathcal{L}(x_4(v))\mathcal{L}(x_4(w)) = \mathcal{L}(x_4(v))\mathcal{L}(x_{24}(w_2))\mathcal{L}(x_4(w_1,0,w_3))\langle X,Y\rangle = \mathcal{L}(x_4(v_1,v_2+w_2,v_3))\mathcal{L}(x_4(w_1,0,w_3))\langle X,Y\rangle = \mathcal{L}(x_4(v+w))\langle X,Y\rangle$ by the previous cases.

5.20 Let us return to 3.30 now. We have proved 3.30. We can forget 3.31 because we have another definition now. Then 3.32 can be deleted too. We proved 3.33 and Lemma 3.34 is obvious from the present definition of $\mathcal{L}(x_4(v))$, $\mathcal{L}(x_4(w))$. (i.e. the definition in 5.18). So we did what we promised in 3.35 and we can enter Section 4. We get stuck again in 4.9. So we have to give a different proof for Proposition 4.9. Then the remainder of Section 4 will apply. So we have to prove:

5.21 <u>Proposition</u> (cf. 4.9). Let $v \in R^3$, $w = (w_2,w_3,w_4)$. Then $\mathcal{L}(x_4(v))\cdot\mathcal{R}(x_1(w)) \approx \mathcal{R}(x_1(w))\cdot\mathcal{L}(x_4(v))$.

<u>Proof</u> So say $\mathcal{L}(x_4(v))\mathcal{R}(x_1(w))\langle X,Y\rangle$ and $\mathcal{R}(x_1(w))\mathcal{L}(x_4(v))\langle X,Y\rangle$ are both defined. We have to show that they are equal. We may assume that $X = x_{41}(q_1)x_{42}(q_2)x_{43}(q_3)$, $Y = x_{12}(r_2)x_{13}(r_3)x_{14}(r_4)$.

Case 1: $v_2 = w_3 = w_4 = 0$. By means of Lemma 3.34 with $T = x_{13}(*)$ we can reduce to the case $1 + v_1 q_1 \in R^*$. One checks that <u>mat</u>$(x_{14}(v_1)XYx_1(w)) \in$ <u>mat</u>(C), using Lemma 3.36. From 5.18 it should be clear that both $\mathcal{L}(x_{14}(v_1))\mathcal{R}(x_1(w))\langle X,Y\rangle$ and $\mathcal{R}(x_1(w))\mathcal{L}(x_{14}(v_1))\langle X,Y\rangle$ can be expressed in the form $\langle P',Q'\rangle$ with $P' \in St(\{3\}^* \times \{4\}^*)$, $Q' \in St(\{3\}^* \times \{1\}^*)$. So they are equal by the squeezing principle, with $i = 3$. One also checks that $\mathcal{L}(x_{34}(v_3))$, $\mathcal{R}(x_1(w))$ slide past each other at $\mathcal{L}(x_{14}(v_1))\langle X,Y\rangle$. The result therefore follows from Proposition 4.6.

Case 2: $v_2 = 0$. We may now assume that $1 + r_4 w_4 \in R^*$. We can apply

Lemma 3.34 to get rid of r_2, r_3. So say $r_2 = r_3 = 0$. Then $\mathcal{R}(x_1(0,w_3,w_4))\langle X,Y\rangle$ has the form $\langle XP',Q'\rangle$ with $P' \in St(\{1,3,4\} \times \{1,3\})$, $Q' \in St(\{1,3,4\} \times \{3,4\})$. From this one sees that the maps $\mathcal{L}(x_4(v))$, $\mathcal{R}(x_{21}(w_2))$ behave like in Case 1 at $\mathcal{R}(x_1(0,w_3,w_4))\langle X,Y\rangle$. So $\mathcal{L}(x_4(v))\mathcal{R}(x_1(w))\langle X,Y\rangle =$ $\mathcal{R}(x_{21}(w_2))\mathcal{L}(x_4(v))\mathcal{R}(x_1(0,w_3,w_4))\langle X,Y\rangle$. The result now follows from the squeezing principle with $i = 2$.

Case 3: General case. We may reduce to the case $v_2 \in Rad(R)$ as follows: For $T \in St(\{2,3\} \times \{2,3\})$ the element $\mathrm{Inn}(T)\langle X,Y\rangle =$ $\mathcal{L}(T)\mathcal{R}(T^{-1})\langle X,Y\rangle$ has the same general form as $\langle X,Y\rangle$. So we may apply $\mathrm{Inn}(T)$ to the problem and move the $\mathrm{Inn}(T)$ over to the $\langle X,Y\rangle$, using results like Lemma 3.34. Once we have $v_2 \in Rad(R)$ we may write $\mathcal{L}(x_4(v))\mathcal{R}(x_1(w))\langle X,Y\rangle$ as $\mathcal{L}(x_{24}(v_2))\mathcal{L}(x_4(v_1,0,v_3))\mathcal{R}(x_1(w))\langle X,Y\rangle$. It follows from Case 2 that $\mathcal{L}(x_4(v))\mathcal{R}(x_1(w))\langle X,Y\rangle =$ $\mathcal{L}(x_{24}(v_2))\mathcal{R}(x_1(w))\mathcal{L}(x_4(v_1,0,v_3))\langle X,Y\rangle$, everything being defined. The element $\mathcal{L}(x_4(v_1,0,v_3))\langle X,Y\rangle$ can be written as $\langle P',Q'\rangle$ with $P' \in St(\{2\}^* \times \{2,4\}^*)$, $Q' \in \underline{Up}$, because $\langle X,Y\rangle$ can be written as $\langle x_{41}(q_1),*\rangle$. Therefore $\mathcal{L}(x_{24}(v_2))$, $\mathcal{R}(x_1(w))$ behave at $\mathcal{L}(x_4(v_1,0,v_3))\langle X,Y\rangle$ as if we are in the case $v_3 = 0$. But that case can be reduced to Case 2.

§6. Multiple Stable Range Conditions for some non-commutative rings.

6.1 In order to apply Theorem 4 (see 3.1) to some non-commutative rings, we want to prove the following generalization of Theorem 2.

THEOREM 6 Let A be a ring, R a subring contained in the center of A. Assume that A is finitely generated as an R-module and that R has a noetherian maximal spectrum of dimension $d < \infty$. Then A satisfies $\widetilde{\widetilde{SR}}_n$ for $n \geq \max(3, d+2)$.

6.2 Similarly we want to apply Theorem 5 (see 5.4) to some non-commutative rings. Recall that $\mathrm{Rad}(R)$ is the Jacobson radical of R. We will prove

THEOREM 7 Let $R/\mathrm{Rad}(R)$ be artinian. Then R satisfies SR_2^*.

6.3 We will prove Theorem 7, but only indicate how to prove Theorem 6. The reason is that the proof of Theorem 6 is not significantly different from the proof of Theorem 2.

6.4 So let $R/\mathrm{Rad}(R)$ be artinian. We have to prove SR_2^*. We may assume $\mathrm{Rad}(R) = 0$ and by the Wedderburn Structure Theorem we may further assume $R = M_q(D)$, the ring of q by q matrices over a division ring D. Let $\binom{a}{b} \in R^2$. So a, b are q by q matrices over D and $\binom{a}{b}$ can be viewed as a $2q$ by q matrix over D. Recall that the rank of a matrix over D is the dimension of the left linear space spanned by its rows, or, equivalently, the dimension of the right linear space spanned by its columns. (See [13], Ch. II, Thm. 9). Viewing a and b as describing left linear maps from D^q to D^q one sees there is $x \in R$ such that the rank of $b + xa$ is the same as the rank of $\binom{a}{b}$. Then there is $y \in R$ with $a = y(b + xa)$ and the matrix $T = \begin{pmatrix} 1 & -y \\ 0 & 1 \end{pmatrix} \begin{pmatrix} 1 & 0 \\ x & 1 \end{pmatrix}$ satisfies $T\binom{a}{b} = \binom{0}{*}$. That proves one of the constituents of SR_2^*. The proof of SR_2 is similar and well known anyway. Remains to prove $SR_4^3(4,4)$ for $R = M_q(D)$.

6.5 So let A_1, A_2, A_3 be 3 by 4 matrices over R, or, in other words, 3q by 4q matrices over D. We assume that A_i is completable to an element of $E_{4q}(D)$, so we assume that A_i has rank 3q. $(i = 1,2,3)$. We have to show that there is a column λ in R^3 such that $A_i\binom{1}{\lambda}$ is unimodular for $i = 1,2,3$. (Here the 1 in $\binom{1}{\lambda}$ is a q by q identity matrix over D). We view λ as a 3q by q matrix over D and construct it column by column. All we have to do is chosing the j-th column of λ in such a way that the j-th column of $A_i\binom{1}{\lambda}$ is right linearly independent from the first $j - 1$ columns of $A_i\binom{1}{\lambda}$. $(i = 1,2,3)$. This amounts to avoiding three right linear varieties of codimension at least $q + 1$. (at least 2). (A "right linear variety" V is a coset of a right linear subspace L. The codimension of V is 3q minus the dimension of L, 3q being the dimension of the space of columns of length 3q). If D is finite one just counts to see that it can be done. If D is infinite one argues as follows: Let V_1, V_2, V_3 be the three varieties to be avoided. Choose v_i outside V_i. Then the line joing v_1 and v_2 intersects $V_1 \cup V_2$ in at most two points. So choose v outside $V_1 \cup V_2$ and consider the line joing v and v_3. On this line there is a point outside $V_1 \cup V_2 \cup V_3$. This proves Theorem 7.

6.6 Now let us consider Theorem 6. By an argument like the one above one proves $\widetilde{\widetilde{SR}}_n$ for A in the case that R is a field, $n \geq 3$. So Theorem 6 will follow from the following analogue of Theorem 3.

<u>THEOREM 8</u> Let A,R,d be as in Theorem 6. Let $c \geq u \geq n - 1 \geq d + 1$ and $p \geq 1$. If $SR_n^p(c,u)$ holds for all $A \underset{R}{\otimes} (R/y)$, $y \in$ maximal spectrum of R, then it holds for A.

6.7 To prove Theorem 8 we want to argue as in Section 2. So for $y \in$ j-spec, let us denote the field of fractions of R/y by k(y) again. Put $A(y) = (A \underset{R}{\otimes} k(y))/\mathrm{Rad}(A \underset{R}{\otimes} k(y))$, or $A(y) = A_y/\mathrm{Rad}(A_y)$, where $A_y = A \underset{R}{\otimes} R_y$, R_y is the localization at (the complement of) y. (Use the Nakayama Lemma to see that these two definitions of A(y) are equivalent). There is a natural map $A^s \to A(y)^s$, denoted by $f \to \bar{f}$ $(s \geq 1)$. Given $a_1,\ldots,a_m$ in A^s we need a notion of rank over A(y) of $\bar{a}_1,\ldots,\bar{a}_m$. Consulting [4], Ch. IV, §1, §2, we see that the proper definition is as follows: **Form the s by m matrix M over** A(y) with columns $\bar{a}_i$. Look for the highest r such that there is an r by s matrix T over A(y) and an m by r matrix U over A(y) with TMU = identity. This highest r is the rank over A(y). (f-rank in terminology of [4]). One can study this rank factor by factor in the Wedderburn decomposition of A(y). Let us look at one factor, so say $A(y) = M_q(D)$. Then the rank of $\bar{a}_1,\ldots,\bar{a}_m$ over A(y) is the highest r such that M, as a matrix over the division ring D, has rank at least qr.

6.8 Given $\bar{a}_1,\ldots,\bar{a}_m$ in $(M_q(D))^s$, of rank r over $M_q(D)$, what can we say about the rank over $M_q(D)$ of $\bar{a}_1 + \bar{a}_m\bar{t}_1,\ldots,\bar{a}_{m-1} + \bar{a}_m\bar{t}_{m-1}$, with $\bar{t}_i \in M_q(D)$? It is at least r - 1, because the matrix rank over D doesn't drop by more than q. (cf. [4], Ch. IV, (1.6)). If $m - 1 \geq r$, how can we avoid a drop from r to r - 1? Let us write it in matrix form. Let M be made up of $\bar{a}_1,\ldots,\bar{a}_m$ again and let L be the m by m - 1 matrix over $M_q(D)$ consisting of an m - 1 by m - 1 identity matrix on top and the row $(\bar{t}_1,\ldots,\bar{t}_{m-1})$ at the bottom. Then it is sufficient to get qr independent columns in ML (over D.)

This can be done by making sure that for each j the first j columns span a right linear space of dimension at least $j - q(m-r-1)$. That again can be done by choosing the columns inductively, just as in 6.5. (cf. 2.7).

6.9 Now suppose $\bar{a}_1,\ldots,\bar{a}_m$ has rank r over $A(y)$ and that we look for $\bar{t}_1,\ldots,\bar{t}_m$ in $A(y)$ such that

(i) $\bar{a}_1 + \bar{a}_m\bar{t}_1,\ldots,\bar{a}_{m-1} + \bar{a}_m\bar{t}_{m-1}$ has rank r over $A(y)$.

(ii) The $\bar{t}_i$ are in the image $\bar{A}$ of A in $A(y)$.

(We no longer assume $A(y) = M_q(D)$). Of course we need $m - 1 \geq r$. If $k(y)$ is finite, one has $\bar{A} = A(y)$ and (ii) is superfluous. So we may assume $k(y)$ is infinite and therefore R/y is infinite. We know how we can solve (i) by constructing the components of the $\bar{t}_i$ in the Wedderburn decomposition, column by column. A matrix can be seen as the sum of matrices representing its columns, each of the summands having all but one of its columns zero. So to satisfy (ii) let us try to choose the columns in such a way that the corresponding summands are in $\bar{A}$. Given a choice of a column which is good for solving (i) there is a non-zero element $\bar{a}$ of R/y so that multiplication by $\bar{a}$ gets the summand into $\bar{A}$. The resulting column may not solve (i) any more, but there is $\bar{b}\in R/y$ so that multiplication by $\bar{b}$ will correct this defect. (R/y is infinite).

6.10 The rank over $A(y)$ of $\bar{a}_1,\ldots,\bar{a}_m$ has the required semi-continuity property when y ranges over j-spec(R). (See [4], Ch. IV (2.3) and use the Nakayama Lemma). We need not say more about the proof of Theorem 8. (Consult [4],[9],[25], if necessary).

§7. Examples of non-stability for K_2.

7.1 We want to give examples where $K_2(n,R) \to K_2(n+1,R)$ is not injective or not surjective. We will sketch how such examples can be obtained from the homotopy theory of real orthogonal groups. Then we will give explicit formulas for a particular case. (starting in 7.17).

7.2 We use the setting of §7 in [20]. So let Λ be a commutative Banach algebra over the reals. Banach algebras tend not to be noetherian and we would like to have noetherian examples too. So let R be a subring of Λ which is dense in Λ and which satisfies $R \cap \Lambda^* = R^*$. Let $n \geq 2$. Sending $x_{ij}(a)$ to the path $t \mapsto \underline{\mathrm{mat}}(x_{ij}(ta)), t\in[0,1]$, we get a homomorphism from $\mathrm{St}(n,\Lambda)$ onto $\tilde{E}$, the (topological) universal covering group of $E(n,\Lambda)$. (See [20], §7 for the case $n \geq 3$. The case $n = 2$ is similar, and also known). Composing with $\mathrm{St}(n,R) \to \mathrm{St}(n,\Lambda)$ we get a homomorphism φ_n: $\mathrm{St}(n,R) \to \tilde{E}$. Inspecting §7 of [20] one sees there is an integer N_1 such that: for each $\varepsilon > 0$ there is a neighborhood V_1 of the identity in $E(n,\Lambda)$ such that any matrix in V_1 can be written as a product of N_1 factors of type $\underline{\mathrm{mat}}(x_{ij}(a))$, $\|a\| < \varepsilon$. (Of course the values of i,j,a vary for the factors). Using the Steinberg relations one can break up the $x_{ij}(a)$ still further. That gives an integer N_2 such that: for each $\varepsilon > 0$ there is a neighborhood V_2 of the identity in $E(n,\Lambda)$ such that any matrix in V_2 can be written as a product of N_2 factors of type $\underline{\mathrm{mat}}(x_{ij}(a))$, $\|a\| < \varepsilon$, $|i-j| = 1$. It can be arranged that the factors are in $E(n,R)$ if the original matrix is. One sees from this and from the density of R in Λ that a loop in $E(n,\Lambda)$, going from 1 to 1, can be approximated by loops coming from $\mathrm{St}(n,R)$, also going from 1 to 1. As $E(n,\Lambda)$ is locally contractible this means that the restriction of φ_n to $K_2(n,R)$ maps onto $\pi_1(E(n,\Lambda))$, the first homotopy group of $E(n,\Lambda)$.

7.3 Lemma If the map $\pi_1(E(n,\Lambda)) \to \pi_1(E(n+1,\Lambda))$ is not surjective, then the map $K_2(n,R) \to K_2(n+1,R)$ is not surjective. $(n \geq 2)$.

Proof By 7.2 the map φ_{n+1} in the depicted commutative square is surjective.

$$\begin{array}{ccc} K_2(n+1,R) & \xrightarrow{\varphi_{n+1}} \!\!\!\!\to & \pi_1(E(n+1,\Lambda)) \\ \uparrow & & \uparrow \\ K_2(n,R) & \xrightarrow{\varphi_n} & \pi_1(E(n,\Lambda)) \end{array}$$

7.4 Now consider an element x of $K_2(n,R)$ that can be written as a product of $4N_2$ factors of type $x_{ij}(a)$, $\|a\| < \varepsilon$, $|i-j| = 1$. (Compare 7.2). Call such an element of ε-type.

Lemma Let ε be sufficiently small. $(\varepsilon > 0)$. Then all elements of ε-type are in the kernel of φ_n and together with their conjugates in $St(n,R)$ they generate this kernel. $(n \geq 2)$.

Sketch of Proof As $E(n,\Lambda)$ is locally contractible it is easy to get them in the kernel. Now suppose $\varphi_n(x) = 1$. Consider the homotopy between the path represented by x and the trivial path. It is a map $[0,1] \times [0,1] \to E(n,\Lambda)$. Choose a fine grid of points in $[0,1] \times [0,1]$, so that neighbors in the grid have images which differ by an element of V_2 (see 7.2). We may assume that the points on the grid are mapped into $E(n,R)$. Given two neighbors in the grid, lift the difference of their images to a product of N_2 factors of type $x_{ij}(a)$, $\|a\| < \varepsilon$, $|i-j| = 1$. Then squares in the grid correspond to elements of ε-type and one can break up x into conjugates of them.

7.5 Lemma Let $n \geq 3$. Let ε be sufficiently small. ($\varepsilon > 0$). Then every element of ε-type is a product of symbols $\{u,v\}$, where $\|u-1\|$, $\|v-1\|$ are smaller than 1/10. In particular, elements of ε-type are central.

Sketch of proof One gradually rearranges the product into a "lower-diagonal-upper" normal form. This one does as in §2 (up to 2.10) of [21]. (For the relevant explicit computations, elements with small norm behave like elements of a Jacobson radical). (1/10 is just a number between zero and one).

7.6 Proposition If the map $\pi_1(E(n,\Lambda)) \to \pi_1(E(n+1,\Lambda))$ is not injective, then the map $K_2(n,R) \to K_2(n+1,R)$ is not injective. ($n \geq 2$).

Proof Let $y \in \pi_1(E(n,\Lambda))$ vanish in $\pi_1(E(n+1,\Lambda))$, $y \neq 1$. Lift y to x in $K_2(n,R)$. (see 7.2). Let z be the image of x in $K_2(n+1,R)$. Then $\varphi_{n+1}(z) = 1$ so z is a product of elements $\{u,v\}$ of the type described in Lemma 7.5. These elements $\{u,v\}$ can be lifted to elements $\{u,v\}_{12}$ in $K_2(n,R)$ and these liftings are in $\ker(\varphi_n)$. So we can modify x such that it has image 1 in $K_2(n+1,R)$, but still image $y(\neq 1)$ in $\pi_1(E(n,\Lambda))$.

7.7 We want to apply Lemma 7.3 and Proposition 7.6 in the following situation: Let S^d be the unit sphere in $\mathbb{R}^{d+1}$; let Λ_d be the Banach algebra of continuous real valued functions on S^d. Let k be a subfield of $\mathbb{R}$ and let $X_1,\ldots,X_{d+1}$ be the co-ordinate functions on S^d. Then $A_d = k[X_1,\ldots,X_{d+1}]$ is a dense subring of Λ_d (Stone-Weierstrass), isomorphic to $k[x_1,\ldots,x_{d+1}]/(x_1^2 + \cdots + x_{d+1}^2 - 1)$.

Put $S = \Lambda_d^* \cap A_d$. Localizing A_d at the multiplicative system S we get the ring $R_d = S^{-1}A_d$. Both A_d and R_d are noetherian rings of dimension d and the above is applicable to $R = R_d$, $\Lambda = \Lambda_d$.

7.8 So we want to have an interpretation of $\pi_1(SL_n(\Lambda_d))$, with Λ_d as in 7.7. An element of this group is a homotopy class of continuous maps $f: [0,1] \times S^d \to SL_n(\mathbb{R})$ with $f(0,v) = f(1,v) = 1$ for $v \in S^d$. Choose a base point v_0 in S^d. Then $t \mapsto f(t,v_0)$ represents an element of $\pi_1(SL_n(\mathbb{R}))$. So we have a map $\rho: \pi_1(SL_n(\Lambda)) \to \pi_1(SL_n(\mathbb{R}))$. An element of $\pi_{d+1}(SL_n(\mathbb{R}))$ can be viewed as a homotopy class of continuous maps $f: [0,1] \times S^d \to SL_n(\mathbb{R})$ with $f(0,v) = 1$, $f(1,v) = f(1,v_0)$ for $v \in S^d$. So there also is a homomorphism $\sigma: \pi_1(SL_n(\Lambda_d)) \to \pi_{d+1}(SL_n(\mathbb{R}))$. On the other hand there is a homomorphism

$$\begin{array}{ccc} \pi_1(SL_n(\mathbb{R})) \times \pi_{d+1}(SL_n(\mathbb{R})) & \xrightarrow{\tau} & \pi_1(SL_n(\Lambda_d)) \text{ given by} \\ (f,g) & \longmapsto & [(t,v) \mapsto g(t,v)g(t,v_0)^{-1}f(t)] \end{array}$$

In $\pi_{d+1}(SL_n(\mathbb{R}))$ the maps $g(t,v)g(t,v_0)^{-1}f(t)$ and $g(t,v)$ represent the same class. Therefore it is easy to see that τ is an isomorphism with inverse $\rho \times \sigma$.

7.9 <u>Proposition</u>

(a) If $\pi_{d+1}(SO_n(\mathbb{R})) \to \pi_{d+1}(SO_{n+1}(\mathbb{R}))$ is not surjective then $K_2(n,R_d) \to K_2(n+1,R_d)$ is not surjective. $(n \geq 2)$.

(b) Same with "not injective."

<u>Proof</u> Combine 7.3, 7.6, 7.8 and recall that $\pi_1(SL_n(\mathbb{R}))$ is canonically isomorphic with $\pi_1(SO_n(\mathbb{R}))$.

7.10 <u>Remark</u> Of course we may replace R_d in the Proposition by any ring between R_d and Λ_d. We may in this context declare that R is between R_d and Λ_d if there are homomorphisms $R_d \to R$ and $R \to \Lambda_d$ whose composition is the natural inclusion $R_d \to \Lambda_d$.

7.11 Data on the map $\pi_{d+1}(SO_n(\mathbb{R})) \to \pi_{d+1}(SO_{n+1}(\mathbb{R}))$ can for instance be obtained from the long exact homotopy sequence

$$\cdots \to \pi_i(SO_n(\mathbb{R})) \to \pi_i(SO_{n+1}(\mathbb{R})) \to \pi_i(S^n) \to \pi_{i-1}(SO_n(\mathbb{R})) \to \cdots$$

In particular, if one knows all the groups in a certain piece of this sequence, one may be able to decide which maps are surjective and which ones are injective. A table for some $\pi_i(SO_n(\mathbb{R}))$ can be found in [17] and some $\pi_i(S^n)$ are listed in [11]. Further information can be found in [18], [26], [3], [1], [30] etc.

7.12 We give a table of examples that are closest to the reach of the stability theorems for K_2, so as to illustrate how sharp the bounds in these stability theorems are. (Of course they need not be sharp at all if one looks at specific rings). We use the following notation: If $\pi_{d+1}(SO_n(\mathbb{R})) \to \pi_{d+1}(SO_{n+1}(\mathbb{R}))$ has a non-trivial cokernel, we put a "c" in the table, at the place corresponding to these values of d and n. If it has a non-trivial kernel, we put a "k" in the table. If the map has both a non-trivial kernel and a non-trivial cokernel, we put "ck" in the table. Not all columns have been pursued to the same depth and we put stars at the bottom where we stopped listing results. (So a star means: this pair of values is left to the reader). The number s is an integer, $s \geq 1$.

Table

$d-n$ \ n	2	3	4	5	6	7	8s	8s+1	8s+2	8s+3	8s+4	8s+5	8s+6	8s+7
-2	k		k	k	k		k	k	k	k	k	k	k	k
-1		c	k	ck		c	k	ck	k	c	k	ck	k	c
0	c	c	k		c	c	k	k		c	k	k		c
1	c	c	k	c	ck	c	k	k	ck	ck	k	k	ck	c
2	c	c	ck	c	c	c		c	c	c	*	c	c	c
3	c	c	k	c	k		*	*	*	*	*	*	*	*
4	c	c	*	*	*	*	*	*	*	*	*	*	*	*
*	*	*	*	*	*	*	*	*	*	*	*	*	*	*

7.13 From the table one immediately sees the following: For $d \neq 1,5$ there is a ring R_d with maximal spectrum of dimension d such that $K_2(d+2,R_d) \to K_2(R_d)$ is not injective. Theorem 1 tells us that $K_2(d+3,R_d) \to K_2(R_d)$ is injective, so this result is sharp for $d \neq 1,5$.

For surjective stability one sees from the table that the Surjective Stability Theorem is sharp for d even. For d odd the table gives less nice results. One can get examples for d odd just by taking the polynomial ring $R_{d-1}[x]$ over the example R_{d-1} which has even dimension. Using 7.10 one thus sees that the bound for d odd can be at most one unit off. It is unlikely that it actually is one unit off. For $d = 1$ one has other examples (certain rings of integers in number fields) which show that the bound is sharp. (See [8]). Similarly one can see that the bound in Theorem 1 can only be one unit off for $d = 1$ or 5. But again the bound is probably sharp in those cases too.

7.14 We can also look at the table for fixed n and see what examples we get. For $n = 3,7$ there are no "k"'s in the table, because SO_3 is topologically a direct factor of SO_4, and SO_7 is one of SO_8. For $n = 2$ the table tells too little: recall that one already has examples of non-surjectivity (see [8] again). For $n \equiv 0 \bmod 4$, $n > 4$, one doesn't see a "c" in the table. However, there is one if one goes further down: n even, $n > 8$, $d = 2n - 2$ gives non-surjective maps (see [3]) and the pairs $n = 8$, $d = 14$; $n = 12$, $d = 14$ also give examples. (Seems to be in range where no tables are available).

7.15 <u>Lemma</u> Let R be a ring, $M_q(R)$ the ring of $q \times q$ matrices over R. Then $K_2(n,M_q(R))$ and $K_2(nq,R)$ are isomorphic, for $n \geq 3$.

<u>Sketch of proof</u> Say e_{ij} is the element of $M_q(R)$ with a one on place (i,j) and zeroes elsewhere. So the e_{ij} form a basis of $M_q(R)$ over R. Sending $x_{ij}(ae_{k\ell})$ to $x_{kn-n+i,\ell n-n+j}(a)$ we get a homomorphism $St(n,M_q(R)) \to St(nq,R)$. One shows that there is a map in the other direction by proving sufficiently many relations between the $x_{ij}(ae_{k\ell})$ so that one can apply Theorem B' of [6]. (For instance, $x_{12}(ae_{11})$ commutes with $x_{21}(be_{22})$ because $x_{21}(be_{22})$ can be written as $[x_{23}(e_{22}), x_{31}(be_{22})]$). One arranges it in such a way that the composite map $St(nq,R) \to St(n,M_q(R)) \to St(nq,R)$ is the identity. (To see that it is the identity one shows that $St(nq,R)$ is generated by the elements which are left fixed by the map). The other composite is then similarly seen to be the identity, and restriction to $K_2(n,M_q(R))$, $K_2(nq,R)$ yields the result.

7.16 From the Lemma one sees that $K_2(n,M_q(R)) \to K_2(n+1,M_q(R))$ is not surjective if $K_2(nq+q-1,R) \to K_2(nq+q,R)$ is not surjective. $(n \geq 3)$. So for instance $K_2(8,M_2(R_{16})) \to K_2(9,M_2(R_{16}))$ is not surjective, where R_{16} means R_d from 7.7 with $d = 16$. And $K_2(n,M_q(R)) \to K_2(n+1,M_q(R))$ is not injective if $K_2(nq,R) \to K_2(nq+1,R)$ is not injective $(n \geq 3)$. So for instance

$K_2(3,M_2(R_4)) \to K_2(4,M_2(R_4))$ is not injective.

7.17 We now work out the case $n = d + 2$, $R = R_d$, $d \neq 1,5$. We will give an element of $\ker(K_2(d+2,R_d) \to K_2(d+3,R_d))$. It will turn out that this element actually lives over $B_d = \mathbb{Z}[x_1,\ldots,x_{d+1}]/(x_1^2+\ldots+x_{d+1}^2-1)$ so that we also find an element of $\ker(K_2(d+2,B_d) \to K_2(d+3,B_d))$, or of $\ker(K_2(d+2,A_d) \to K_2(d+3,A_d))$, $d \neq 1,5$. The form of this element suggests various problems.

7.18 From now on $d \neq 1,5$; $d \geq 0$. The non-injectivity of $\pi_{d+1}(SL_{d+2}(\mathbb{R})) \to \pi_{d+1}(SL_{d+3}(\mathbb{R}))$ can be made more explicit: By [1] the sphere S^{d+2} is not parallelizable, so by [30], Corollary to Theorem 10, the "canonical map" represents a non-trivial element of $\pi_{d+1}(SL_{d+2}(\mathbb{R}))$. (This element is also known to generate $\ker \pi_{d+1}(SL_{d+2}(\mathbb{R})) \to \pi_{d+1}(SL_{d+3}(\mathbb{R}))$, see [30] again). In the setting of 7.8 the canonical map can be described as the map $[0,1] \times S^d \to SL_{d+2}(\mathbb{R})$ given by the formula

$$\left[I_{d+2} - 2\begin{pmatrix} \cos \pi t \\ X_1 \sin \pi t \\ \vdots \\ X_{d+1} \sin \pi t \end{pmatrix} (\cos \pi t, X_1 \sin \pi t, \ldots, X_{d+1} \sin \pi t) \right] \times$$

$$\begin{pmatrix} -1 & & & & 0 \\ & 1 & & & \\ & & 1 & & \\ & & & \ddots & \\ 0 & & & & 1 \end{pmatrix}$$

where I_{d+2} is the identity of $SL_{d+2}(\mathbb{R})$. (Compare [30], §6. We permuted the indices, but that doesn't matter. Note that the product of a column of length d + 2 and a row of length d + 2 is a d + 2 by d + 2 matrix indeed. Also note that the resulting matrix is actually an orthogonal matrix of determinant 1, because $\cos^2\pi t + X_1^2 \sin^2\pi t + \cdots + X_{d+1}^2 \sin^2 \pi t = \cos^2\pi t + \sin^2 \pi t = 1$, while the determinant is 1 for t = 0.)

7.19 So now we look for an element of $K_2(d + 2, A_d)$ which is mapped to the class of the canonical map in $\pi_{d+1}(SL_{d+2}(\mathbb{R}))$. We claim the following element will do the trick:

$$\tau = \sigma^4, \text{ where } \sigma = \Big(\prod_{i=2}^{d+2} x_{1i}(-X_{i-1})\Big)\Big(\prod_{i=2}^{d+2} x_{i1}(X_{i-1})\Big)\Big(\prod_{i=2}^{d+2} x_{1i}(-X_{i-1})\Big).$$

Note that σ is analogous to the elements $w_{12}(u) = x_{12}(u)x_{21}(-u^{-1})x_{12}(u)$. Here the unimodular column $(-X_1,\ldots,-X_{d+1})$ is the analogue of u, and the unimodular row $(-X_1,\ldots,-X_{d+1})$ is the analogue of u^{-1}, as $X_1^2 + \cdots + X_{d+1}^2 = 1$. Anyway, let us prove the claim. For simplicity of notations we take d = 0. (The general case goes the same way). The image of σ in $\pi_{d+1}(SL_{d+2}(\mathbb{R}))$ is represented by $\begin{pmatrix} 1-t^2 & X_1(t^3-2t) \\ X_1 t & 1-X_1^2t^2 \end{pmatrix}$, which is homotopic to $\begin{pmatrix} 1-2\sin^2(\frac{\pi}{4}t) & \text{something} \\ X_1\sqrt{2}\sin(\frac{\pi}{4}t) & 1-2X_1^2\sin^2(\frac{\pi}{4}t) \end{pmatrix}$, because t is homotopic to $\sqrt{2}\sin(\frac{\pi}{4}t)$ on [0,1]. Now conjugate by the matrix $\begin{pmatrix} (\sqrt{2}\cos(\frac{\pi}{4}t))^u & 0 \\ 0 & 1 \end{pmatrix}$, $u\in[0,1]$, to get a homotopy with a matrix $\begin{pmatrix} 1-2\sin^2(\frac{\pi}{4}t) & \text{something} \\ X_1\sin(\frac{\pi}{2}t) & 1-2X_1^2\sin^2(\frac{\pi}{4}t) \end{pmatrix}$. This last matrix still has determinant 1 and is therefore homotopic to the matrix which one gets from it by "pulling the top row straight", so as to get the top row orthogonal to the other one. (other ones, if $d > 0$). Using $X_1^2 = 1$ we see that the image of σ is represented by the orthogonal matrix

$$\begin{pmatrix} 1-2\sin^2(\frac{\pi}{4}t) & -X_1\sin(\frac{\pi}{2}t) \\ X_1\sin(\frac{\pi}{2}t) & 1-2X_1^2\sin^2(\frac{\pi}{4}t) \end{pmatrix}.$$

(Actually this matrix equals the previous one). The image of τ is represented by the fourth power of this orthogonal matrix, and that happens to be the canonical map.

7.20 One may have guessed that τ is the element in the kernel, but that is wrong. The image of τ in $St(d+3, A_d)$ is central, so let us conjugate it by a suitable element, transforming $\prod_{i=2}^{d+2} x_{1i}(-X_{i-1})$ to $x_{12}(-1)$ and $\prod_{i=2}^{d+2} x_{i1}(X_{i-1})$ to $x_{21}(1)$. (Use that the column $(-X_1, \ldots, -X_{d+1})$ is unimodular, and, more in particular, that $X_1^2 + \cdots + X_{d+1}^2 = 1$). So the image of τ in $K_2(d+3, A_d)$ is $w_{12}(-1)^4$. This element is also known under the names $h_{12}(-1)^2$ or $\{-1,-1\}$. Its square is 1 in $K_2(d+3, \mathbf{Z})$, so certainly also in $K_2(d+3, A_d)$. And $h_{12}(-1)^2$ also makes sense in $K_2(d+2, A_d)$.

7.21 <u>Lemma</u> $h_{12}(-1)^2\tau$ is a non-trivial element of the kernel of $K_2(d+2, A_d) \to K_2(d+3, A_d)$ for $d \geq 0$, $d \neq 1,5$. Similarly for B_d, Λ_d, and rings in between, like R_d. (cf. 7.10).

<u>Proof</u> Clearly $h_{12}(-1)^2\tau$ is in the kernel. The image of $h_{12}(-1)^2$ in $\pi_{d+1}(SL_{d+2}(\mathbf{R}))$ is trivial because its representative only depends on t and not on v, in the notations of 7.8. So $h_{12}(-1)^2\tau$ is mapped to the class of the canonical map.

7.22 <u>Remark</u> The case $d = 0$ is well known. It suggested the other cases. (Compare [20]).

7.23 Let C_d be the ring

$\mathbb{Z}[u_1,\ldots,u_{d+1},v_1,\ldots,v_{d+1}]/(u_1v_1 + \cdots + u_{d+1}v_{d+1} + 1)$. Define τ_1 to be σ_1^4 with

$$\sigma_1 = \left(\prod_{i=2}^{d+2} x_{1i}(u_{i-1})\right)\left(\prod_{i=2}^{d+2} x_{i1}(v_{i-1})\right)\left(\prod_{i=2}^{d+2} x_{1i}(u_{i-1})\right) \text{ in } St(d + 2, C_d).$$

The homomorphism $C_d \to A_d$ given by $u_i \to -X_i$, $v_i \to X_i$ induces a map $K_2(d + 2, C_d) \to K_2(d + 2, A_d)$ sending τ_1 to τ. One concludes as in 7.20, 7.21 that $h_{21}(-1)^2\tau_1$ is a non-trivial element of the kernel of $K_2(d + 2, C_d) \to K_2(d + 3, C_d)$, for $d \geq 0$, $d \neq 1,5$. Similarly for rings between C_d and Λ_d. (cf. 7.10).

7.24 <u>Question</u> Are $d = 1$, $d = 5$ really exceptions in 7.21, 7.23?

§8. Comparison with $\pi_2(BGL(n,R)^+)$

8.1 Let us recall the connection between classical non-stable K-groups and Quillen's non-stable groups: If the natural map $K_1(n,R) \to K_1(R)$ is injective, then $E(n,R)$ is the commutator subgroup of $GL(n,R)$. ($n \geq 3$, R any ring). And if $E(n,R)$ is the commutator subgroup of $GL(n,R)$, then Quillen's group $\pi_1(BGL_n(R)^+)$, as described in [10] problem 2, is just $K_1(n,R)$. Thus injective stability for K_1, proved for the $K_1(n,R)$, carries over to Quillen's groups. In general one has a sequence of natural maps $K_1(n,R) \twoheadrightarrow \pi_1(BGL(n,R)^+) \to \pi_1(BGL(R)^+) \xrightarrow{\cong} K_1(R)$, so that surjective stability for K_1 carries over too. (in both directions). But for K_2 the situation is not so nice. Let us assume that the map $K_2(n,R) \to K_2(R)$ is injective. Then $K_2(n,R)$ is central in $St(n,R)$ and therefore there is a surjective map $H_2(E(n,R)) \to K_2(n,R)$, if $n \geq 3$. (As usual $H_2(E(n,R))$ means the second homology group $H_2(E(n,R),\mathbb{Z})$, also called the "Schur Multiplier" of $E(n,R)$). The map $H_2(E(n,R)) \to K_2(n,R)$ is an isomorphism precisely when $H_2(St(n,R)) = 0$. This condition is satisfied if $n \geq 5$. If $n = 4$ then $H_2(St(n,R)) = 0$ precisely then when there is no ring homomorphism from R onto the field with 2 elements. (to appear). If $n = 3$, a sufficient condition is that $\{u^3 - 1 | u \text{ is central unit}\}$ generates the unit ideal. (see [23]). Anyway, $K_2(n,R)$ and $H_2(E(n,R))$ may fail to be isomorphic, even "in the stable range", if n is small. On the other hand, if $E(n,R)$ is the commutator subgroup of $GL(n,R)$, then $H_2(E(n,R)) = \pi_2(BGL(n,R)^+)$. So for $n \leq 5$ the stability results for the $K_2(n,R)$ don't carry over to Quillen's groups, unless R satisfies additional conditions.

8.2 <u>Example</u> Take $R = \mathbb{Z}$. The sequence $K_2(2,\mathbb{Z}) \to K_2(3,\mathbb{Z}) \to K_2(4,\mathbb{Z}) \to \cdots$ starts with a surjective map and has isomorphisms beyond. (see [20]). This behavior is better than predicted by the dimension of the ring, which is 1. (From d = 1 it follows that $K_2(3,\mathbb{Z}) \to K_2(4,\mathbb{Z})$ is surjective and that $K_2(4,\mathbb{Z}) \to K_2(\mathbb{Z})$ is an isomorphism). But look at the $\pi_2(BGL(n,\mathbb{Z})^+)$. They are equal to the $H_2(E(n,\mathbb{Z}))$, which take much longer to stabilize:

$H_2(E(3,\mathbb{Z})) = (\mathbb{Z}/2\mathbb{Z}) \oplus (\mathbb{Z}/2\mathbb{Z})$

$H_2(E(4,\mathbb{Z})) = (\mathbb{Z}/2\mathbb{Z}) \oplus (\mathbb{Z}/2\mathbb{Z})$

$H_2(E(5,\mathbb{Z})) = K_2(\mathbb{Z}) = \mathbb{Z}/2\mathbb{Z}$. (From 5 on it is stable).

The natural map $H_2(E(3,\mathbb{Z})) \to H_2(E(4,\mathbb{Z}))$ is not an isomorphism (see [14]) and thus not even surjective. In other words, the map $\pi_2(BGL(3,\mathbb{Z})^+) \to \pi_2(BGL(4,\mathbb{Z})^+)$ fails to be surjective, contrary to the statement in [10], problem 2. The map $\pi_2(BGL(4,\mathbb{Z})^+) \to \pi_2(BGL(5,\mathbb{Z})^+)$ also defies the stability conjecture for the $\pi_i(BGL(n,R)^+)$, as it is not injective. (Compare [10], problem 2). So some additional hypothesis is needed on the ring, as is well known. In the case of $\mathbb{Z}$ the source of trouble is the anomalous behavior of $H_2(E(3,\mathbb{F}_4))$ and $H_2(E(4,\mathbb{F}_2))$, where $\mathbb{F}_q$ denotes the field with q elements. (The trouble is carried over to $\mathbb{Z}$ via the ring homomorphisms $\mathbb{Z} \to \mathbb{F}_4$, $\mathbb{Z} \to \mathbb{F}_2$). Note the similarity with the technical difficulties in this paper for semi-local rings with at least one small residue field.

Key to notations and terminology

We give a loose description and/or a reference.

C The chunk. (1.5), (3.6).

$E(m) = E(m,R) = E_m(R)$. The subgroup of $GL(m,R)$ generated by elementary matrices. Often embedded in $GL(n+2,R)$. (3.4).

$E(I \times J) = \underline{mat}(St(I \times J))$. (3.4).

$K_1(n,R)$ The set of cosets $GL(n,R)/E(n,R)$. It can be a group.

$K_1(R)$ The limit of the $K_1(n,R)$, $n \to \infty$. It is a group. [20].

$K_2(n,R)$ The kernel of $St(n,R) \to E(n,R)$.

$K_2(R)$ The limit of the $K_2(n,R)$, $n \to \infty$. [20].

$St(n) = St(n,R) = St_n(R)$. The Steinberg group over R, on n row indices and n column indices. (3.4).

$St(I \times J) = St(I \times J,R)$. The Steinberg group on row index set I and column index set J. (3.4).

SR_n Stable range condition. (Given a unimodular column of length n,...). (2.1).

$\widetilde{\widetilde{SR}}_n$ Shorthand for list of stable range conditions. (2.3).

SR_2^* Shorthand for similar list. (5.2).

$SR_n^p(c,u)$ Multiple stable range condition. (2.2).

π Embeds the chunk into $St(n + 2)$. (1.5), (3.8).

id Identity map.

<u>inv</u> Interchanges left and right. (4.1).

<u>left</u> The set of "left translations" of C. (1.3),(1.5),(4.10).

<u>Low</u> = St([n + 2] × {n + 2}*). (3.6), (3.4).

<u>mat</u> Associates a matrix to a more abstract entity. (3.4),(3.8).

<u>Med</u> = St([n + 2] × {1,n + 2}*). (3.6), (3.4).

<u>Up</u> = St([n + 2] × {1}*). (3.6), (3.4).

L(x) Describes $\mathcal{L}(x)$ in terms of elements of <u>Low</u> × <u>Up</u>. (3.9), (3.13), (3.15), (5.9).

$\mathcal{L}(x)$ An element of <u>left</u>, eventually. (1.5), (3.9), (3.13), (3.18), (3.31), (5.7), (5.14), (5.18).

R(x) The analogue at the right of L(x). (3.9).

$\mathcal{R}(x)$ The analogue at the right of $\mathcal{L}(x)$. (3.9), (4.1).

$x_{ij}(r)$ Can be sent to a generator of St(n + 2). (3.4).

$w_{ij}(t)$ $= x_{ij}(t)x_{ji}(-t^{-1})x_{ij}(t)$.

$h_{ij}(t)$ $= w_{ij}(t)w_{ij}(-1)$.

$x_1(w)$ Product of the $x_{i,1}(w_i)$. (3.12).

$x_{n+2}(v), x_4(v)$. Product of the $x_{i,n+2}(v_i)$. (3.12), (5.6).

S^d The d-sphere in real d + 1-space. (7.7).

$A_d, B_d, C_d, R_d, \Lambda_d$. Ring of continuous functions on S^d and variations. (§7).

Rad Jacobson radical.

R* Group of units in R.

~ $(X,Y) \sim (X',Y')$ iff $\langle X,Y\rangle = \langle X',Y'\rangle$. (3.6).

$\simeq$ $f \simeq g$ if $f(x) = g(x)$ holds whenever both sides are defined. (4.3).

$f \circ g$ Composition of f and g. (1.5), (4.3).

$\mathcal{L} * \mathcal{L}_1$ Composition of $\mathcal{L}$ and $\mathcal{L}_1$ in left. (1.5), (4.13).

$\{1,n+2\}^*$ Complement of $\{1,n+2\}$ in $[n+2]$. (3.4).

$[k]$ $= \{1,\ldots,k\}$. (3.4).

(X,Y) Representative in Low x Up of $\langle X,Y\rangle \in C$. (3.6).

$\langle X,Y\rangle$ Element of chunk. (3.6).

additivity. $\mathcal{L}(x_{n+2}(v+w)) = \mathcal{L}(x_{n+2}(v))\mathcal{L}(x_{n+2}(w))$, and same for $\mathcal{R}$'s. (3.25),(3.27),(3.30),(3.33),(4.1),(5.19).

Chunk. Imitates big piece of group. (1.5), (3.6).

defined at the bottom. After change of basis the bottom two rows carry the relevant data. (3.18).

j-spec. A subspace of the prime spectrum. (2.5).

semi-direct product. The quotient homomorphism splits. (3.5).

slide past each other. Evaluation of $\mathcal{L}$ can be done in harmony with evaluation of other $\mathcal{L}$ or $\mathcal{R}$. (3.26), (4.5).

squeezing principle. If a problem can be handled inside $St(n+1)$, don't use $St(n+2)$. (3.10), (3.11).

trivial row or column. As in the identity matrix.

unimodular One can multiply by a row and get 1. (2.1).

References

1. J. F. Adams, On the non-existence of elements of Hopf invariant one, Ann. of Math. 72 (1960), 20-103.

2. M. Artin, Théorème de Weil sur la construction d'un groupe à partir d'une loie rationelle, Schémas en Groupes II, SGA 3, Lecture Notes in Math., vol. 152, Springer, Berlin, 1970, 632-653.

3. M. G. Barratt and M. E. Mahowald, The metastable homotopy of $O(n)$, Bull. Amer. Math. Soc. 70(1964), 758-760.

4. H. Bass, Algebraic K-Theory, Benjamin, New York, 1968.

5. R. K. Dennis, Stability for K_2, Procedings of the Conference on Orders and Group Rings held at Ohio State University, Columbus, Ohio, May 12-15, 1972, Lecture Notes in Math. 353, Springer, Berlin, 1973, 85-94.

6. R. K. Dennis and M. R. Stein, Injective Stability for K_2 of Local Rings, Bull. Amer. Math. Soc. 80, 1974, 1010-1013.

7. ________________, The functor K_2: A survey of computations and problems, Algebraic K-Theory II, Lecture Notes in Math. 342, Springer, Berlin, 1973, 243-280.

8. ________________, K_2 of discrete valuation rings, Advances in Math. 18, 1975, 182-238.

9. D. Eisenbud and E. G. Evans, Jr., Generating Modules Efficiently: Theorems from Algebraic K-Theory, J. Algebra 27, 1973, 278-305.

10. S. M. Gersten, Problems about higher K-functors, Algebraic K-Theory I, Lecture Notes in Math. 341, Springer, Berlin, 1973, 43-56 or 41-54.

11. S. Hu, Homotopy Theory, Academic Press, New York and London, 1959.

12. N. Jacobson, Structure of Rings, Amer. Math. Soc. Colloquium Publications, XXXVII, Providence, Rhode Island, 1964.

13. N. Jacobson, Lectures in Abstract Algebra, II, Van Nostrand, Princeton, 1953.

14. W. van der Kallen, The Schur multipliers of SL(3,$\mathbb{Z}$) and SL(4,$\mathbb{Z}$), Math. Ann. 212, 1974, 47-49.

15. _______________, The K_2 of rings with many units. (in preparation).

16. W. van der Kallen, H. Maazen and J. Stienstra, A presentation for some $K_2(n,R)$, Bull. Amer. Math. Soc. 81, 1975, 934-936.

17. M. A. Kervaire, Some non-stable homotopy groups of Lie Groups, Illinois J. Math. 4, 1960, 161-169.

18. M. Mahowald, The metastable homotopy of S^n, Memoirs of the Amer. Math. Soc. 72, Providence, Rhode Island, 1967.

19. H. Matsumoto, Sur les sous-groupes arithmétiques des groupes semi-simples deployés, Ann. Scient. Ec. Norm. Sup(4) 2, 1969, 1-62.

20. J. Milnor, Introduction to Algebraic K-Theory, Annals of Math. Studies 72, Princeton University Press, Princeton, 1971.

21. M. R. Stein, Surjective stability in dimension 0 for K_2 and related functors, Trans. Amer. Math. Soc. 178, 1973, 165-191.

22. __________, Stability theorems for K_1, K_2 and related functors modelled on Chevalley groups (to appear).

23. J. R. Strooker, The fundamental group of the general linear group over a ring (preprint).

24. R. G. Swan, The number of generators of a module, Math. Zeitschrift 102, 1967, 318-322.

25. __________, Serre's Problem, Conference on Commutative Algebra 1975, Queen's Papers in Pure and Applied Math. 42, Queen's University, Kingston, Ontario, 1975, 1-60.

26. H. Toda, Composition methods in homotopy groups of spheres, Annals of Math. Studies 49, Princeton University Press, Princeton, 1962.

27. L. N. Vaserstein, Stable rank of rings and dimensionality of topological spaces, Funkcional. Anal. i Prilozen (2) 5, 1971, 17-27. (Consultants Bureau Translation, 102-110).

28. _______________, On the stabilization of Milnor's K_2-functor (Russian), Uspehi Mat. Nauk 30, 1, 1975, 224.

29. _______________, On the stabilization of the general linear group over a ring, Math. USSR Sbornik 8, 1969, 383-400.

30. G. D. Whitehead, Homotopy properties of the real orthogonal groups, Ann. of Math. 43, 1942, 132-146.

Rijksuniversiteit Utrecht, Utrecht, The Netherlands.

LES MATRICES MONOMIALES ET LE GROUPE DE WHITEHEAD Wh_2

Jean-Louis LODAY

On donne une démonstration algébrique de l'exactitude de la suite

$$\pi_2^s(B\pi \cup pt) \to K_2(\mathbb{Z}[\pi]) \to Wh_2(\pi) \to 0$$

en utilisant le groupe des matrices monomiales à coefficients dans π.

Dans le dernier paragraphe, on étudie le groupe des matrices monomiales à coefficients ± 1, plus précisément on identifie l'image de $\pi_n^s(RP^\infty) \to K_n(\mathbb{Z})$.

1. - Le groupe des matrices monomiales.

Soit π un groupe, $\mathbb{Z}[\pi]$ l'anneau de π sur $\mathbb{Z}$ et $GL_n(\mathbb{Z}[\pi])$ le groupe linéaire des $n \times n$-matrices inversibles à coefficients dans $\mathbb{Z}[\pi]$.

DEFINITION 1.1. - Le groupe monomial $M_n(\pi)$ est le sous-groupe de $GL_n(\mathbb{Z}[\pi])$ formé des matrices de la forme $P.D$ où P est une matrice de permutation et D une matrice diagonale à coefficients dans π.

Le groupe $M_n(\pi)$ s'identifie au produit semi-direct du groupe symétrique Σ_n et du produit π^n. L'inclusion naturelle $M_n(\pi) \to M_{n+1}(\pi)$ déduite du groupe linéaire nous amène à poser

$$M(\pi) = \varinjlim M_n(\pi) \subset GL(\mathbb{Z}[\pi]) .$$

Mots-clés : Matrice monomiale, homotopie stable, groupe de Whitehead, K- théorie algébrique.

Soit $EM_n(\pi)$ le sous-groupe de $M_n(\pi)$ formé des matrices $P.\,\mathrm{Diag}(g_1,\ldots,g_n)$ telles que P appartient au sous-groupe alterné A_n et le produit $g_1 g_2 \ldots g_n$ appartient au sous-groupe des commutateurs $[\pi,\pi]$. On pose $EM(\pi) = \underrightarrow{\lim}\, EM_n(\pi)$.

On rappelle qu'un groupe G est dit <u>parfait</u> (resp. <u>quasi-parfait</u>) s'il est égal à son sous-groupe des commutateurs : $G = [G,G]$ (resp. si son sous-groupe des commutateurs est parfait).

LEMME 1.2. - <u>Le groupe $EM_n(\pi)$ est parfait dès que</u> $n \geq 5$. <u>Le groupe monomial $M(\pi)$ est quasi-parfait.</u>

<u>Démonstration.</u> - Puisque le groupe A_n est parfait pour $n \geq 5$, il suffit de montrer que toute matrice diagonale dont le produit des éléments diagonaux est dans $[\pi,\pi]$ s'écrit comme un produit de commutateurs dans $EM_n(\pi)$. La matrice $\mathrm{Diag}(g_1,\ldots,g_n)$ est égal au produit

$$\mathrm{Diag}(g_1,g_1^{-1},1,1,\ldots,1).\mathrm{Diag}(1,g_1 g_2,(g_1 g_2)^{-1},1,\ldots,1).\ \ldots$$
$$\ldots\mathrm{Diag}(1,\ldots,1,1,g_1\ldots g_{n-1},(g_1\ldots g_{n-1})^{-1}).\mathrm{Diag}(1,\ldots 1,g_1\ldots g_n)\ .$$

Puisque le produit $g_1\ldots g_n$ est un produit de commutateurs $\prod_i [h_i,h_i']$, la matrice $\mathrm{Diag}(1,\ldots 1,1,g_1\ldots g_n)$ est égale au produit des commutateurs $\prod_i [\mathrm{Diag}(1,\ldots 1,h_i^{-1},h_i)\ ,\ \mathrm{Diag}(1,\ldots,h_i'^{-1},1,h_i')]$. Finalement, la première assertion résulte de la formule

$$\begin{pmatrix} g &&&& \\ & g^{-1} &&& \\ && 1 && \\ &&& 1 & \\ &&&& 1 \end{pmatrix} = \left[\begin{pmatrix} g &&&& \\ & 1 &&& \\ && g^{-1} && \\ &&& 1 & \\ &&&& 1 \end{pmatrix} , \begin{pmatrix} 0 & 1 &&& \\ 1 & 0 &&& \\ && 1 && \\ &&& 0 & 1 \\ &&& 1 & 0 \end{pmatrix}\right]$$

L'application $M(\pi) \xrightarrow{\det} \{+1,-1\} \times \pi/[\pi,\pi]$ qui associe à $P.D$ la signature de P et la classe du produit des éléments diagonaux de D a pour noyau $EM(\pi)$. Puisque $\{+1,-1\} \times \pi/[\pi,\pi]$ est abélien et $EM(\pi)$ parfait, on a $EM(\pi) = [EM(\pi),EM(\pi)] = [M(\pi),M(\pi)]$. $\square$

On peut alors appliquer la construction "+" de Quillen au classifiant $BM(\pi)$ du groupe discret $M(\pi)$ relativement à $EM(\pi)$. (cf. par exemple [L, ch.I]).

PROPOSITION 1.3. - <u>L'espace</u> $\mathbb{Z} \times BM(\pi)^+$ <u>a le type d'homotopie de</u> $\Omega^\infty S^\infty(B\pi \cup pt)$ <u>et par suite</u> $\pi_n(BM(\pi)^+) = \pi_n^s(B\pi \cup pt)$.

Pour tout espace X, la réunion disjointe de X et d'un point sera noté $X^\cdot = X \cup pt$; ce point supplémentaire étant le point base de $X^\cdot$. On rappelle les notations usuelles suivantes $\Omega^\infty S^\infty X = \varinjlim_k \Omega^k S^k X$ et $\pi_n^s(X) = \varinjlim_k \pi_{n+k}(S^k X) = \pi_n(\Omega^\infty S^\infty X)$.

Cette proposition est un cas particulier du théorème ci-dessous dû à Barratt, Priddy et Quillen.

Soit X un CW-complexe connexe, pointé. Le groupe Σ_n opère sur le produit X^n et on pose $M_n(X) = E\Sigma_n \times_{\Sigma_n} X^n$ où $E\Sigma_n$ est le revêtement universel de l'espace classifiant $B\Sigma_n$. Le groupe fondamental de $M_n(X)$ est $M_n(\pi_1(X))$. Les inclusions $\Sigma_n \to \Sigma_{n+1}$ et $X^n \to X^{n+1}$, $(x_1,\dots,x_n) \mapsto (x_1,\dots,x_n,*)$, induisent une inclusion $M_n(X) \to M_{n+1}(X)$. On pose $M(X) = \varinjlim M_n(X)$ et on a $\pi_1(M(X)) = M(\pi_1(X))$.

THEOREME 1.4. - <u>L'espace</u> $\mathbb{Z} \times M(X)^+$ <u>construit relativement au sous-groupe parfait</u> $EM(\pi_1(X))$ <u>a le type d'homotopie de</u> $\Omega^\infty S^\infty(X^\cdot)$.

<u>Démonstration</u>. - La juxtaposition $\Sigma_n \times \Sigma_m \to \Sigma_{n+m}$ et l'identité $X^n \times X^m \to X^{n+m}$ munissent la réunion infinie $\coprod_{n\geq 0} M_n(X)$ d'une structure de monoïde topologique. La "complétion en groupe" de ce monoïde à savoir $\Omega B(\coprod_{n\geq 0} M_n(X))$ a le type d'homotopie de $\Omega^\infty S^\infty(X^\cdot)$ [S]. D'autre part, l'espace M_∞ construit dans [M-S] à partir du monoïde $M = \coprod_{n\geq 0} M_n(X)$ est l'espace $\mathbb{Z} \times M(X)$. Donc il existe une application $\mathbb{Z} \times M(X) \to \Omega^\infty S^\infty(X^\cdot)$ qui est une équivalence d'homologie à coefficients locaux. La propriété universelle de la construction "+" nous permet de conclure. □

1.5. - <u>Remarque</u>. - Une autre démonstration, due à Vogel, peut être déduite de [V]. Le classifiant $\wedge X$ des "étalements de type X" [V, ch.I, 4.1] est un espace analogue à $M(X)$ qui est homotopiquement équivalent à $\Omega^{\infty}S^{\infty}X$ lorsque X est connexe [V, Ch.II, thm. 4] . On peut construire des applications f et g qui rendent le diagramme ci-dessous commutatif

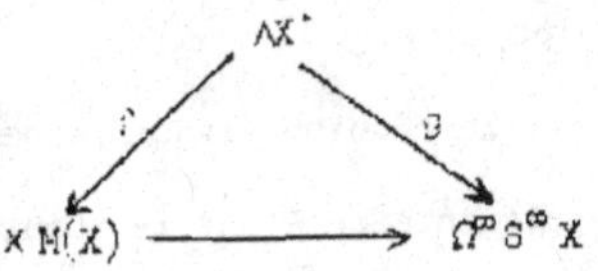

De plus, on montre que, pour tout corps k , on a des isomorphismes

$f_* : H_*(\Lambda X^{\cdot};k) \underset{k[e]}{\otimes} k[e,e^{-1}] \to H_*(\mathbb{Z}\times M(X);k)$ et

$g_* : H_*(\Lambda X^{\cdot};k) \underset{k[e]}{\otimes} k[e,e^{-1}] \to H_*(\Omega^{\infty}S^{\infty}X;k)$ où e est l'élément de $H_0(\Lambda X^{\cdot};k)$ induit par $S^0 \to X^{\cdot} \to \Lambda X^{\cdot}$. Par conséquent, $\mathbb{Z}\times M(X) \to \Omega^{\infty}S^{\infty}X$ est bien une équivalence d'homologie.

1.6. - <u>Homotopie stable et K-théorie algébrique.</u> - Les groupes d'homotopie $\pi_n(BGL(\mathbb{Z}[\pi])^+)$ sont, par définition, les groupes de K-théorie algébrique $K_n(\mathbb{Z}[\pi])$, $n\geq 1$. L'inclusion $M(\pi) \to GL(\mathbb{Z}[\pi])$ induit donc un homomorphisme $\lambda_n(\pi) : \pi_n^s(B\pi^{\cdot}) = \pi_n(BM(\pi)^+) \to \pi_n(BGL(\mathbb{Z}[\pi])^+) = K_n(\mathbb{Z}[\pi])$.

Dans [L] , on a montré que si $h_n(-;\underline{\underline{K}}_{\mathbb{Z}})$ désigne la théorie d'homologie généralisée associée au spectre de la K-théorie de $\mathbb{Z}$ alors λ_n se factorise à travers $h_n(B\pi;\underline{\underline{K}}_{\mathbb{Z}})$.

2. - <u>Le groupe de Whitehead supérieur</u> $Wh_2(\pi)$.

Le groupe de Steinberg $St(A)$ d'un anneau A est l'extension centrale universelle de $E(A)$, sous-groupe des commutateurs de $GL(A)$. Le noyau de $\Phi : St(A) \to E(A)$ est $K_2(A)$. On note x_{ij}^a les générateurs usuels de $St(A)$. Pour tout élément inversible $a\in A^*$, on note $w_{ij}(a) = x_{ij}^a\, x_{ji}^{-a^{-1}}\, x_{ij}^a$ et lorsque $A = \mathbb{Z}[\pi]$, on note $W(\pm\pi)$ le sous-groupe de $St(\mathbb{Z}[\pi])$ engendré par les $w_{ij}(\pm g)$, $g\in\pi$. Hatcher et Wagoner [H-W] ont défini $Wh_2(\pi)$ comme le quotient $K_2(\mathbb{Z}[\pi]) \,/\, W(\pm\pi)\cap K_2(\mathbb{Z}[\pi])$.

Dans [L] , nous avons montré que la suite

$h_2(B\pi;\, \underline{\underline{K}}_{\mathbb{Z}}) \to K_2(\mathbb{Z}[\pi]) \to Wh_2(\pi) \to 0$ est exacte par une méthode topologique. On en déduit la

PROPOSITION. - <u>La suite</u> $\pi_2^s(B\pi') \to K_2(\mathbb{Z}[\pi]) \to Wh_2(\pi) \to 0$ <u>est exacte.</u>

En effet, pour $n=2$, l'homomorphisme $\pi_2^s(B\pi') \to h_2(B\pi;\, \underline{\underline{K}}_{\mathbb{Z}})$ est un isomorphisme. On se propose de donner une démonstration algébrique de cette proposition, c'est-à-dire sans passer par l'intermédiaire de $h_2(B\pi;\, \underline{\underline{K}}_{\mathbb{Z}})$ mais en utilisans l'extension centrale universelle de $EM(\pi)$.

LEMME. - <u>Soit</u> $U(\pi)$ <u>l'extension centrale universelle de</u> $EM(\pi)$ <u>et</u> $\theta : U(\pi) \to St(\mathbb{Z}[\pi])$ <u>l'homomorphisme couvrant l'inclusion</u> $EM(\pi) \to E(\mathbb{Z}[\pi])$. <u>On a alors</u> $\operatorname{Im}\theta \subset W(\pm\pi)$.

<u>Démonstration.</u> - Soit $s : F_G \twoheadrightarrow G$ un épimorphisme avec F_G libre. Si le groupe G est parfait, il existe alors un épimorphisme canonique s' du groupe des commutateurs F_G' sur l'extension centrale universelle de G [M,p.46] .

Puisque $Ker(St(\mathbb{Z}) \to E(\mathbb{Z}))$ est égal à $Ker(U(1) \to M(1))$, on peut considérer $U(1)$ comme un sous-groupe de $St(\mathbb{Z}) \subset St(\mathbb{Z}[\pi])$. On a même $U(1) \subset W(\pm 1)$. Soit $H(\pi)$ le sous-groupe de $St(\mathbb{Z}[\pi])$ engendré par les $h_{ij}(g) = w_{ij}(g)\, w_{ij}(-1)$, $g \in \pi$. L'image par Φ du sous-groupe $\langle U(1), H(\pi)\rangle$ engendré par $U(1)$ et $H(\pi)$ est $EM(\pi)$ (cf. démonstration du lemme 1.2) Soit $\mathfrak{H}(\pi)$ le groupe libre de générateurs $k_{ij}(g)$, $1 \leq i \neq j \geq 1$, $g \in \pi$. L'homomorphisme $\mathfrak{H}(\pi) \to H(\pi)$, $k_{ij}(g) \mapsto h_{ij}(g)$ permet de définir un homomorphisme ψ du produit libre $F_{St(\mathbb{Z})} * \mathfrak{H}(\pi)$ dans $St(\mathbb{Z}[\pi])$, qui est surjectif. En effet, de [M, Cor. 9.4.], on déduit $x_{ij}^g = h_{i1}(g)\, x_{ij}^1\, h_{i1}(g)^{-1}$, $1 \neq j$; d'où $St(\mathbb{Z})$ et $H(\pi)$ engendrent $St(\mathbb{Z}[\pi])$.

La surjectivité de $\Phi\circ\psi$ et la commutativité du diagramme

$$\begin{array}{ccc} F_{U(1)} * \mathfrak{H}(\pi) & \longrightarrow & EM(\pi) \\ \big\downarrow & & \big\downarrow \\ F_{St(\mathbb{Z})} * \mathfrak{H}(\pi) & \xrightarrow{\ \Phi\circ\psi\ } & E(\mathbb{Z}[\pi]) \end{array}$$

implique la commutativité du diagramme

$$\begin{array}{ccc} (F_{U(1)} * H(\pi))' & \xrightarrow{\nu'} & U(\pi) \\ \downarrow & & \downarrow \theta \\ (F_{St(\mathbf{Z})} * H(\pi))' & \xrightarrow{\psi| = (\Phi \circ \psi)'} & St(\mathbf{Z}[\pi]) \end{array}$$

où $\psi|$ est la restriction de ψ à $(F_{St(\mathbf{Z})} * H(\pi))'$. On en déduit les inclusions $\operatorname{Im}\theta = \operatorname{Im}(\theta \circ \nu') \subset \psi(F_{U(1)} * H(\pi)) \subset \langle U(1), H(\pi)\rangle \subset \langle W(\pm 1), H(\pi)\rangle = W(\pm\pi)$. □

Démonstration de la proposition. - Dans le diagramme commutatif

$$\begin{array}{ccccccccc} 1 & \longrightarrow & \pi_2^s(B\pi') & \longrightarrow & U(\pi) & \longrightarrow & EM(\pi) & \longrightarrow & 1 \\ & & \downarrow \theta| & & \downarrow \theta & & \downarrow & & \\ 1 & \longrightarrow & K_2(\mathbf{Z}[\pi]) & \longrightarrow & St(\mathbf{Z}[\pi]) & \longrightarrow & E(\mathbf{Z}[\pi]) & \longrightarrow & 1 \end{array}$$

$\theta|$ est la restriction de θ à $\pi_2^s(B\pi')$. D'après le lemme précédent, on a $\operatorname{Im}\theta| \subset W(\pm\pi) \cap K_2(\mathbf{Z}[\pi])$. D'autre part, un calcul explicite dans le groupe de Steinberg [L, lemmes 4.2.3. et 4.3.5.] montre que $\operatorname{Im}(\theta|) \supset W(\pm\pi) \cap K_2(\mathbf{Z}[\pi])$. Il s'en suit $\operatorname{Coker}(\theta|) = Wh_2(\pi)$. □

3. - Sur l'homomorphisme $\pi_n^s(RP^\infty) \to K_n(\mathbf{Z})$. (*)

Le classifiant $B\mathbf{Z}/2$ du groupe discret $\mathbf{Z}/2$ est l'espace projectif réel RP^∞. Soit $M_p(\pm 1)$ le sous-groupe de $GL(\mathbf{Z})$ des matrices de permutations à coefficients +1 ou -1 :

$$M_p(\pm 1) = \{P \times D \mid P = \text{matrice de permutation},\ D = \operatorname{diag}(\varepsilon_1, \dots, \varepsilon_p) \text{ avec } \varepsilon_i = +1 \text{ ou } -1\}.$$

Le groupe $M_p(\pm 1)$ est clairement isomorphe au groupe monomial $M(\mathbf{Z}/2)$, donc l'inclusion $M(\pm 1)$ dans $GL(\mathbf{Z})$ induit un homomorphisme

$$\zeta_n : \pi_n^s(S^0) \oplus \pi_n^s(B\mathbf{Z}/2) = \pi_n^s(B\mathbf{Z}/2^\cdot) = \pi_n(BM(\pm 1)^+) \to \pi_n(BGL(\mathbf{Z})^+) = K_n(\mathbf{Z}).$$

(*) Ce paragraphe résulte d'une conversation avec S. Priddy.

Nous nous proposons d'identifier l'image de la restriction ζ'_n de ζ_n à $\pi^s_n(B\mathbf{Z}/2)$.

PROPOSITION. - <u>L'image de</u> $\zeta'_n : \pi^s_n(B\mathbf{Z}/2) \to K_n\mathbf{Z}$ <u>est contenue dans l'image de</u> $\lambda_n : \pi^s_n(S^0) \to K_n\mathbf{Z}$.

<u>Démonstration</u>. - Le groupe $\pi^s_n(B\mathbf{Z}/2)$ est de 2-torsion, donc il suffit de raisonner sur la 2-torsion uniquement. Le théorème de localisation [Q 1] et les calculs de la K-théorie des corps finis [Q 2] de Quillen permettent d'affirmer que l'application $K_n(\mathbf{Z}) \to K_n(\mathbf{Z}[\frac{1}{2}])$ est un isomorphisme sur la 2-torsion. Par conséquent, il nous suffit de montrer que l'image de $\pi_n(BM(\pm 1)^+) \to K_n(\mathbf{Z}[\frac{1}{2}])$ est dans l'image de $\pi_n(BM(1)^+) \to K_n(\mathbf{Z}[\frac{1}{2}])$.

On utilise les homomorphismes suivants :

$$j_p : M_p(\pm 1) \hookrightarrow GL_p(\mathbf{Z}) \hookrightarrow GL_p(\mathbf{Z}[\tfrac{1}{2}])$$
$$i_p : M_p(1) \hookrightarrow GL_p(\mathbf{Z}) \hookrightarrow GL_p(\mathbf{Z}[\tfrac{1}{2}])$$
$$\mu_p : M_p(\pm 1) \longrightarrow M_p(1) \quad , \quad P \times D \longmapsto P$$
$$\omega_p : M_p(\pm 1) \longrightarrow M_{2p}(1) \quad , \quad \sigma \longmapsto \bar{\sigma}$$

où la $2p \times 2p$-matrice $\bar{\sigma}$ est déduite de σ en remplaçant chaque coefficent +1 (resp. -1 , resp. 0) par la 2×2-matrice $\begin{pmatrix}1&0\\0&1\end{pmatrix}$ (resp. $\begin{pmatrix}0&1\\1&0\end{pmatrix}$, resp. $\begin{pmatrix}0&0\\0&0\end{pmatrix}$).

LEMME. - <u>Les homomorphismes</u> $i_{2p}\circ\omega_p$ <u>et</u> $(i_p\circ u_p)\oplus j_p$ <u>de</u> $M_p(\pm 1)$ <u>dans</u> $GL_{2p}(\mathbf{Z}[\frac{1}{2}])$ <u>sont conjugués par un élément de</u> $GL_{2p}(\mathbf{Z}[\frac{1}{2}])$.

<u>Démonstration</u>. - L'homomorphisme $(i_p\circ u_p)\oplus j_p$ envoie P.D sur $\begin{pmatrix}P&0\\0&P.D\end{pmatrix}$.

Soit ρ_p l'homomorphisme qui associe à la matrice $\sigma = P.D \in M_p(\pm 1)$ la $2p\times 2p$-matrice σ' déduite de P en remplaçant le 1 de la ligne i par la 2×2-matrice $\begin{pmatrix}1&0\\0&\varepsilon_i\end{pmatrix}$ avec $D = \mathrm{Diag}(\varepsilon_1,\dots,\varepsilon_n)$. L'homomorphisme ρ_p est conjugué de $(i_p\circ u_p)\oplus j_p$ par une matrice de permutation. Posons $\alpha = \begin{pmatrix}\frac{1}{2}&\frac{1}{2}\\1&-1\end{pmatrix}$ on a

$$\alpha\begin{pmatrix}1&0\\0&1\end{pmatrix}\alpha^{-1} = \begin{pmatrix}1&0\\0&1\end{pmatrix} \quad \text{et} \quad \alpha\begin{pmatrix}1&0\\0&-1\end{pmatrix}\alpha^{-1} = \begin{pmatrix}0&1\\1&0\end{pmatrix} .$$

Par conséquent, la matrice conjuguée de σ' par $\alpha \oplus \ldots \oplus \alpha \in GL_{2p}(Z[\frac{1}{2}])$ est $\bar{\sigma}$; donc les homomorphismes ρ_p et $i_{2p} \circ \omega_p$ sont conjugués. □

En définitive, les homomrophismes $(i \circ \omega)_*$ et $(i \circ u)_* + j_*$ de $\pi_n(BM(\pm 1)^+)$ dans $K_n(Z[\frac{1}{2}])$ sont égaux (cf. [L.,Prop. 1.1.9.]), et donc on a $\mathrm{Im}\ j_* \subset \mathrm{Im}\ i_*$. Il s'en suit $\mathrm{Im}\ \zeta_n \subset \mathrm{Im}\ \lambda_n$, d'où $\mathrm{Im}\ \zeta'_n \subset \mathrm{Im}\ \lambda_n$.

<u>Remarque.</u> - On peut même affirmer que l'image de ζ'_n est <u>exactement</u> la 2-torsion de l'image de λ_n car Kahn et Priddy [K-P] ont montré que l'application $\pi_n^s(RP^\infty) \to \pi_n^s(S^0)$ déduite de $\omega : M(\pm 1) \to M(1) = \Sigma_\infty$ est surjective sur la 2-torsion.

BIBLIOGRAPHIE

[H-W] HATCHER A. and WAGONER J. Pseudo-isotopies of compact manifolds. Astérisque 6 , Soc. Math. de France (1973) .

[K-P] KAHN D. and PRIDDY S. Applications of the transfer to stable homotopy. Bull. A.M.S. 78 (1972) 981-987 .

[L] LODAY J.L. K-théorie algébrique et représentations de groupes. Ann. Sc. Ec. Norm. Sup., 4ème série, 9 , n° 3, (1976) .

[M-S] MAC DUFF D. and SEGAL G. Homotopy fibration and the "group completion" theorem. Invent. Math. 31 (1976) 279-284 .

[M] MILNOR J. Introduction to algebraic K-theory. Annals of Math. Studies Princeton, 72 (1971).

[Q1] QUILLEN D. Higher algebraic K-theory I . Springer Lecture Note in Math. 341 (1973) 85-147 .

[Q2] QUILLEN D. On the cohomology and K-theory of the general linear group over a finite field, Ann. of Math. 96 (1972), 552-586 .

[S] SEGAL G.B. Categories and cohomology theories. Topology 13 (1974) 293-312 .

[V] VOGEL P. Cobordisme d'immersions. Ann. Sc. Ec. Norm. Sup., 4ème série, 7 (1974), 317-358 .

FINITELY PRESENTED GROUPS OF MATRICES

U. Rehmann, C. Soulé

1. Let A be a commutative ring with 1 and n a positive integer. The aim of this text is to find conditions on A and n under which the group $GL_n(A)$ is finitely presented.

First of all, the ring A must be finitely generated as a $\mathbb{Z}$-algebra, so that it can be written as a quotient $\mathbb{Z}[t_1,\ldots,t_g]/(P_\rho)_{1\leq\rho\leq r}$, where $(P_\rho)_{1\leq\rho\leq r}$ is the ideal generated by polynomials P_ρ, $1\leq\rho\leq r$. The ring A is then noetherian with finite Krull dimension.

THEOREM 1: *Assume A is a finitely generated commutative $\mathbb{Z}$-algebra with 1, with Krull dimension d, and n an integer greater or equal to $d+3$. If furthermore $K_1(A)$ and $K_2(A)$ are finitely generated, then $GL_n(A)$ is finitely presented.*

Examples and remarks:

- Finite fields, the integers of global fields satisfy the hypotheses of the theorem. If a regular ring A satisfies the hypotheses of the theorem, so do $A[t]$, $A[t,t^{-1}]$.

- If F_q is a finite field, $A = F_q[t]$ (resp. $F_q[t,t^{-1}]$) and G is a simple Chevalley group over $\mathbb{Z}$ and rank $G \geq 3$ (resp. rank $G \geq 2$ and $G \neq G_2$), then one can also prove that $G(A)$ is finitely presented ([3], [4]).

2. Let $E_n(A)$ (resp. $E(A)$) be the group of elementary matrices and $St_n(A)$ (resp. $St(A)$) the Steinberg group corresponding to $GL_n(A)$ (resp. $GL(A)$, the infinite linear group). We know from Bass [1] and from Dennis, Stein and Van der Kallen [5] that, for $n \geq d+3$, the morphisms $GL_n(A)/E_n(A) \to GL(A)/E(A) = K_1(A)$ and $\mathrm{Ker}(St_n(A) \to E_n(A)) \to \mathrm{Ker}(St(A) \to E(A)) = K_2(A)$ are isomorphisms.

The following Lemma is easy:

LEMMA: *If $1 \to G_1 \to G_2 \to G_3 \to 1$ is an exact sequence of groups such that G_1 is finitely presented, then G_2 is finitely presented if and only if G_3 is.*

So, to prove theorem 1, we only need to show that $St_n(A)$ is finitely presented.

THEOREM 2: *Let* A *be as in Th.* 1, Φ *a reduced irreducible root system, and* $St(\Phi,A)$ *the Steinberg group associated to* Φ *with coefficients in* A. *Then, if the rank of* Φ *is greater or equal to* 3, $St(\Phi,A)$ *is finitely presented.*

3. The proof of theorem 2:

For shortness, we shall assume that all the roots in Φ have the same length (for the other cases, see [4]).

3.1 Preliminaries:

3.1.1 LEMMA: *Let* x,y,z *be elements in any group, and* ${}^{y}x = yxy^{-1}$, $[x,y] = xyx^{-1}y^{-1}$. *Then,*

i) $[y,z][x,z] = [x,[y,z]]^{-1}[xy,z]$

ii) *if* $[x,z] = 1$, *then* $[x,[y,z]] = [[x,y],{}^{y}z]$.

3.1.2 LEMMA (cf. [2]): *Let* $\alpha,\alpha' \in \Phi$, $\alpha \neq \alpha'$. *There exists a Dynkin diagram of type* α — β — γ *with* $\beta,\gamma \in \Phi$ *and*

i) $\alpha' = \beta$ *when* $\alpha + \alpha' \in \Phi$

ii) $\alpha' = \alpha + \beta$ *or* $\alpha' = \gamma$ *when* $\alpha + \alpha' \notin \Phi \cup \{0\}$.

3.1.3 LEMMA (cf. [2]): *Let* $\alpha,\alpha',\beta,\beta' \in \Phi$ *be such that* $\alpha + \beta = \alpha' + \beta' \in \Phi$, *and* $\psi = \langle\alpha,\beta,\alpha',\beta'\rangle$ *the* $\mathbb{Z}$*-module generated by these roots.*

i) *If rank* $\psi = 3$, *then* (*up to a change of* α *and* β) $\alpha - \alpha', \beta' - \beta \in \Phi$, *and* $\alpha + \alpha', \beta + \beta', \alpha \pm \beta', \alpha' \pm \beta \notin \Phi \cup \{0\}$.

ii) *If rank* $\psi = 2$, *there exist* $\alpha'',\beta'' \in \Phi$ *such that* $\alpha + \beta = \alpha'' + \beta''$ *and* $\alpha,\beta,\alpha'',\beta''$ *generate a* $\mathbb{Z}$*-module of rank* 3.

3.1.4 The Steinberg group $St(\Phi,A)$ is generated by symbols $x_\alpha(a)$, $\alpha \in \Phi$, $a \in A$, submitted to the relations

A) $x_\alpha(a + b) = x_\alpha(a)x_\alpha(b)$

B) $[x_\alpha(a), x_\beta(b)] = x_{\alpha+\beta}(N_{\alpha,\beta}ab)$, $\alpha + \beta \neq 0$

where $N_{\alpha,\beta} = -N_{\beta,\alpha} = \pm 1$, if $\alpha,\beta,\alpha+\beta \in \Phi$, and $N_{\alpha,\beta} = 0$ if $\alpha,\beta \in \Phi$, $\alpha + \beta \notin \Phi$. We also have, by Jacobi's identity:

LEMMA: *If* $\alpha,\beta,\gamma \in \Phi$ *are pairwise linearly independent, and* $\alpha + \beta + \gamma \neq 0$, *then*

$$N_{\alpha,\beta+\gamma}N_{\beta,\gamma} + N_{\beta,\gamma+\alpha}N_{\gamma,\alpha} + N_{\gamma,\alpha+\beta}N_{\alpha,\beta} = 0.$$

3.2 The case $A = \mathbb{Z}[t_1,\dots,t_g]$

3.2.1 If $\mu = (\mu_1,\dots,\mu_g) \in \mathbb{N}^g$ is a multi-integer and $t = (t_1,\dots,t_g)$, we write $t^\mu = t_1^{\mu_1} \dots t_g^{\mu_g}$, $|\mu| = \max(\mu_i)_{1\le i\le g}$, and $\varepsilon(\mu) = (\varepsilon_1,\dots,\varepsilon_g)$ where $\varepsilon_i = 1$ if $\mu_i = |\mu| > 0$, $\varepsilon_i = 0$ if not.

Let m be a positiv integer and G_m the group generated by symbols $x_\alpha(at^\mu)$, $\alpha \in \Phi$, $a = \pm 1$, $\mu \in \mathbb{N}^g$, $|\mu| \le m$, with the relations

a_m) $x_\alpha(t^\mu)x_\alpha(-t^\mu) = 1$

b_m) $[x_\alpha(t^\mu), x_{\alpha'}(t^\nu)] = 1$ when $\alpha + \alpha' \notin \Phi \cup \{0\}$

b'_m) $[x_\alpha(at^\mu), x_{\alpha'}(bt^\nu)] = x_{\alpha+\alpha'}(N_{\alpha,\alpha'}abt^{\mu+\nu})$ when $\alpha,\alpha',\alpha+\alpha' \in \Phi$.

Because of b'_{m+1}), there exists a canonical epimorphism $\varphi_m : G_m \to G_{m+1}$. We shall prove that this is an isomorphism for $m \ge 2$, and the theorem will follow since G_2 is finitely presented and $\varinjlim G_m$ is equal to $St(\Phi,A)$ (using relations A), $x_\alpha(a)$ can be defined without any ambiguity for all a in A).

3.2.2 PROPOSITION: Let $\alpha,\beta,\alpha+\beta \in \Phi$; $|\mu|,|\nu| \le m$, $|\mu+\nu| = m+1$ and $a,b = \pm 1$. In G_m the following is true:

i) If $\alpha',\beta' \in \Phi$, $\alpha'+\beta' = \alpha+\beta$; $|\mu'|,|\nu'| \le m$, $\mu'+\nu' = \mu+\nu$; $a',b' = \pm 1$, and $N_{\alpha,\beta}ab = N_{\alpha',\beta'}a'b'$, then

$$[x_\alpha(at^\mu), x_\beta(bt^\nu)] = [x_{\alpha'}(a't^{\mu'}), x_{\beta'}(b't^{\nu'})].$$

ii) If $\gamma \in \{\alpha,\beta,\alpha+\beta\}$, $|\mu'| \le m$, $c = \pm 1$, then

$$[[x_\alpha(at^\mu), x_\beta(bt^\nu)], x_\gamma(ct^{\mu'})] = 1.$$

Proof: i) According to 3.1.3, we first consider the case rank $\Psi = 3$ and $\alpha - \alpha' = \beta' - \beta \in \Phi$. Then if $\varepsilon := \varepsilon(\mu+\nu)$, we have $\tau = \nu - \varepsilon \in \mathbb{N}^g$, $|\tau|$, $|\mu+\tau| \le m$ and

$$[x_\alpha(at^\mu), x_\beta(bt^\nu)] = [x_\alpha(at^\mu),[x_{\beta-\beta'}(ct^\tau), x_{\beta'}(t^\varepsilon)]]$$

$$= [x_{\alpha'}(c't^{\mu+\tau}), x_{\beta'}(t^\varepsilon)x_\beta(bt^\nu)], \text{ by 3.1.1 ii), and } \alpha+\beta' \in \Phi \cup \{0\}$$

$$= [x_{\alpha'}(c't^{\mu+\nu-\varepsilon}), x_{\beta'}(t^\varepsilon)], \text{ because } \alpha'+\beta \notin \Phi \cup \{0\},$$

where $c = \pm 1$ suitable and $c' = N_{\alpha,\beta-\beta'}N_{\beta-\beta',\beta'}ab = a'b'$ by 3.1.4. By symmetry we get the result. In the case $\beta - \alpha' = \beta' - \alpha \in \Phi$ one uses $N_{\alpha,\beta} = -N_{\beta,\alpha}$ and gets

$$[x_\alpha(at^\mu), x_\beta(bt^\nu)] = {}^{x_\alpha(at^\mu)}[x_\beta(bt^\nu), x_\alpha(-at^\mu)]$$

$$= {}^{x_\alpha(at^\mu)}[x_{\alpha'}(t^{\mu'}), x_{\beta'}(b't^{\nu'})]$$

from the above, and $\alpha + \alpha'$, $\alpha + \beta' \notin \Phi \cup \{0\}$ yields the formula.
When rank $\psi = 2$, we write

$$[x_\alpha(at^\mu), x_\beta(bt^\nu)] = [x_{\alpha''}(N_{\alpha,\beta}N_{\alpha'',\beta''}abt^{\mu+\nu-\varepsilon(\mu+\nu)}), x_{\beta''}(t^{\varepsilon(\mu+\nu)})]$$
$$= [x_{\alpha''}(N_{\alpha',\beta'}N_{\alpha'',\beta''}a'b't^{\mu'+\nu'-\varepsilon(\mu'+\nu')}), x_{\beta''}(t^{\varepsilon(\mu'+\nu')})]$$
$$= [x_{\alpha'}(a't^{\mu'}), x_{\beta'}(b't^{\nu'})].$$

Part ii) uses α'', β'' in Φ such that $\alpha'' + \gamma$, $\beta'' + \gamma \notin \Phi \cup \{0\}$:

$$[[x_\alpha(at^\mu), x_\beta(bt^\nu)], x_\gamma(ct^{\mu'})]$$
$$= [[x_{\alpha''}(c't^{\mu+\nu-\varepsilon(\mu+\nu)}), x_{\beta''}(t^{\varepsilon(\mu+\nu)})], x_\gamma(ct^{\mu'})] = 1 \qquad \text{q.e.d.}$$

<u>3.2.3</u> Now let us define in G_m, when $\alpha \in \Phi$, $|\mu| = m + 1$ and $a = \pm 1$,

$$x_\alpha(at^\mu) = [x_{\alpha-\alpha'}(N_{\alpha-\alpha',\alpha}at^{\mu-\varepsilon(\mu)}), x_{\alpha'}(t^{\varepsilon(\mu)})]$$

for a choice (and then any choice by 3.2.2 i)) of $\alpha' \in \Phi$ such that $\alpha - \alpha' \in \Phi$ (possible by irreducibility).

We have to show that these elements verify $a_{m+1})$, $b_{m+1})$, and $b'_{m+1})$.
The relations $a_{m+1})$ come from 3.1.1. i) and 3.2.2 ii).
We show relations $b_{m+1})$, $b'_{m+1})$ by induction on $k = |\mu| + |\nu|$, and we may assume $k > m \geq 2$. For $b'_{m+1})$, by symmetry and with respect to 3.2.2 i), we have to consider only the case $|\mu| \leq |\mu + \nu| = |\nu| = m + 1$. Then it is clear that $|\mu + 2\varepsilon(\nu)| \leq \max\{|\mu|, 2\}$, and by 3.1.2 i) we have

$$[x_\alpha(at^\mu), x_\beta(bt^\nu)]$$
$$= [x_\alpha(at^\mu), [x_{\beta+\gamma}(t^{\varepsilon(\nu)}), x_{-\gamma}(b't^{\nu-\varepsilon(\nu)})]]$$
$$= [x_{\alpha+\beta+\gamma}(a't^{\mu+\varepsilon(\nu)}), {}^{x_{\beta+\gamma}(t^{\varepsilon(\nu)})}x_{-\gamma}(b't^{\nu-\varepsilon(\nu)})] \quad \text{from 3.1.1 ii)}$$
$$= {}^{x_{\beta+\gamma}(t^{\varepsilon(\nu)})}[x_{\alpha+\beta+\gamma}(a't^{\mu+\varepsilon(\nu)}), x_{-\gamma}(b't^{\nu-\varepsilon(\nu)})] \quad \text{from induction hypothesis}$$
$$= {}^{x_{\beta+\gamma}(t^{\varepsilon(\nu)})}[x_{\alpha+\beta+\gamma}(a't^{\mu+2\varepsilon(\nu)}), x_{-\gamma}(b't^{\nu-2\varepsilon(\nu)})] \quad \text{from 3.2.2 i)}$$
$$= [x_{\alpha+\beta+\gamma}(a't^{\mu+2\varepsilon(\nu)}), {}^{x_{\beta+\gamma}(t^{\varepsilon(\nu)})}x_{-\gamma}(b't^{\nu-2\varepsilon(\nu)})]$$
$$= [x_\alpha(at^{\mu+\varepsilon(\nu)}), x_\beta(bt^{\nu-\varepsilon(\nu)})]$$

using again 3.1.1 ii) and induction hypothesis. This proves $b'_{m+1})$ if $|\mu| + |\nu| = k$, since $|\nu - \varepsilon(\nu)| \leq m$.

The relations $b_{m+1})$ will follow from 3.2.2 ii), when $\alpha = \alpha'$. If $\alpha \neq \alpha'$, by 3.1.2 ii), we may assume $|\nu| = m+1$, and then we have two cases.

In the first one $\alpha - \alpha' = \beta \in \Phi$, and we can write

$$(*) \quad [x_\alpha(at^\mu), x_{\alpha+\beta}(bt^\nu)] = [x_\alpha(at^\mu), [x_{\alpha+\beta+\gamma}(b't^{\nu-\varepsilon(\nu)}), x_{-\gamma}(t^{\varepsilon(\nu)})]] = 1,$$

by $b'_{m+1})$ and induction hypothesis.

In the second case, we have $\alpha' = \gamma$, and we first assume $\mu \leq \nu$ (i.e. $\mu_i \leq \nu_i$ for all i), $\epsilon = \varepsilon(\nu - \mu)$. Then, with suitable constants,

$$\begin{aligned}
&{}^{x_\alpha(at^\mu)}x_\gamma(bt^\nu) \\
&= {}^{x_\alpha(at^\mu)}[x_{-\beta}(b't^{\nu-\varepsilon}), x_{\beta+\gamma}(t^\varepsilon)] && \text{from } b'_{m+1}) \\
&= [x_{-\beta}(b't^{\nu-\varepsilon}), {}^{x_\alpha(at^\mu)}x_{\beta+\gamma}(t^\varepsilon)] && \text{from } (*) \\
&= [[x_{-\alpha-\beta}(b''t^{\nu-\mu-\varepsilon}), x_\alpha(at^\mu)], {}^{x_\alpha(at^\mu)}x_{\beta+\gamma}(t^\varepsilon)] && \text{from } b'_{m+1}) \\
&= [x_{-\alpha-\beta}(b''t^{\nu-\mu-\varepsilon}), [x_\alpha(at^\mu), x_{\beta+\gamma}(t^\varepsilon)]] && \text{from 3.1.1 ii) and } b_m) \\
&= x_\gamma(b''' t^\nu),
\end{aligned}$$

and computing b''' using the Lemma in 3.1.4 we get

$$b''' = N_{-\alpha-\beta,\alpha+\beta+\gamma} N_{\alpha,\beta+\gamma} N_{-\alpha-\beta,\alpha} N_{-\beta,\beta+\gamma}\, b = b.$$

Now, in general, let $\mu' = (\max\{\mu_i - \nu_i, 0\})_{1\leq i\leq g}$. Then $\mu - \mu' \leq \mu \leq \nu + \mu'$, and from the above we get, for $a', c = \pm 1$ suitably chosen,

$$\begin{aligned}
x_{\alpha+\beta+\gamma}(bt^{\nu+\mu'}) &= {}^{x_{-\beta}(a't^{\mu-\mu'})}x_{\alpha+\beta+\gamma}(bt^{\nu+\mu'}) \\
&= {}^{x_{-\beta}(a't^{\mu-\mu'})}[x_{\alpha+\beta}(ct^{\mu'}), x_\gamma(bt^\nu)] && \text{from } b'_{m+1}) \\
&= [x_\alpha(at^\mu)x_{\alpha+\beta}(ct^{\mu'}), x_\gamma(bt^\nu)] && \text{from } (*) \\
&= {}^{x_\alpha(at^\mu)}x_{\alpha+\beta+\gamma}(bt^{\mu'+\nu})[x_\alpha(at^\mu), x_\gamma(bt^\nu)],
\end{aligned}$$

and this together with $(*)$ implies $[x_\alpha(at^\mu), x_\gamma(bt^\nu)] = 1$, and the proof is finished.

3.3 The general case:

Let $A = \mathbb{Z}[t_1,\ldots,t_g]/(P_\rho)_{1\leq\rho\leq r}$ and p be the epimorphism $St(\Phi,\mathbb{Z}[t_1,\ldots,t_g]) \to St(\Phi,A)$. We claim that Ker p is equal to the subgroup K generated, as a normal subgroup, by the elements $x_\alpha(P_\rho)$, $\alpha \in \Phi$, $1\leq\rho\leq r$. Actually, if $p(x_\alpha(P)) = p(x_\alpha(Q))$, then $P - Q = \sum_\rho P_\rho Q_\rho$, and

$$x_\alpha(P)x_\alpha(Q)^{-1} = \prod_\rho [x_\beta(P_\rho), x_\gamma(Q_\rho)] \in K$$

($\beta,\gamma \in \Phi$ chosen such that $\alpha = \beta + \gamma$). So $St(\Phi,Z[t_1,\dots,t_g])/K$ has a set of generators in bijection with $\{x_\alpha(a) | \alpha \in \Phi, a \in A\}$ and the relations they satisfy are just the same as those in $St(\Phi,A)$. q.e.d.

References

1. Bass, H.: Algebraic K-Theory. Benjamin, 1968
2. Chevalley, C.: Sur certains groupes simples. Tôhoku Math. J. 7 14-66 (1955)
3. Hurrelbrink, J.: Endlich präsentierte arithmetische Gruppen im Funktionenkörper-Fall, Preprint. 1975
4. Rehmann, U.: Präsentationen von Chevalley-Gruppen über $k[t]$. Preprint, 1975
5. Van der Kallen, W.: Stability for K_2. These proceedings.

HOMOLOGY SPHERE BORDISM AND QUILLEN PLUS CONSTRUCTION

Jean-Claude Hausmann[(1)]

Let X be a CW-complex and N a perfect normal subgroup of $\pi_1(X)$. The Quillen plus construction $X \longrightarrow X^+$ with respect to N is then defined. In this paper, an interpretation of $\pi_k(X^+)$ $(k \geq 5)$ as certain bordism groups of X is given. Some mild assumptions on the 2-type of X seem to be necessary.

1. The group $\pi_k^N(X)$.

Definition. $\pi_k^N(X)$ denotes the set of equivalence classes of pairs (M,f) where:

1) M is a k-dimensional oriented topological manifold such that $H_*(M;\mathbb{Z}) \simeq H_*(S^k;\mathbb{Z})$ (one says that M is a k-homology sphere).

2) $f : M \longrightarrow X$ is a continuous map such that $\operatorname{Im} \pi_1 f = N$ and $\operatorname{Ker} \pi_1 f$ is a central subgroup in $\pi_1 M$.

Two pairs (M_1,f_1) and (M_2,f_2) are considered as equivalent if there exists a pair (W,F) such that:

i) W is a $(k+1)$-dimensional oriented manifold such that $\operatorname{Bd} W = M_1 + (-M_2)$ and $H_*(W,M_i;\mathbb{Z}) = 0$,

ii) $F : W \longrightarrow X$ is a continuous map such that $F|M_i = f_i$, $\operatorname{Im} \pi_1 F = N$ and $\operatorname{Ker} \pi_1 F$ is central in $\pi_1(W)$. (W has a base arc joining the base points of M_i and sent by F to the base point of X).

Lemma 1.1. Let $0 \longrightarrow H_2(N) \longrightarrow \tilde{N} \xrightarrow{p} N \longrightarrow 1$ be the universal central extension of $N([K])$. Let (M,f) represent a class in $\pi_k^N(X) k \geq 3$. Then, there exists a unique isomorphism $\varphi : \pi_1(\pi) \longrightarrow \tilde{N}$ such that the following diagram commutes:

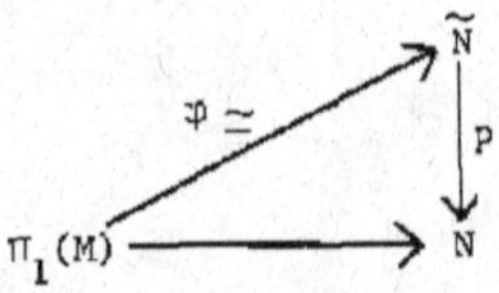

Proof. By condition (2), $\pi_1 f$ is a central extension of N. Since M is a k-

homology sphere, $k \geq 3$, one has $H_1(\pi_1(M)) = H_2(\pi_1(M)) = 0$. These conditions characterize the universal central extension up to a unique isomorphism ([K]).

Since $\pi_1(M)$ is finitely presented, one has:

<u>Corollary 1.2</u>. If $\tilde{N}$ is not finitely presented, then $\pi_k^N(X) = \emptyset$.

Note that if N is finitely presented, so is $\tilde{N}$. For, $H_2(N)$ is finitely presented.

<u>2. Addition laws on</u> $\pi_k^N(X)$.

<u>Case 1</u>. One assumes that:

i) $\tilde{N}$ is finitely presented and $k \geq 5$,

ii) N acts trivially on $\pi_2(X)$,

iii) $k_1(X) \in \ker p^* \circ i^*$, where $k_1(X) \in H^3(\pi_1(X);\pi_2(X))$ is the first Postnikov invariant of X and p^* and i^* are the homomorphisms induced in cohomology by $p : \tilde{N} \longrightarrow N$ and $i : N \subset \pi_1(X)$ respectively.

The hypothesis (iii) can be translated in the following way: Let

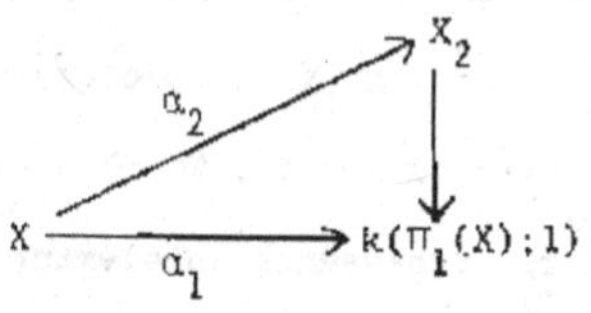

be the first two stages of the Postnikov tower of X. Then the map $i \circ p : k(\tilde{N};1) \longrightarrow k(\pi_1(X);1)$ lifts in:

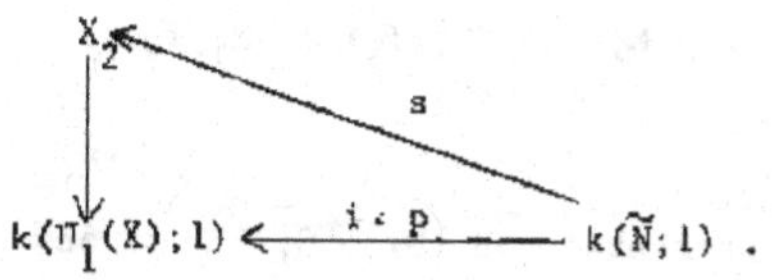

We are now able to define the addition $(M_1,f_1) + (M_2,f_2)$. Take first the connected sum $M^0 = M_1 \# M_2$, $f^0 = f_1 \# f_2 : M^0 \longrightarrow X$. M^0 is a homology sphere and

$\mathrm{Im}\,\pi_1 f^0 = N$. One has the lifting:

$$\begin{array}{ccc} & & k(\widetilde{N};1) \\ & \overset{\varphi_2 \,\#\, \varphi_2}{\nearrow} & \big\downarrow{\scriptstyle i \circ p} \\ M^0 & \xrightarrow[\alpha_1 \circ f^0]{} & k(\pi_1(X);1) \end{array}$$

given by liftings φ_1 and φ_2 of $\alpha_1 \circ f_1$ and $\alpha_2 \circ f_2$ respectively. By obstruction theory and condition (ii), $\alpha_2 \circ f^0$ is homotopic to $s \circ (\varphi_1 \# \varphi_2)$ and then to $s \circ \alpha_1 \circ f^0$.

By attaching finitely many 2-handles to $M^0 \times I$, one can build up a cobordism (W^0, F^0) from (M^0, f^0) to (M^1, f^1) such that $\mathrm{Im}\,\pi_1 f^1 = N$, $\ker \pi_1 f^1$ is central in $\pi_1(M^1) \simeq \pi_1(W^0)$ (or, equivalently, $\pi_1(M^1) = \widetilde{N}$). Here one uses that $\widetilde{N}$ is finitely presented. By obstruction theory, F^0 can be chosen such that $\alpha_2 \circ F^0 = s \circ \alpha_1 \circ F^0$.

Now $H_2(M^1;\mathbb{Z}) \simeq H_2(W^0;\mathbb{Z})$ is free abelian of finite rank. The Hurewicz homomorphism $\pi_2(M^1) \longrightarrow H_2(M^1)$ is surjective since $H_2(\widetilde{N}) = 0$. Therefore, one can add finitely many 3-handles to W^0 on M^1 (use $n \geq 5$) to obtain a cobordism (W,F) from (M^0, f^0) to (M,f) such that M is a homology sphere and $\pi_2(M) \simeq \pi_1(W) \simeq N$. By our assumption on X, F^0 extends to $F : W \longrightarrow X$ leading by restriction to $f : M \longrightarrow X$. Therefore, (M,f) represents an element of $\pi_k^N(X)$. If (W',F') is another cobordism from (M^0, f^0) to (M',f') constructed as above, one can use the cobordism $(W \cup W', F \cup F')$ to show that (M,f) and (M',f') represent the same class in $\pi_k^N(X)$. It is thus possible to define

$$(M,f) = (M_1,f_1) + (M_2,f_2)\ .$$

<u>Case 2</u>. N is finitely presented, $H_2(N) = 0$, $k \geq 5$ and N acts trivially on $\pi_2(X)$.

Our hypothesis $H_2(N) = 0$ implies that $N = \widetilde{N}$.

In this case, we will use the following consequence of our hypothesis on X:

Let $f : L \longrightarrow X$ be a continuous map inducing an isomorphism on the fundamental groups and such that $H_2(L) \longrightarrow H_2(X)$ is the zero homomorphism. Suppose $H_2(L)$ is free of finite rank. Then, there exists a complex K obtained by adding finitely many 3-cells to L together with an extension $F : K \longrightarrow X$ of f and such that $H_2(L) = 0$. Indeed, one has the diagram

$$\begin{array}{ccccccc}
 & & \pi_3(X;L) & & & & \\
 & & \downarrow & & & & \\
H_3(X,L;\mathbb{Z}N) & \longrightarrow & H_3(X;L) & \longrightarrow & \mathrm{Tor}_1^{\mathbb{Z}N}(H_2(X,L;\mathbb{Z}N);\mathbb{Z}) & \longrightarrow & 0 \\
 & & \downarrow & & & & \\
 & & H_2(L) & & & &
\end{array}$$

where the horizontal sequence is exact and comes from the universal coefficients spectral sequence applied to $C_*(\widetilde{X};\widetilde{L})$. $H_2(X;L;\mathbb{Z}N)$ is a quotient of $\pi_2(X)$ on which $\widetilde{N}$ acts trivially. Therefore, $\mathrm{Tor}_1^{\mathbb{Z}N}(H_2(X,L;\mathbb{Z}N);\mathbb{Z}) = H_1(N;H_2(X,L;\mathbb{Z}N)) = 0$. Then $\pi_3(X;L) \longrightarrow H_2(X)$ is onto and the assertion follows.

Now we can define $(M_1,f_1) + (M_2,f_2)$ in the same way as in Case 1. One first constructs a cobordism (W^0,F^0) from $(M_1,f_1) \# (M_2,f_2) = (M^0,f^0)$ to (M^1,f^1) such that $\pi_1 f^1$ is an isomorphism onto $N = \widetilde{N}$. W^0 is obtained by attaching a finite collection of 2-handles on $M^0 \times \{1\} \subset M^0 \times I$. Since $H_2(N) = 0$, the Hurewicz homomorphism $\pi_2(X) \longrightarrow H_2(X)$ is surjective and we may choose F_0 such that the induced homomorphism $H_2(W^0) \longrightarrow H_2(X)$ is zero. By using the consequence of our hypothesis, one may construct a cobordism (W,F) from (M^0,f^0) to (M,f) and one is able to define $(M,f) = (M_1,f_1) + (M_2,f_2)$ as in Case 1.

__Proposition 2.1.__ Suppose $k \geq 5$, $\widetilde{N}$ finitely presented and N acts trivially on $\pi_2(X)$. If X satisfies the hypothesis of Case (1) or (2) above, then the additions defined in these cases endow $\pi_k^N(X)$ with an abelian group structure. A pair (M,f) represents the zero element of $\pi_k^N(X)$ if and only if there exists an acyclic manifold V^{k+1} such that $\mathrm{Bd}\, V = M$, the inclusion $M \subset V$ inducing an isomorphism on fundamental groups and f extending to $F : V \longrightarrow X$. The inverse element $-(M,f)$ of (M,f) is the class of $(-M,f)$.

<u>Proof</u>. The associativity of the addition is clear. So is the characterization of the zero element. But one has to prove the existence of this zero element. Let L be a finite 2-dimensional complex such that $\pi_1(L) = \tilde{N}$. Find a map $f^L : L \longrightarrow X$ inducing the composition $\tilde{N} \xrightarrow{P} N \longrightarrow \pi_1(X)$ on fundamental groups. In Case (1), one can choose f^L such that $\alpha_1 \circ f^L$ is homotopic to $s \circ \alpha_2 \circ f^L$. In Case (2), since $H_2(N) = 0$, $\pi_2(\tilde{X}_N) \longrightarrow H_2(\tilde{X}_N)$ is surjective where $\tilde{X}_N$ is the covering of X with fundamental group N. Using a lifting $\tilde{f}^L : L \longrightarrow \tilde{X}_N$ of f^L, one can change f^L such that $f^L : H_2(L) \longrightarrow H_2(X)$ is the zero homomorphism. By the hypothesis in Case (1) or (2), there exists a 3-dimensional acyclic complex K whose 2-skeleton is L and a map $f : K \longrightarrow X$ such that $f|L = f^L$.

By [W] Theorem 1, there exists a PL-embedding of K in $\mathbb{R}^{k+1}$ if $k + 1 \geq 6$. Let E^{k+1} be a regular neighborhood of K in R^{k+1} and $F : E^{k+1} \longrightarrow X$ such that $F|K = f$. $(\mathrm{Bd}\, E^{k+1}, F|\mathrm{Bd}\, E^{k+1})$ represents the neuter element of $\pi_k^N(X)$.

To prove that $(M,f) + (-M,f) = 0$, observe that $M \# (-M) = \mathrm{Bd}((M - \mathring{D}^k) \times I)$ where $\mathring{D}^k$ is an open k-disk in M. $f \# f$ extends to $F : (M - \mathring{D}^k) \times I \longrightarrow X$. Then, if $(\bar{M},\bar{f}) = (M,f) + (-M;f)$, $\bar{M}$ will bound an acyclic manifold A^{k+1} such that $\pi_1(\bar{M}) \longrightarrow \pi_2(A)$ is an isomorphism and $\bar{f}$ extends to $\bar{F} : A \longrightarrow X$. Then $(\bar{M},\bar{f}) = 0$.

<u>Remark</u>. $\pi_k^{\{0\}}(X)$ is simply $\pi_k(X)$.

3. The main theorem.

Let $X \longrightarrow X^+$ be the plus construction of Quillen with respect to N. Let (M,f) represent an element of $\pi_k^N(X)$. By functoriality of the plus construction, one gets a homotopy commutative diagram:

$$\begin{array}{ccc} M & \xrightarrow{f} & X \\ \downarrow & & \downarrow \\ S^k \simeq M^+ & \xrightarrow{f^+} & X^+ \end{array}$$

The reader will check that the correspondence $(M,f) \longmapsto f^+ : S^k \longrightarrow X^+$ gives rise to a map $T_k : \pi_k^N(X) \longrightarrow \pi_k(X^+)$; if, moreover, the assumptions of Case (1) or (2) are fulfilled, T_k is a homomorphism.

Theorem 3.1. Suppose that $k \geq 5$, $\tilde{N}$ is finitely presented and N acts trivially on $\pi_2(X)$. Then

$$T_k : \pi_k^N(X) \longrightarrow \pi_k(X^+)$$

is a bijection. In particular, if X satisfies the hypothesis of Case (1) or (2), T_k is an isomorphism.

Proof. The basic ingredients are the results of homological surgery developed in [H2], [V] and [H3].

Proof of surjectivity. Let $\alpha : S^k \longrightarrow X^+$ represent a class in $\pi_k(X^+)$. Consider the pull-back diagram:

$$\begin{array}{ccc} Z & \longrightarrow & X \\ \downarrow & & \downarrow \\ S^k & \xrightarrow[\alpha]{} & X^+ \end{array}$$

Let A be the common fiber of $X \longrightarrow X^+$ and of $Z \longrightarrow S^k$. A is acyclic and $\pi_1(A) = \tilde{N}$. Therefore, $H_*(Z) \simeq H_*(S^k)$, $\pi_1(Z) \simeq \tilde{N}$ and $\pi_2(Z) = \pi_2(A)$. Moreover, A is the Drov acyclic functor [D] of $\tilde{X}_N$, where $\tilde{X}_N$ is the covering of X of fundamental group N. For a proof of this, see the argument of Quillen ([G] proof of 2.22) which holds in general for a map $Z \longrightarrow Z^+$ when $\pi_1(Z)$ is perfect. By [D] theorem 2.1 (iv), $\pi_1(A)$ acts trivially on $\pi_2(A)$ since the same holds for N and $\pi_2(\tilde{X}_N) = \pi_2(X)$. One deduces that $\pi_1(Z) = \tilde{N}$ acts trivially on $\pi_2(Z)$.

To prove the surjectivity of T_k, it is enough to find a manifold M^k together with a map $g : M \longrightarrow Z$ inducing an isomorphism on integral homology. The identity map of S^k gives a normal map of degree one:

$$\begin{array}{ccc} \nu(V^k) & \xrightarrow{b_0} & \xi \\ \downarrow & & \downarrow \\ V^k & \xrightarrow[g_0]{} & Z \end{array}$$

where $\nu(V^k)$ is the stable normal bundle of the manifold V^k and ξ a stable vector bundle over Z. (For this step, see [H1, §5].)

By Theorem 3.1 of [H2], (g_0,b_0) is normally cobordant to (g,b), $g : M^k \longrightarrow Z$ such that g induces an isomorphism on homology. Thus surjectivity of T_k is proved. One could also use the argument of Vogel ([V], 1.7).

<u>Proof of injectivity</u>. Let (M_1,f_1) and (M_2,f_2) be such that $T_k((M_1,f_1)) = T_k((M_2,f_2)) = [\alpha]$, for $\alpha : S^k \longrightarrow X^+$. One has a factorization $\bar{f}_i$ of f_i as follows:

$$\begin{array}{ccccc} M_i & & & & \\ & \searrow^{\bar{f}_i} & & \searrow^{f_i} & \\ & & Z & \longrightarrow & X \\ & & \downarrow & \text{pull-back} & \downarrow \\ & & S^k & \longrightarrow & X^+ \end{array} \qquad (i = 1,2) \; .$$

$\bar{f}_i$ induces an isomorphism on integral homology, and then an isomorphism in KO-theory. Therefore, $\bar{f}_i$ is covered by a stable bundle map

$$\begin{array}{ccc} \nu_{M_i} & \xrightarrow{\bar{b}_i} & \xi \\ \downarrow & & \downarrow \\ M_i & \xrightarrow[\bar{f}_i]{} & Z \end{array}$$

and $(\bar{f}_i,\bar{b}_i)$ is a normal map of degree one. The injectivity of T_k will be proved when we will have established that $(\bar{f}_1,\bar{b}_1)$ and $(\bar{f}_2,\bar{b}_2)$ are joined by a normal cobordism (W,F) such that $H_*(W,M_i) = 0$ and $\pi_1 F$ is an isomorphism. This step is given by [H2] §4.3.

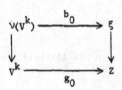

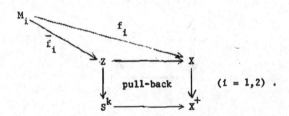

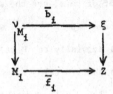

Remark. If the conditions of Case (1) or (2) are realized, one knows that T_k is a homomorphism. Then it suffices to prove that $T_k^{-1}(0) = 0$ for proving the injectivity of T_k. If $T_k((M,f)) = 0$, f lifts in $g : M \longrightarrow A$, and A is acyclic. Then, one can use the argument of [V] 1.8 to prove that $(M,f) = 0$.

4. Homology spheres with a given fundamental group.

Let G be a finitely presented group satisfying $H_1(G) = H_2(G) = 0$. Denote by BG the Eilenberg-MacLane space $K(G;1)$. Lemma 1.1 implies that an element of $\pi_i^G(BG)$ is represented by an i-homology sphere M^i together with an isomorphism $\pi_1(M) \xrightarrow{\simeq} G$. Roughly speaking, $\pi_i^G(BG)$ is the set of i-dimensional homology spheres with fundamental group identified to G, up to (homology-and-fundamental-group)-cobordism.

By Proposition 3.1, $\pi_i^G(BG)$ is a group for $i \geq 5$, which is isomorphic to $\pi_i(BG^+)$ by Theorem 3.1.

A pair (M^i,f) represents zero in $\pi_i^G(BG)$ if and only if $M = \mathrm{Bd}\, A^{i+1}$, where A^{i+1} is an acyclic compact manifold such that $\pi_1(M) \longrightarrow \pi_1(A)$ is an isomorphism. Indeed, there is no obstruction for extending $f : M \longrightarrow BG$ to $F : A \longrightarrow BG$ and the assertion follows from Theorem 3.1. The classical way for constructing homology spheres of dimension $k \geq 5$, i.e., taking the boundary of a regular neighborhood of an acyclic subpolyhedron of $\mathbb{R}^{k+1}$, cannot be used to produce non-zero elements of $\pi_k^G(BG)$.

We restrict ourselves here to examples offering connections with algebraic K-theory. Other computations are given in [H2].

Let $St_n(\Lambda)$ be the Steinberg groups of a ring with unit Λ. One has $H_1(St_n(\Lambda)) = H_2(St_n(\Lambda)) = 0$ if $n \geq 5$ ([K]). The condition for $St_n(\Lambda)$ to be finitely presented leads to a condition on Λ, which is fulfilled if, for instance:

-- Λ is a finite ring. Then $St_n(\Lambda)$ is finite.

-- Λ is the ring of integers in a finite extension field of the rational numbers. For $S\ell_n(\Lambda)$ is finitely presented ([B1]) and so is $St_n(\Lambda)$.(2)

For all these cases, $\pi_k(BSt_n(\Lambda)^+)$ classifies homology spheres of dimension k ($k \geq 5$) with fundamental group identified to $St_n(\Lambda)$. Using the homomorphism $i_k : \pi_k(BSt_n(\Lambda)^+) \longrightarrow \pi_k(BSt(\Lambda)^+) \simeq K_k(\Lambda)$, the last isomorphism holding for $k \geq 3$

[L, p. 28], elements of $K_k(\Lambda)$ thus appear as obstructions for homology spheres with fundamental group $St_n(\Lambda)$ to bound acyclic manifold with the same fundamental group. If $K_k(\Lambda)$ is of finite type, any $x \in K_k(\Lambda)$ appears as such an obstruction when n is large enough.

Example. By [B2], one has $K_5(\mathbb{Z}) \otimes \mathbb{Q} \simeq \mathbb{Q}$. Then, for n large enough, there are infinitely many 5-dimensional homology spheres with fundamental group $St_n(\mathbb{Z})$ such that any two of them are not cobordant by a homology cobordism with fundamental group $St_n(\mathbb{Z})$. In order to compute $\pi_k(BSt_n(\Lambda))$ the following question arises naturally:

Problem. For a given integer n, can one find an integer $s(n)$ such that $s(n) \longrightarrow \infty$ when $n \longrightarrow \infty$ and $i_k : \pi_k(St_n(\Lambda)) \longrightarrow K_k(\Lambda)$ is an isomorphism for $k \leq s(n)$?

The last condition is equivalent to:

-- $H_k(E_n(\Lambda)) \longrightarrow H_k(E(\Lambda))$ is an isomorphism for $k \leq s(n)$,

or, in the case where Λ is commutative and $E_n(\Lambda) = Sl_n(\Lambda)$:

(*) -- $H_k(Gl_n(\Lambda)) \longrightarrow H_k(Gl(\Lambda))$ is an isomorphism and $\Lambda^{\bullet}$ acts trivially on $H_k(Sl_n(\Lambda))$ for $k \leq s(n)$.

Although some general methods were proposed ([Q]) very little is known about the homology of $Sl_n(\Lambda)$ (except results of [B2]). However, there are some stability theorems for $H_k(Gl_n(\Lambda)) \longrightarrow H_k(Gl(\Lambda))$ (Quillen for fields, Wagoner for local ring, both to appear). Then one can use condition (*) and get some partial results. For the example of a finite field $\mathbb{F}_q$, $\mathbb{F}_q^{\bullet}$ acts trivially on $H_*(Sl_n(\mathbb{F}_q))$ when $n \equiv 1$ (mod $g - 1$); for, $a \in \mathbb{F}_q^{\bullet}$ lifts in $Gl_n(\mathbb{F}_q)$ by Diag(a). For other such elementary facts, see [H3].

5. Homology sphere bordism.

Definition. Let X be a space. The i-th homology sphere bordism group of X, denoted by $\Omega_i^{HS}(X)$ is the set of pairs (M,f) where:

-- M is a i-dimensional oriented topological homology sphere,

-- $f : M \longrightarrow X$ is a pointed continuous map

under the equivalence of homology bordism (i.e. $(M_1,f_1) \sim (M_2,f_2)$ if there exists a cobordism (W^{i+1},M_0,M_1) together with an extension $F : W \longrightarrow X$ of both f_i and such that $H_*(W,M_1) = H_*(W,M_2) = 0.$)

$\Omega_i^{HS}(X)$ is a group for the connected sum, its zero is $(S^i$, constant) and $(-M,f) = -(M,f)$. $\Omega_i^{HS}(-)$ is a covariant functor from the category of topological spaces and maps to the category of abelian group and homomorphisms.

For $i > 1$, $\Omega_i^{HS}(X) \xleftarrow{\cong} \Omega_i^{HS}(\tilde{X}_N)$ where $\tilde{X}_N$ is the covering of X whose fundamental group is the maximal perfect subgroup N of $\pi_1(X)$. Then one can assume that $\pi_1(X)$ is perfect and consider $X \longrightarrow X^+$, the plus map with respect to $\pi_1(X)$. As in §3, one gets a homomorphism $\psi_i : \Omega_i^{HS}(X) \longrightarrow \pi_i(X^+)$.

__Theorem 5.1.__ Suppose that X is an inductive limit of spaces X_n such that $\pi_1(X_n)$ is a finitely presented perfect group acting trivially on $\pi_2(X_n)$. Then ψ_i is an isomorphism for $i \geq 5$.

__Proof.__ Let $G(n) = \pi_1(X_n)$ and $X_n \longrightarrow X^+$ be the plus map with respect to $G(n)$. By Theorem 3.1, one has $\pi_i^{G(n)}(X_n) \cong \pi_i(X_n^+)$. There is a commutative diagram for all n:

$$\begin{array}{ccc}
\pi_i^{G(n)}(X_n) & \longrightarrow & \Omega_i^{HS}(X) \\
\downarrow \cong & & \downarrow \psi_i \\
\pi_i(X_n^+) & \longrightarrow & \pi_i(X^+)
\end{array}$$

Since $\pi_i(X^+) = \varinjlim \pi_i(X_n^+)$, ψ_i is surjective. If $\psi_i(M^i,f) = 0$, one has a lifting:

$$\begin{array}{ccc}
 & \overset{\bar{f}}{\nearrow} & F \\
 & & \downarrow \\
M & \xrightarrow{\ f\ } & X \\
 & & \downarrow \\
 & & X^+
\end{array}$$

and F is acyclic. By the argument of [V, 1.8], $\bar{f} : M \longrightarrow F$ extends to $\bar{g} : A \longrightarrow F$ where A^{i+1} is an acyclic manifold. Then $(M,f) = 0$ in $\Omega_i^{HS}(X)$ and

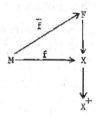

ψ_i is injective.

Corollary 5.2. Let G be a group which is an inductive limit of finitely generated perfect groups (whence G is itself perfect). Then, for $i \geq 5$,

$$\Omega_i^{HS}(BG) \xrightarrow{\simeq} \pi_i(BG^+)$$

is an isomorphism.

Proof. The hypothesis implies that G is an inductive limit of finitely presented perfect groups [V, Lemma 3.5]. Then 5.2 follows from 5.1.

Corollary 5.3. Let Λ be any ring. Then, for $i \geq 5$, one has

$$K_i(\Lambda) \simeq \Omega_i^{HS}(BSt\ \Lambda) \simeq \Omega_i^{HS}(BE(\Lambda)).$$

Analogous results were found by P. Vogel.

Proof. Λ is the inductive limit of its sub-rings Λ_j which are finitely generated as Z-algebras. For each such Λ_j, the groups $St_n(\Lambda)$ (or $E_n(\Lambda)$) are finitely generated and perfect for $n \geq 5$. Then 5.3 follows from 5.2.

Footnotes:

(1) Supported in part by NSF grant MPS72-05055 A03.

(2) Recently, C. Soulé proved that $St_n(\Lambda)$ is finitely presented for every finitely generated commutative Z-algebra Λ and $n \geq 5$.

References

[B1] Borel A., Arithmetic properties of linear groups, Proc. I. C. M. Stockholm (1962), 10-22.

[B2] ________, Stable real cohomology of arithmetic groups, Ann. Sc. Ec. norm. sup. 4e serie, t. 7 (1974), 235-272.

[D] Dror, E., Acyclic Spaces, Topology 11 (1972), 339-348.

[G] Gersten, S. M., Higher K-theory of rings, Algebraic K-theory I, Springer Lecture Notes 341, 1-41.

[H1] Hausmann, J-Cl., Homological surgery, to appear.

[H2] ______________, Manifolds with a given homology and fundamental group, in preparation.

[H3] ______________, Stabilité partielle pour l'homologie des groupes speciaux linéaires, C. R. Ac. Sc. Paris 281 (1975), 687-690.

[K] Kervaire, M., Multiplicateurs de Schur et K-theorie, Essays on topology (Memoire dedié á G. de Rham), Springer (1970), 212-225.

[L] Loday, J. L., Thesis, 1975, University or Strasbourg.

[Q] Quillen, D., Cohomology of groups, Actes I. M. C. Nice (1970), vol. 2, 47-51.

[V] Vogel, P., Un theorem de Hurewicz homologique, to appear.

[W] Weber, Cl., Deux remarques sur les plongements d'un AR dans un espace euclidien, Bull. Ac. Polonaise des Sci. XVI, Nb. 11 (1968), 851-855.

Institute for Advanced Study, Princeton, New Jersey.

University of Geneva, Switzerland.

Letter from Quillen to Milnor on $\mathrm{Im}(\pi_i O \longrightarrow \pi_i^s \longrightarrow K_i \mathbb{Z})$

July 26, 1972

Dear Jack,

As I wrote you earlier, the assertion in your note that I can prove the injectivity of the map

$$J(\pi_i O) \subset \pi_i^s \longrightarrow K_i Z$$

is inaccurate with respect to the 2-torsion. Unfortunately, the corrections I sent are also incorrect. Since Kervaire has requested some details, I am sending the following account of what I know about the above map, in order to clear the confusion.

1. First consider the dimensions $i = 8k$, $8k+1$, where $J(\pi_i O) = Z/2$. I do not know whether this group injects into $K_* \mathbb{Z}$, and suspect that it does not, except of course when $k = 0$.

However, Adams has produced elements of order 2, $\eta_j \in \pi_j^s$, $j = 8k+1$, $8k+2$, closely related to the image of J in the preceding dimensions, which do map non-trivially into $K_* \mathbb{Z}$. To see this, consider the square

$$(1) \qquad \begin{array}{ccc} B\Sigma_\infty^+ & \longrightarrow & BO \\ \downarrow & & \downarrow \wr \\ BGL(\mathbb{Z})^+ & \longrightarrow & BGL(R) \end{array}$$

induced by the various group inclusions. Passing to homotopy groups, we obtain homomorphisms $\pi_j^s = \pi_j B\Sigma_\infty^+ \longrightarrow K_j \mathbb{Z} \longrightarrow \pi_j BO$ whose composition is the degree map for KO-theory. Since Adams has shown that the degree map carries η_j to the generator of $\pi_j BO = \mathbb{Z}/2$, the image of η_j in $K_j\mathbb{Z}$ is non-trivial. In fact, we have

$$K_j\mathbb{Z} = \mathbb{Z}/2 \oplus \; ?, \qquad j = 8k+1,\ 8k+2.$$

I should mention that this observation appears already in one of Gersten's papers.

2. Next consider the dimension $i = 4s-1$, where $J(\pi_i O)$ is cyclic of order denom$(B_s/4s)$. I shall prove the injectivity:

$$J(\pi_{4s-1}O) \longrightarrow K_{4s-1}\mathbb{Z}$$

by showing that the Adams e-invariant on π^s_{4s-1}, which detects $J(\pi_{4s-1}O)$, comes from an invariant defined on $K_{4s-1}\mathbb{Z}$.

Following Sullivan, consider the fibration

$$F \longrightarrow BO \xrightarrow{(ch_{4i})} \prod_{i\geq 1} K(\mathbb{Q},4i)$$

where $K(\mathbb{Q},j)$ is an Eilenberg-Maclane space and ch_j represents the j-th component of the Chern character. Since $B\Sigma^+_\infty$ has trivial rational cohomology, the degree map $B\Sigma^+_\infty \longrightarrow BO$ lifts by obstruction theory, uniquely up to homotopy, to a map

(2) $$B\Sigma^+_\infty \longrightarrow F$$

which induces a homomorphism

$$\pi^s_{4s-1} \longrightarrow \pi_{4s-1}F \simeq \mathbb{Q}/a_s\mathbb{Z}$$

where a_s is 1 or 2 depending on whether s is even or odd.

I claim this homomorphism is the negative of the Adams e-invariant. Assuming this for the moment, consider the diagram

Now Adams defines the e-invariant of $\overline{f}$ by choosing an element z of $\widetilde{KO}(\text{Cone } f)$ restricting to the generator of $\widetilde{KO}(S^{8k})$, and forming

$$ch_{8k+4s}(z) \quad \varepsilon \; H^{8k+4s}(\text{Cone } f, \mathbb{Q}) \cong H^{8k+4s}(S^{8k+4s}, Q) \cong \mathbf{Q} .$$

The image of this rational number in $\mathbb{Q}/a_s\mathbf{Z}$ is then $e(\overline{f})$. Clearly z and $ch_{8k+4s}(z)$ may be identified with the maps x and y in the diagram, hence we have the formula

$$e(\overline{f}) \quad = \quad \{c,b,f\} .$$

On the other hand, from the theory of Toda brackets one knows that the map u in the diagram represents the negative of $\{c,b,f\}$. Thus we have the formula

(3) $$e(\overline{f}) \quad = \quad -f^*(v_k) \; \varepsilon \; \pi_{8k+4s-1}F(8k) \cong \mathbf{Q}/a_s\mathbf{Z}$$

where $v_k = v$ is the unique element of $\pi_{8k}F(8k)$ mapping to the generator of $\pi_{8k}BO(8k)$. Now by periodicity we have $\Omega^{8k}F(8k) \simeq \mathbb{Z} \times F$. The maps v_k fit together to induce a map

$$\overline{v} : \lim_{k} \Omega^{8k}_{o}S^{8k} \longrightarrow F$$

which covers the degree map into BO. Thus $\overline{v}$ is the map (2). The formula (3) shows that its effect on homotopy groups is the negative of the e-invariant, which proves the claim.

4. Additional information on the image of $J(\pi_{4s-1}O)$ in $K_{4s-1}\mathbf{Z}$ can be obtained from the computation of the K-groups of finite fields as follows. Let p be a prime number and F_p the field with p elements, and consider the obvious homomorphisms

$$\pi^s_{4s-1} \longrightarrow K_{4s-1}\mathbf{Z} \longrightarrow K_{4s-1}\mathbf{F}_p .$$

I will show below that this composition is essentially the part of the complex e-invariant which is prime to p. More precisely, there is a commutative diagram

(4) $$\begin{array}{ccc} \pi^s_{4s-1} & \longrightarrow & K_{4s-1}\mathbb{F}_p \cong Z/(p^{2s}-1)\mathbf{Z} \\ {\scriptstyle -e}\downarrow & & \downarrow{\scriptstyle e} \\ Q/a_s\mathbb{Z} & \longrightarrow & \mathbb{Q}/\mathbb{Z}[p^{-1}] \end{array}$$

$$\begin{array}{ccccc} B\Sigma_\infty^+ & \longrightarrow & BGL(\mathbb{Z})^+ & & \\ & \swarrow^{\text{(dotted)}} & \downarrow w & & \\ F & \longrightarrow & BO & \xrightarrow{ch} & \Pi K(Q,4i) \end{array}$$

with the map w obtained from (1). Since the Chern classes of representations of discrete groups are torsion classes, the map (ch)w is null-homotopic, and the dotted arrow exists. The induced map from $B\Sigma_\infty^+$ to F must be (2). Thus we obtain a commutative diagram

$$\begin{array}{ccc} \pi_{4s-1}^s & \longrightarrow & K_{4s-1}\mathbb{Z} \\ {\scriptstyle -e}\downarrow & \swarrow^{\text{(dotted)}} & \\ \mathbb{Q}/a_s\mathbb{Z} & & \end{array}$$

as desired.

3. To prove the claim about the e-invariant, consider the map

$$BO(8k) \longrightarrow \prod_{i\geq 1} K(Q,\ 8k+4i)$$

with components ch_{8k+4i}, where BO(8k) is the (8k-1)-connected covering of BO. Denote this map briefly by $c : BO(8k) \longrightarrow E(8k)$ and let F(8k) be its fibre. Let $b\colon S^{8k} \longrightarrow BO(8k)$ represent the generator of $\pi_{8k}BO(8k) = \pi_{8k}BO$ provided by Bott periodicity.

Now suppose given a map $f : S^{8k+4s-1} \longrightarrow S^{8k}$ representing an element $\bar{f}$ of π_{4s-1}^s. We compute the Toda bracket $\{c,b,f\}$ by forming the diagram

$$\begin{array}{ccccccc} S^{8k+4s-1} & \xrightarrow{f} & S^{8k} & \longrightarrow & \text{Cone } f & \longrightarrow & S^{8k+4s} \\ \downarrow u & & \downarrow v & \searrow^{b} & \downarrow x & & \downarrow y \\ \Omega E(8k) & \longrightarrow & F(8k) & \longrightarrow & BO(8k) & \xrightarrow{c} & E(8k) \end{array}$$

in which the arrows x,y and v,u can be filled in as bf and cb are null-homotopic. By definition, the Toda bracket is the element represented by y in

$$\pi_{8k+4s}E(8k)\ /\ c_*\pi_{8k+4s}BO(8k) + f^*\pi_{8k}\Omega E(8k) = \mathbb{Q}/a_s\mathbb{Z}\ .$$

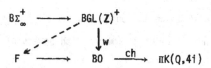

where θ is injective with image the unique subgroup of order $p^{2s}-1$. Here $\mathbb{Z}[p^{-1}]$ denotes the ring of rational numbers with powers of p in the denominator.

Assuming this, let ℓ be an odd prime, and choose p to be a topological generator of the group $\mathbb{Z}_\ell^*$ of ℓ-adic units. According to Adams, the e-invariant is injective on $J(\pi_{4s-1}0)$, and the ℓ-primary component $J(\pi_{4s-1}0)_{(\ell)}$ is cyclic of order ℓ^n, $n = v_\ell(p^{2s}-1)$, v_ℓ = ℓ-adic valuation. We have therefore an isomorphism

$$J(\pi_{4s-1}0)_{(\ell)} \xrightarrow{\sim} (K_{4s-1}\mathbf{F}_p)_{(\ell)}.$$

It follows that the odd part of $J(\pi_{4s-1}0)$ is isomorphic to a direct summand of $K_{4s-1}\mathbb{Z}$.

Suppose now that $\ell = 2$ and take p = 3. Using Adams work, both the source and target of the map

$$J(\pi_{4s-1}0)_{(2)} \longrightarrow (K_{4s-1}\mathbb{F}_3)_{(2)}$$

are cyclic of order 2^n, $n = v_2(3^{2s}-1)$; and the map is essentially multiplication by a_s. It follows that for s even, when $a_s = 1$, $J(\pi_{4s-1}0)_{(2)}$ is isomorphic to a direct summand of $K_{4s-1}\mathbb{Z}$.

Finally, observe that the diagram (4) shows the unique element of order 2 of $J(\pi_{4s-1}0)$, when s is odd, goes to zero in $K_{4s-1}\mathbf{F}_p$ for all p.

Summarizing:

Proposition: The homomorphism $\pi_{4s-1}^s \longrightarrow K_{4s-1}\mathbf{Z}$ induces an injection of $J(\pi_{4s-1}0)$ into $K_{4s-1}\mathbf{Z}$. For even s, the image of $J(\pi_{4s-1}0)$ is a direct summand. For odd s, the odd-torsion part of the image is a direct summand. For odd s, the unique element of order 2 of the image is in the kernel of the homomorphism $K_{4s-1}\mathbb{Z} \longrightarrow K_{4s-1}\mathbf{F}_p$ for all primes p.

I do not know whether or not the image of $J(\pi_{4s-1}0)_{(2)}$ is a direct summand of $K_{4s-1}\mathbf{Z}$ when s is odd. The first case is s=1, where $\mathbb{Z}/24 = J(\pi_3 0) = \pi_3^s \subset K_3\mathbb{Z} = H_3(St(\mathbf{Z}),\mathbf{Z})$. Here $K_3\mathbf{F}_3 = \mathbf{Z}/8$ and the map $J(\pi_3 0) \longrightarrow K_3F_3$ has a kernel of order 6.

5. It remains to construct the diagram (4). Consider the diagram

$$\begin{array}{ccccc}
F & \longrightarrow & BO & \xrightarrow{ch} & \prod_{i\geq 1} K(\mathbb{Q},4i). \\
\downarrow & & \downarrow & & \downarrow \\
F' & \longrightarrow & BU[p^{-1}] & \xrightarrow{ch} & \prod_{j\geq 1} K(\mathbb{Q},2j) \\
\uparrow & & \uparrow & & \uparrow \scriptstyle ((p^j-1)^{-1}ch_{2j}) \\
F\Psi^p & \longrightarrow & BU & \xrightarrow{\psi^p-1} & BU
\end{array}$$

where F' and $F\Psi^p$ are defined so that the rows are fibrations. Here BU $[p^{-1}]$ is the localization of BU which represents the functor $K(?)\otimes Z[p^{-1}]$. Examining the homotopy sequences of these fibrations, we obtain isomorphisms

$$(5)\qquad \begin{array}{ccc}
\pi_{4s-1}F & \approx & Q/a_s\mathbb{Z} \\
\downarrow & & \downarrow \\
\pi_{4s-1}F' & \approx & Q/Z[p^{-1}] \\
\uparrow & & \cup \\
\pi_{4s-1}F\Psi^p & \approx & (p^{2s}-1)^{-1}\mathbb{Z}/Z
\end{array}$$

where the maps at the right are the obvious ones.

From the computation of the K-groups of a finite field, there is a homotopy equivalence

$$BGL(\mathbb{F}_p)^+ \approx F\Psi^p$$

induced by lifting representations of finite groups over F_p to virtual complex representations by means of the Brauer theory. I claim that the diagram

$$\begin{array}{ccccc}
B\Sigma_\infty^+ & \longrightarrow & BGL(\mathbb{F}_p)^+ & \approx & F\Phi^p \\
\downarrow & & & & \downarrow \\
BO & & \longrightarrow & & BU[p^{-1}]
\end{array}$$

is commutative. The upper right path is obtained by lifting the obvious representation of Σ_n on $\mathbb{F}_p^n$ to a virtual complex representation, while the lower right path comes from the obvious action of Σ_n on $\mathbb{C}^n$.

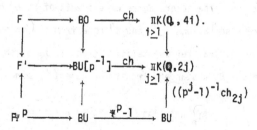

These two virtual representations are not the same in general. However, it is known that their characters agree on elements of Σ_n of order prime to p, because both the representations $\mathbb{F}_p{}^n$ and $\mathbb{C}^n$ come from the integral representation $\mathbb{Z}^n$. Thus the two virtual representations agree on the Sylow ℓ-subgroups Σ_n^ℓ for all primes $\ell \neq p$. By a standard transfer argument, one has

$$[B\Sigma_n, BU[p^{-1}]] \hookrightarrow \prod_{\ell \neq p} [B\Sigma_n^\ell, BU[p^{-1}]] .$$

Consequently, the above diagram commutes as claimed.

Since $B\Sigma_\infty^+$ has trivial rational cohomology, it follows by obstruction theory that the diagram

$$\begin{array}{ccc} B\Sigma_\infty^+ & \longrightarrow & BGL(\mathbb{F}_p)^+ \simeq F\Psi^p \\ \downarrow & & \downarrow \\ F & \longrightarrow & F' \end{array}$$

is commutative, where the vertical arrow at the left is the one inducing minus the e-invariant. The desired commutative diagram (4) now results by taking homotopy groups, and using the isomorphisms (5).

This concludes the account of the map $J(\pi_* O) \longrightarrow K_*\mathbb{Z}$. To the best of my knowledge, nothing more is known about $K_i\mathbb{Z}$ for $i>2$ beyond what this and Borel's theorem provide.

Best regards,

Dan Quillen

Characteristic classes of representations

by Daniel Quillen *

Let A be a ring with identity. For any group B, let Is(G,A) be the set of isomorphism classes of representation of G over A, that is, finitely generated projective A-modules on which G acts linearly. This paper is concerned with characteristic classes of such representations, that is, natural transformations from Is(G,A) to group cohomology. These bear the same relation to the algebraic K-theory of the ring A as characteristic classes of vector bundles do in topological K-theory.

The first part of the paper begins with the definitions and shows how characteristic classes can be described in terms of the homology of the infinite general linear group GL(A), at least when the coefficients are over a field k. I have tried to explain a point of view I have found very useful, namely, that the "points" (as in algebraic geometry) of the algebra $H_*(GL(A),k)$ may be identified with stable exponential characteristic classes, that is, characteristic classes which transform direct sums into products like the total Chern class.

The second part is devoted to an important theorem (Th.2',§5) saying roughly that any exact sequence of representations splits as far as stable characteristic classes are concerned. As a corollary one obtains that for any finite field F of characteristic p, GL(F) has trivial mod p cohomology; in fact, the proof of the theorem is a generalization of this special case. In another paper devoted to algebraic K-theory, I plan to use this result to show that the exterior power operations on representations over a commutative ring induce a λ-ring structure on the corresponding K-theory.

---

*Supported by The Institute for Advanced Study and National Science Foundation.

First part: Generalities about characteristic classes of representations

§1. Representations. Let $\underline{\underline{A}}$ be an additive category whose isomorphism classes form a set S. The direct sum operation in $\underline{\underline{A}}$ provides an abelian monoid structure on S; we denote by $K_o\underline{\underline{A}}$ the associated abelian group. By a representation of a group G in $\underline{\underline{A}}$ we mean an object of $\underline{\underline{A}}$ endowed with an action of G. We denote by $Is(G, \underline{\underline{A}})$ the set of isomorphism classes of these representations; it is an abelian monoid in a natural way, and we let $R'(G, \underline{\underline{A}})$ be the associated abelian group.

Let $\underline{\underline{Is}}\ \underline{\underline{A}}$ be the groupoid whose morphisms are the isomorphisms in $\underline{\underline{A}}$. A representation of G in $\underline{\underline{A}}$ may be viewed as a functor

$$E : G \longrightarrow \underline{\underline{Is}}\ \underline{\underline{A}}$$

where the group G is considered as a category with one object. Let $[\underline{\underline{B}}, \underline{\underline{C}}]$ denote the set of isomorphism classes of functors from $\underline{\underline{B}}$ to $\underline{\underline{C}}$, so that on viewing groups as categories, $[G,G']$ is the set of equivalence classes of homomorphisms from G to G', two homomorphisms being equivalent if they are conjugate by an element of G'. Then

$$Is(G, \underline{\underline{A}}) = [G, \underline{\underline{Is}}\ \underline{\underline{A}}] = \coprod_s [G, Aut(P_s)]$$

where s runs over S and P_s is an object in the isomorphism class S.

We say that two representations E and E' of G are stably isomorphic if $E \oplus \varepsilon \simeq E' \oplus \varepsilon'$ for two trivial representations ε and ε'. This is an equivalence relation on $Is(G, \underline{\underline{A}})$, and we denote the quotient by $St(G, \underline{\underline{A}})$. Let (S,S) be the category obtained by letting the monoid S act on itself: S is the set of objects, and an arrow $s \longrightarrow s'$ is a pair (s,t) such that $s + t = s'$. We define a functor from (S,S) to sets by assigning to s the set $[G,Aut(P_s)]$, and to the arrow (s,t): $s \longrightarrow s'$ the map induced by the homomorphism

$$Aut(P_s) \xrightarrow{u \mapsto u \oplus id} Aut(P_s \oplus P_t) \simeq Aut(P_{s'})$$

where the last map is induced by any isomorphism of $P_s \oplus P_t$ with $P_{s'}$. Then clearly we have

$$St(G, \underline{\underline{A}}) = \lim_{\substack{(S,S)}} \text{ind.} \ [G, Aut(P_s)].$$

Example 1. Let A be a ring (always with identity, but not necessarily commutative) and let $\underline{\underline{A}}$ be the category of its finitely generated projective modules. In this case representations in $\underline{\underline{A}}$ will be called representations over A, and frequently we replace $\underline{\underline{A}}$ by A in the notation, e.g. $R'(G,A)$, K_oA. The group $R'(G,A)$ is the Grothendieck group of representations in which the relations are of the form $[E' \oplus E''] = [E'] + [E'']$; it therefore has as a quotient the Grothendieck group in which the relations come from exact sequences.

Let $GL_n(A) = Aut(A^n)$, and let $GL(A)$ be the limit of the $GL_n(A)$ with respect to the standard inclusions. We consider the monoid homomorphism $\mathbb{N} \longrightarrow S$ sending n to A^n. Since every object of $\underline{\underline{A}}$ is a direct summand of A^n for some n, the induced functor $(\mathbb{N},\mathbb{N}) \longrightarrow (S,S)$ is cofinal ([1], p.149), hence

$$St(G,A) = \lim_{n} \text{ind.} \ [G, GL_n(A)].$$

As the limit maps to $[G,GL(A)]$, it follows that any representation E of G over A gives rise to a canonical element

$$\rho_E \in [G, GL(A)]$$

depending only on the stable isomorphism class of E. In concrete terms, one chooses a trivial representation ε and an isomorphism of the underlying A-module of $E \oplus \varepsilon$ with A^n, thus obtaining a homorphism from G to $GL_n(A)$; composing with the embedding of the latter in $GL(A)$ yields a homomorphism from G to $GL(A)$ representing ρ_E.

Example 2. Suppose in the preceding discussion we replace "representation of G in $\underline{\underline{A}}$ " by "(complex) vector bundle over X". Then $Is(G,\underline{\underline{A}})$ becomes the isomorphism classes of vector bundles:

$$Vect(X) = [X, \coprod_n BU_n]$$

while $St(G, \underline{A})$ and $R'(G, \underline{A})$ become respectively

$$\widetilde{K}(X) = [X, BU] \ , \ K(X) = [X, \mathbb{Z} \times BU]$$

provided X is a finite-dimensional CW complex.

§2. Characteristic classes. For any graded abelian group

$$M = \bigoplus_{i \geq 0} M_i$$

put

$$H^o(G, M_i) = \prod_{i \geq 0} H^i(G, M_i)$$

where $H^i(G, M_i)$ is the cohomology of the group G with coefficients in M_i regarded as a trivial G-module. By a characteristic class for representations in $\underline{A}$, or simply characteristic class of $\underline{A}$, we shall mean a morphism of contravariant functors from groups to sets of the form

$$\theta : Is(G, \underline{A}) \longrightarrow H^o(G, M)$$

for some M; we say that θ has coefficients in M. We call θ stable if $\theta(E \oplus \varepsilon) = \theta(E)$ when ε is a trivial representation, and we call θ additive if it is a monoid homomorphism:

$$\theta(0) = 0 \ , \ \theta(E' \oplus E'') = \theta(E') + \theta(E'').$$

If θ is additive, it extends to a homomorphism from $R'(G, \underline{A})$ to $H^o(G, M)$.

Supposing that M is a graded anti-commutative ring, $H^o(G, M)$ is a commutative ring with multiplication derived from the cup-product:

$$H^i(G, M_i) \otimes H^j(G, M_j) \longrightarrow H^{i+j}(G, M_{i+j}).$$

A characteristic class θ with coefficients in M will be called an exponential characteristic class if

$$\theta(0) = 1 \ , \ \theta(E' \oplus E'') = \theta(E')\theta(E'').$$

Such a θ will be called invertible if $\theta(E)$ is a unit for every representation E. In this case θ extends uniquely to a homomorphism from $R'(G,\underline{A})$ to the group of units $H^o(G,M)^*$.

Examples. 1. In the vector bundle situation (Ex. 2, §1), the total Chern class

$$c(E) = \Sigma c_i(E) \in \prod_{i\geq 0} H^{2i}(X,\mathbb{Z})$$

is a stable exponential characteristic class with coefficients in $\mathbb{Z}[t]$ with t of degree 2. The Chern character is an additive characteristic class with coefficients in $Q[t]$.

2. The various species of Chern classes considered in [4] provide characteristic classes for representations over a commutative ring.

3. We call a characteristic class inessential if it is independent of the action of the group and depends only on the underlying object of the representation. Inessential characteristic classes with coefficients in M are in one-one correspondence with maps $S \longrightarrow M_o$, the additive (resp. exponential) characteristic classes corresponding to monoid homomorphisms for the addition (resp. multiplication) of M_o.

Remarks. In the following M will be a fixed anti-commutative ring.

1. The set of exponential characteristic classes with coefficients in M is an abelian monoid with product: $(\theta_1\theta_2)(E) = \theta_1(E)\theta_2(E)$. The invertible exponential characteristic classes are the invertible elements of this monoid.

2. An element of $H^o(G,M)$ is invertible if and only if it is carried to an invertible element by the augmentation: $H^o(G,M) \longrightarrow M_o$. Consequently an exponential characteristic class θ is invertible if and only if for each object P, considered as a representation of the one element group e, we have that $\theta(P)$ is invertible in $H^o(e,M) = M^o$. In particular, any stable exponential characteristic class is

invertible.

3. The group of invertible exponential characteristic classes with coefficients in M is the direct product of the subgroup of inessential classes, which is isomorphic to $\mathrm{Hom}(K_0\,\underline{\underline{A}}, M_0{}^*)$, and the group of stable exponential characteristic classes.

4. Supposing that M is an algebra over Q, the exponential series

$$\exp(x) = \Sigma x^n/n!$$

converges for $x \in H^0(G,M)$ of augmentation zero, and gives rise to an isomorphism of the additive subgroup of elements of augmentation zero with the multiplicative group of elements of augmentation one. Consequently there is a one-one correspondence between stable additive characteristic classes α and stable exponential characteristic classes θ given by the formula $\theta(E) = \exp(\alpha(E))$.

5. Suppose M is of the form $k \oplus N$, where k is a commutative ring located in degree zero and N is an ideal of square zero. Then the formula $\theta(E) = 1 + \alpha(E)$ gives a one-one correspondence between additive characteristic classes α with coefficients in N and exponential characteristic classes θ with coefficients in M which become identically one under the augmentation $M \longrightarrow k$.

§3. <u>Classification of characteristic classes</u>. Let k be a field, and let $H_*(\underline{\underline{C}})$ denote the homology of the category $\underline{\underline{C}}$ with coefficients in k. The homology of a category is defined to be that of the simplicial set sometimes called the nerve of the category. In the case of groups one obtains group homology. One knows that two functors induce the same map on homology when they are connected by a morphism of functors.

Given a representation $E: G \longrightarrow \underline{\underline{I}}\mathrm{s}\ \underline{\underline{A}}$, let

$$E_*: H_*(G) \longrightarrow H_*(\underline{\underline{I}}\mathrm{s}\ \underline{\underline{A}})$$

be the induced map on homology; it depends only on the isomorphism class of E. If P_s, $s \in S$, are representatives for the isomorphism classes of $\underline{\underline{A}}$, we have

$$(*) \qquad \oplus_s \; H_*(\mathrm{Aut}(P_s)) \xrightarrow{\sim} H_*(\underline{\underline{I}}s\ \underline{\underline{A}})$$

where the map has components $(\sigma_s)_*$, σ_s denoting the obvious representation of $\mathrm{Aut}(P_s)$ on P_s.

Since k is a field, there are canonical isomorphisms

$$H^i(G,V) = \mathrm{Hom}_k(H_i(G),V)$$

for any k-module V. Hence for any graded k-module M there is a canonical isomorphism

$$H^o(G,M) = \mathrm{Hom}_k^{(0)}(H_*(G),M)$$

where the '(0)' denotes homomorphisms of graded k-modules (homogeneous of zero). We say that a cohmology class $\alpha \in H^o(G,M)$ corresponds to a homomorphism $f\colon H_*(G) \longrightarrow M$; and conversely, if α and f are related by this isomorphism.

Given $E\colon G \longrightarrow \underline{\underline{I}}s\ \underline{\underline{A}}$, let

$$\theta(E) \in H^o(G,H_*(\underline{\underline{I}}s\ \underline{\underline{A}}))$$

denote the class corresponding to E_*. It is clear that θ is a characteristic class of $\underline{\underline{A}}$ with coefficients in $H_*(\underline{\underline{I}}s\ \underline{\underline{A}})$.

Proposition 1. If θ is a characteristic class of $\underline{\underline{A}}$ with coefficients in a graded k-module M, then there is a unique homomorphism $u\colon H_*(\underline{\underline{I}}s\ \underline{\underline{A}}) \longrightarrow M$ of graded k-modules such that $u(\Theta) = \theta$, i.e. such-that $\theta(E)$ corresponds to $u\,E_*$ for all representations E.

Proof. By naturality, it suffices to show that there is a unique u such that $u(\theta)$ and θ coincide on the representation σ_s, i.e. such that $\theta(\sigma_s)$ and $u(\sigma_s)_*$ correspond for all s. But the existence and uniqueness of such a u is clear from the decomposition (*).

Remark. This proposition says that Θ is the "universal" characteristic class of $\underline{\underline{A}}$ with coefficients in a graded k-module.

We shall say that a characteristic class θ corresponds to $U: H_*(\underline{\underline{Is}}\ \underline{\underline{A}}) \longrightarrow M$, and conversely, when they are related as in proposition 1.

Abbreviating $\underline{\underline{Is}}\ \underline{\underline{A}}$ to $\underline{\underline{I}}$, let μ be the composition

$$H_*(\underline{\underline{I}}) \otimes H_*(\underline{\underline{I}}) \longrightarrow H_*(\underline{\underline{I}} \times \underline{\underline{I}}) \longrightarrow H_*(\underline{\underline{I}})$$

where the first map is the cartesian product for homology classes and the second is induced by the direct sum in $\underline{\underline{A}}$. Then μ makes $H_*(\underline{\underline{Is}}\ \underline{\underline{A}})$ a graded anti-commutative algebra over k.

Proposition 2. Let θ correspond to $u: H_*(\underline{\underline{Is}}\ \underline{\underline{A}}) \longrightarrow M$, and suppose M is a graded algebra over k. Then θ is an exponential characteristic class if and only if u is a ring homomorphism.

Proof. ($\Longleftarrow$). Since $u(\Theta) = \theta$ it suffices to show that Θ is an exponential characteristic class. The following assertions are immediate from the definitions.

i) If $c, c' \in H^0(G,M)$ correspond to $f, f': H_*(G) \longrightarrow M$, then the cup product cc' corresponds to $\mu_M(f \otimes f')\Delta_G$ where $\mu_M: M \otimes M \longrightarrow M$ is the product of M, and where Δ_G, is the homomorphism

$$H_*(G) \longrightarrow H_*(G \times G) \xrightarrow{\sim} H_*(G) \otimes H_*(G)$$

induced by the diagonal map of G.

ii) The identity of $H_*(\underline{\underline{Is}}\ \underline{\underline{A}})$ is the distinguished generator of $H_o(\underline{\underline{Is}}\ \underline{\underline{A}})$ associated to the component of $\underline{\underline{Is}}\ \underline{\underline{A}}$ consisting of the zero objects.

iii) If E and E' are representations of G, then

$$(E \oplus E')_* = \mu(E_* \otimes E'_*)\Delta_G.$$

Assertion ii) implies that $\Theta(0) = 1$, whereas i) and iii) imply that $\Theta(E \oplus E') = \Theta(E)\Theta(E')$, so Θ is an exponential characteristic class.

($\Longrightarrow$). Assuming that θ is an exponential characteristic class we wish to show that $u(1) = 1$ and that $\mu_M(u \otimes u) = u\mu$. The former follows from $\theta(0) = 1$ and ii). For the latter it will suffice to establish the formula

$$u\mu(E_* \otimes E'_*)\Delta_G = \mu_M(u \otimes u)\,(E_* \otimes E'_*)\Delta_G$$

for all pairs of representation of the same group, because $H_*(\underline{\underline{I}}s\ \underline{\underline{A}}) \otimes H_*(\underline{\underline{I}}s\ \underline{\underline{A}})$ is generated by the images of the maps $(E_* \otimes E'_*)\Delta_G$, with E and E' the two obvious representations of $G = \mathrm{Aut}(P_s) \times \mathrm{Aut}(P_{s'})$, as s and s' run over S. But the left side corresponds to $\theta(E \oplus E')$ by iii), and the right to $\theta(E)\theta(E')$ by i), and these are equal as θ is an exponential characteristic class. q.e.d.

Corollary. *Let θ correspond to $u\colon H_*(\underline{\underline{I}}s\ \underline{\underline{A}}) \longrightarrow M$ where M is a graded k-module. Then θ is additive if and only if u is a derivation, i.e.*

$$u(xy) = \varepsilon(x)\,u(y) + \varepsilon(y)\,u(x)$$

where $\varepsilon\colon H_*(\underline{\underline{I}}s\ \underline{\underline{A}}) \longrightarrow k$ is the augmentation.

This follows from the proposition and remark 5 of §2, using the fact that an algebra homorphism to $k \oplus M$ with $(M)^2 = 0$ is the same as a derivation.

Remark. Proposition 2 says that $H_*(\underline{\underline{I}}s\ \underline{\underline{A}})$ represents the functor from the category of graded anti-commutative k-algebras to sets assigning to M the set of exponential characteristic classes of $\underline{\underline{A}}$ with coefficients in M. In other terms, the *points* of $H_*(\underline{\underline{I}}s\ \underline{\underline{A}})$ with values in M may be identified with exponential characteristic classes with coefficients in M. This means that exponential characteristic classes determine $H_*(\underline{\underline{I}}s\ \underline{\underline{A}})$ up to isomorphism. Moreover, properties of exponential characteristic classes will be reflected in properties of $H_*(\underline{\underline{I}}s\ \underline{\underline{A}})$ and conversely. In the rest of this section we give some examples to illustrate this.

Examples. 1. Consider characteristic classes of vector bundles. In this situation the analogue of the ring $H_*(\text{Is } A)$ is $\otimes H_*(BU_n)$. Now it is an easy consequence of the computation of the cohomology of the projective bundle associated to a vector bundle that for any sequence $\underline{m} = \{m_o, m_1, \ldots,\}$ of elements of a graded anti-commutative ring M, with $m_i \in M_{2i}$, there is an unique exponential characteristic class $\phi_{\underline{m}}$ for complex vector bundles with coefficients in M such that

$$\phi_{\underline{m}}(L) = \sum_{i \geq 0} c_1(L)^i m_i$$

for all line bundles L. It follows that if we take M to be the polynomial ring $k[t_o, t_1, \ldots]$ with $\deg(t_i) = 2i$, and $\underline{t} = \{t_o, t_1, \ldots,\}$, then $\phi_{\underline{t}}$ is a universal exponential characteristic class. Therefore by the universal property expressed in Proposition 2, we obtain an isomorphism

$$\otimes_n H_*(BU_n) \simeq k[t_o, t_1, \ldots]$$

carrying θ to $\phi_{\underline{t}}$, that is, such that t_i corresponds to the generator of $H_{2i}(BU_1)$ dual to c_i^i.

2. We consider the behavior of the algebra $H_*(\underline{\text{Is}}\ \underline{A})$ corresponding to that of exponential characteristic classes described in the remarks of §2. First of all, to the product of exponential characteristic classes corresponds a coproduct on this algebra making it a bicommutative Hopf algebra. By remark 2 invertible exponential characteristic classes may be identified with points of the Hopf algebra

$$T(\underline{A}) = k[K_o\ \underline{A}] \otimes_{k[S]} H_*(\underline{\text{Is}}\ \underline{A})$$

where $k[?]$ dennotes the monoid algebra functor. To the splitting of invertible exponential characteristic classes into inessential and stable components corresponds a Hopf algebra decomposition

$$T(\underline{A}) = k[K_o\ \underline{A}] \otimes_k T^{st}(\underline{A})$$

where $T^{st}(\underline{A})$ is the Hopf algebra whose points are the stable exponential characteristic classes. Finally when char(k) = 0, the one-one correspondence between stable exponential and stable additive characteristic classes is reflected in the well-known Hopf algebra isomorphism

$$\tilde{S}(\underline{Q}(T^{st}(\underline{A}))) \simeq T^{st}(\underline{A})$$

where $\tilde{S}$ is the analogue of the symmetric algebra for anti-commutative rings.

Remark. It can be proved that all of the preceding holds when k is any commutative ring such that $H_*(\mathrm{Aut}(P_s),k)$ is a projective k-module for all s.

§4. Homology of GL(A). According to §1, example 1, the stable isomorphism classes of representations of G over A may be identified with elements of lim.ind.$[G,GL_n(A)]$. Applying Yoneda's lemma, one sees that stable characteristic classes for representations over A with coefficients in M are classified by elements of

$$\text{lim.proj. } H^o(GL_n(A),M).$$

Supposing M is a graded k-module where k is a field, we have

$$H^o(GL_n(A),M) = \mathrm{Hom}_k^{(0)}(H_*(GL_n(A)),M).$$

Taking the limit and using the isomorphism

$$\text{lim.ind. } H_*(GL_n(A)) = H_*(GL(A)),$$

we see that stable characteristic classes for representations over A with coefficients in M are in one-one correspondence with elements of

$$H^o(GL(A),M) = \mathrm{Hom}_k^{(0)}(H_*(GL(A)),M).$$

Specifically, for a characteristic class θ to correspond to $u\colon H_*(GL(A)) \longrightarrow M$ means that for every representation E of a group G, $\theta(E) \in H^0(G,M)$ corresponds to the homomorphism $u(\rho_E)_*$, where

$$(\rho_E)_*: \ H_*(G) \longrightarrow H_*(GL(A))$$

is the homomorphism induced by the element ρ_E of example 1, §1.

The homomorphisms

$$H_*(GL_m(A)) \otimes H_*(GL_n(A)) \longrightarrow H_*(GL_{m+n}(A))$$

induced by direct sum provide in the limit a product μ on $H_*(GL(A))$, making it a graded anti-commutative k-algebra. This product is characterized by the formula

$$(\rho_{E \oplus E'})_* = u((\rho_E)_* \otimes (\rho_{E'})_*)\Delta_G$$

for any two representations of the same group. As in the proof of proposition 2, this entails that a stable characteristic class with coefficients in a graded k-algebra M is an exponential characteristic class if and only if the corresponding map $u: \ H_*(GL(A)) \longrightarrow M$ is an algebra homomorphism. Therefore <u>points of $H_*(GL(A))$ are stable exponential characteristic classes of representations over</u> A, i.e.

$$T^{st}(A) = H_*(GL(A))$$

in the notation of example 2, §3.

<u>Remark</u>. The preceding arguments show in general that

$$T^{st}(\underline{A}) = \lim_{(S,S)}\text{.ind. } H_*(\text{Aut}(P_s))$$

(notation of §1), hence that stable chacteristic classes correspond with graded k-module homomorphisms with source $T^{st}(\underline{A})$.

Second part: Cohomological splitting of exact sequences of representations

§5. Statements of the theorems and some applications. We begin with a discussion of the problem in order to motivate the results and set the notation. Unless stated otherwise, all characteristic classes have coefficients in graded k-modules, where k is a given field.

Let θ be a characteristic class for representations over a ring A. We wish to know when, for any exact sequence

$$(*) \qquad 0 \longrightarrow E' \longrightarrow E \longrightarrow E'' \longrightarrow 0$$

of representations, it is true that

$$\theta(E) = \theta(E' \oplus E''),$$

that is, whether all exact sequences of representations split in so far as the characteristic class θ is concerned. Let $\underline{A}$ be the category of finitely generated projective A-modules, and let $\underline{A}^{(2)}$ denote the additive category of short exact sequences in $\underline{A}$, so that an exact sequence of representations of G over A is the same as a representation of G in $\underline{A}^{(2)}$. Then more generally, given a characteristic class ϕ for representations in $\underline{A}^{(2)}$, we wish to know when ϕ has the same value for any exact sequence and for the associated split sequence:

$$0 \longrightarrow E' \longrightarrow E' \oplus E'' \longrightarrow E'' \longrightarrow 0.$$

Let $GL_{m,n}(A)$ be the group of automorphisms of the exact sequence

$$0 \longrightarrow A^m \longrightarrow A^{m+n} \longrightarrow A^n \longrightarrow 0$$

and let $\pi_{m,n}$ be the evident homomorphism

$$\pi_{m,n}: \ GL_{m,n}(A) \longrightarrow GL_m(A) \times GL_n(A).$$

Note that $\pi_{m,n}$ admits a section which is an homomorphism, hence $GL_{m,n}(A)$ is a semi-direct product

$$Gl_{m,n}(A) = (GL_m(A) \times GL_n(A)) \tilde{\times} Hom_A(A^n, A^m).$$

Up to isomorphism, an exact sequence of representations of G with $E' \simeq A^m$ and $E'' \simeq A^n$ may be identified with an element of $[G, GL_{m,n}(A)]$. Moreover, the set of stable isomorphism classes of exact sequences of representations is

$$St(G, \underline{\underline{A}}^{(2)}) = \underset{m,n}{\text{lim.ind.}}\ [G, GL_{m,n}(A)]$$

(compare Ex. 1, §1).

Now let ϕ be a characteristic class for representations in $\underline{\underline{A}}^{(2)}$, and for simplicity, let us concentrate only on those representations with $E' \simeq A^m$ and $E'' \simeq A^n$, where m and n are fixed. Then ϕ corresponds in the sense of §3 to a homomorphism

$$u: \ H_*(GL_{m,n}(A)) \longrightarrow M,$$

and for ϕ to have the same value on any exact sequence and its associated split sequence means precisely that u factors through the surjection

$$(\pi_{m,n})_*: \ H_*(GL_{m,n}(A)) \longrightarrow H_*(GL_m(A) \times GL_n(A)).$$

Therefore, when $(\pi_{m,n})_*$ is an isomorphism, every exact sequence of representations of the type under consideration splits in so far as characteristic classes over k are concerned. On the other hand, when $\text{Ker}\ (\pi_{m,n})_* \neq 0$, there exists a non-trivial ϕ which vanishes for all split exact sequences of representations.

<u>Example</u> 1. Let $m = n = 1$, and let A be a finite field of characteristic p, where $k = \mathbb{Z}/p\mathbb{Z}$. The group $GL_{1,1}(A)$ is finite and has elements of order p, hence its mod p homology is non-zero in infinitely many degrees [5]. But $GL_1(A)$ is of order prime to p, hence it has trivial homology, so $(\pi_{1,1})_*$ is not an isomorphism.

<u>Example</u> 2. Suppose either that $\text{char}(k) = \ell > 0$ is invertible

in A, or that char(k) = 0 and $A \otimes_{\mathbb{Z}} k = 0$. Then the abelian group $Hom_A(A^n,A^m)$ has trivial homology with coefficients in k hence by considering the spectral sequence of the extension $GL_{m,n}(A)$, we see that $(\pi_{m,n})_*$ is an isomorphism in this case for all m,n.

We can now state our two theorems in the context of representations over a ring. The first is an improvement of example 2 when char(k) = 0.

Theorem 1'. Assume there exists a prime number ℓ invertible in A which divides char(k), (hence either char(k) = ℓ or char(k) = 0). Then $(\pi_{m,n})_*$ is an isomorphism for all m,n

Remarks. 1. When char(k) = 0, the theorem applies to any ring in which some prime number is invertible. It would be nice to know if the theorem holds for all rings if char(k) = 0, e.g. for $A = \mathbb{Z}$.

2. Let L be the set of prime numbers invertible in A. Assuming this is non-empty, the theorem says that $\pi_{m,n}$ induces isomorphisms on homology with coefficients in $\mathbb{Q}$ and $\mathbb{Z}/\ell\mathbb{Z}$ for all ℓ in L. By standard universal coefficient arguments, it follows that $\pi_{m,n}$ induces isomorphisms on integral homology when A is an algebra over $\mathbb{Q}$

The second result, by far the more significant, may be formulated as follows.

Theorem 2'. For any k, the maps $\pi_{m,n}$ induce an isomorphism in the limit:

$$\underset{m,n}{\text{lim.ind.}}\ H_*(GL_{m,n}(A)) \xrightarrow{\sim} H_*(GL(A) \times GL(A)).$$

A homomorphism from this inductive limit to a graded k-module corresponds to a stable characteristic class for representations in $\underline{\underline{A}}^{(2)}$. Consequently, this theorem says that all exact sequences of representations split as far as stable characteristic classes are concerned.

Corollary 1. Let θ be a stable characteristic class for representations over A. Then for any exact sequence (*) of representations we have $\theta(E) = \theta(E' \oplus E'')$.

This is clear.

Corollary 2. Let F be a finite field of characteristic p. Then

$$H_n(GL(F), \mathbb{Z}/p\mathbb{Z}) = 0 \quad \text{for } n > 0.$$

Proof. Let θ be the stable characteristic class for representations over F obtained from an element of $H^n(GL(F), \mathbb{Z}/p\mathbb{Z})$ with $n > 0$; we must show that $\theta(E) = 0$ for any representation E of a group G. Since the groups $GL_n(F)$ are finite, we may suppose by naturality that G is finite. One knows from transfer theory that the restriction homomorphism on mod p cohomology from G to a Sylow p-subgroup is injective, so we may assume G is a p-group. But then E admits a composition series whose quotients are trivial representations, so $\theta(E) = 0$, by corollary 1 and induction on the dimension of E.

Remarks. 3. Passing to the limit, it follwos that GL(F) has no mod p homology for any field F algebraic over $\mathbb{Z}/p\mathbb{Z}$.

4. A refinement of the preceding argument shows that any stable mod p characteristic class for representations over a regular commutative noetherian ring of characteristic p vanishes for all representations of finite groups. This lends support to the conjecture that for such a ring A, GL(A) has no mod p homology. Gersten [3] has proved this for A=F[T] with F as in remark 3.

We shall prove theorems 1' and 2' in a more general form. Let $\underline{A}$ be a additive category whose isomorphism classes form a set, let n be an integer ≥ 1, and let $\underline{\underline{A}}^n$ be the n-fold product category. Let $\underline{\underline{A}}^{(n)}$ be the category of objects of $\underline{A}$ endowed with an n-stage filtration

$$0 \subset P_1 \subset \ldots \subset P_n = P$$

such that for $1 \leq i \leq n$, P_{i-1} is a direct summand of P_i, i.e. there exists a subobject Q_i of P_i with $P_{i-1} \oplus Q_i = P_i$. With morphisms defined to be filtration-preserving morphisms, $\underline{\underline{A}}^{(n)}$ is an additive category. Note that $\underline{\underline{A}}^{(2)}$ may be identified with the category of short exact sequences in $\underline{\underline{A}}$ which split.

The quotient object $gr_i P = P_i/P_{i-1}$ exists, hence we obtain a functor

$$gr: \quad \underline{\underline{A}}^{(n)} \longrightarrow \underline{\underline{A}}^n, \; P \mapsto (gr_i P)$$

which is unique up to canonical isomorphism. There is also a functor

$$i: \quad \underline{\underline{A}}^n \longrightarrow \underline{\underline{A}}^{(n)}$$

sending (Q_i) to the filtered object with

$$P_i = \bigoplus_{j \leq i} Q_j,$$

and the composition $gr.\, i$ is canonically isomorphic to the identity. Note that $\underline{\underline{A}}^{(n)}$ and $\underline{\underline{A}}^n$ have the same isomorphism classes, hence

$$K_o \; \underline{\underline{A}}^{(n)} = K_o \; \underline{\underline{A}}^n = (K_o A)^{\oplus n}$$

Theorem 1. *Suppose there exists a prime number* ℓ *dividing* char(k) *such that* is $\mathbb{Z}[\ell^{-1}]$*-linear, that is,* $\ell . id_P$ *is an automorphism for every object* P *of* $\underline{\underline{A}}$. *Then* gr *induces an isomorphism*

$$H_*(\underline{\underline{Is}}\; \underline{\underline{A}}^{(n)}) \xrightarrow{\sim} H_*(\underline{\underline{Is}}\; \underline{\underline{A}}^n).$$

Theorem 2. *For any* k, gr *induces an isomorphism*

$$T(\underline{\underline{A}}^{(n)}) \xrightarrow{\sim} T(\underline{\underline{A}}^n).$$

The second theorem says that $\underline{\underline{A}}^{(n)}$ and $\underline{\underline{A}}^n$ have the same invertible exponential characteristic classes, which implies they have the same stable characteristic classes (Remark, §4: Ex. 2, §3). The first theorem says they have the same characteristic classes under suitable conditions on char(k). It is clear that theorems 1 and 2 imply theorems 1' and 2' respectively.

§6. Proof of theorem 1. Arguing by induction on n, starting from the case $n = 1$ which is trivial, we may assume $n \geq 2$ and that the theorem holds for smaller values of n. Let P be an object of $\underline{A}^{(n)}$, let G be its group of automorphisms, and let G'' be the group of automorphisms of the object of $\underline{A}^{(n-1)}$ given by P/P_1 with the induced filtration. We must show that the evident homomorphism

$$G \longrightarrow \prod_{i=1}^{n} \mathrm{Aut}_{\underline{A}} (P_i/P_{i-1})$$

induces isomorphisms on homology with coefficients in k. Since this is true for the similar map from G'' to the product with $2 \leq i \leq n$ by the induction hypothesis, it suffices to show that the homomorphism

$$\pi: \ G \longrightarrow \mathrm{Aut}_{\underline{A}} (P_1) \times G''$$

associating to an automorphism of P the induced automorphisms of P_1 and P/P_1, induces isomorphisms on homology with coefficients in k. Let N be the kernel of π, choose a complement for P_1 in P, and let G' be the subgroup of G consisting of elements preserving the complement. The G is the semi-direct product of G' and N, and π may be identified with the projection $G \longrightarrow G'$ with kernel N. We consider the spectral sequence

$$E^2_{st} = H_s(G', H_t(N,k)) \Longrightarrow H_{s+t}(G,k)$$

associated to the extension G of G' by N.

The group N is abelian, canonically isomorphic to $\mathrm{Hom}\ \underline{A}\ (P/P_1, P_1)$. By hypothesis, the action $n \mapsto n.\mathrm{id}$ of $\mathbf{Z}$ on P_1 extends to an action of the ring $D = \mathbf{Z}\ [\ell^{-1}]$ as endomorphisms of P_1, hence N is a module over D. If $\mathrm{char}(k) = \ell$, one has $H_*(N,k) = k$, so the spectral sequence degenerates, yielding the desired result.

If $\mathrm{char}(k) = 0$ on the other hand, let the group of units D^* act on P_1 by restricting the D-action, and extend this to an action of D^* on P by making it trivial on the given complement of P_1. This gives a homomorphism $h: \ D^* \longrightarrow G'$, and we make D^* act on

G by conjugation: $u(g) = h(u).g.h(u)^{-1}$. Then D^* acts trivially on G', and on N it acts by multiplying with respect to the D-module structure. Thus an element u of D^* acts on

$$H_t(N,k) = \Lambda^t(N \otimes_{\mathbb{Z}} k)$$

by multiplying by u^t, where we identify D with a subring of k in the unique way. Since D^* acts trivially on G', u acts on E^2_{st} by multiplying by u^t. The differentials commute with the D^*-action, and as $u^t \neq u^{t'}$ for $t \neq t'$ when $u = \ell$, all the differentials are zero. Finally, D^* acts trivially on the abutment, because inner automorphisms of a group induce the identity on its homology. Thus E^2_{st} N 0 for $t > 0$, and the spectral sequence degenerates yielding the desired result. q.e.d.

§7. Proof of theorem 2. Using induction on n, we may assume $n \geq 2$ and that the theorem is true for smaller values of n. Observe that any invertible exponential characteristic class θ of $\underline{\underline{A}}^{(n)}$ may be written

$$\theta(E) = \theta'(gr\ E)\theta''(E)$$

where θ' is an invertible exponential characteristic class of $\underline{\underline{A}}^n$, and θ'' is an invertible exponential characteristic class of $\underline{\underline{A}}^{(n)}$ which is trivial on $\underline{\underline{A}}^n$, i.e. such that $\theta''(E) = 1$ if E possesses an invariant splitting. To this decomposition corresponds a Hopf algebra decomposition

$$T(\underline{\underline{A}}^{(n)}) \simeq T(\underline{\underline{A}}^n)\otimes T$$

where the points of T are such classes θ''. We have to prove that $T = k$. Since $K_o\underline{\underline{A}}^{(n)} = K_o\underline{\underline{A}}^n$, T is connected, and it suffices to prove that the indecomposable space $\underline{Q}T$ is zero. Let u be any homogeneous element of the dual space of $\underline{Q}T$, and put $t = \deg(u)$. Then u may be interpreted as an additive characteristic class: $E \mapsto u(E) \in H^t(G,k)$ of $\underline{\underline{A}}^{(n)}$ trivial on $\underline{\underline{A}}^n$, and we must show such a class is identically zero.

If $1 \leq i \leq n$, the full subcategory of P in $\underline{\underline{A}}^{(n)}$ with $gr_i P = 0$ is equivalent to $\underline{\underline{A}}^{(n-1)}$. Applying the induction hypothesis one sees that u vanishes on representations in this full subcategory. It follows that $u(E) = 0$ if E is a representation in $\underline{\underline{A}}^{(n)}$ such that the subrepresentation E_1 admits an invariant complement, for then E is the direct sum of two representations of the former kind.

There is no loss in generality in assuming that k is algebraically closed. We admit for the moment the following.

Lemma. Let k be an algebraically closed field and d an integer > 0. Then there exists an order D in a number field of degree d over $\mathbb{Q}$ with the following properties: Given any D-module N, let the group of units D^* act on it by multiplication, and let the group homology $H_*(N,k)$ be endowed with the induced action of D^*. Then for each t, $H_t(N,k)$ is a direct sum of one-dimensional representations of D^* over k. Furthermore, $H_t(N,k)$ does not contain the trivial representation for $0 < t < d$.

Choose $d > \deg(u)$ and such that $d \neq 0$ in k, and let D be as in the lemma. Let E be a representation of G on an object of $\underline{\underline{A}}^{(n)}$, and let E^d be its d-fold direct sum. Then $u(E^d) = d.u(E)$, and as $d \neq 0$, to prove $u(E) = 0$, it suffices to show that $u(E^d) = 0$. Now the ring D acts as endomorphisms of the representation E^d. Indeed, one has

$$\mathrm{End}(M^d) = \mathrm{End}(M) \otimes_{\mathbb{Z}} \mathrm{End}_{\mathbb{Z}}(\mathbb{Z}^d)$$

for any object M of an additive category, and D can be embedded as a subring of $\mathrm{End}_{\mathbb{Z}}(\mathbb{Z}^d)$ by choosing a $\mathbb{Z}$-basis of D. Therefore, replacing E by E^d, we may suppose that the ring D acts as endomorphisms of the subrepresentation E_1.

By functorality of u, we need only consider the following situation. Let P be an object of $\underline{\underline{A}}^{(n)}$ endowed with an action of D as endomorphisms of P_1, and let G be the full group of automorphisms of P compatible with the filtration and the D-action. If E is the obvious representation of G on P, we must show that $u(E) = 0$.

Choose a complement for P_1 in P, and let G' be the subgroup of G consisting of elements preserving the complement. By what has already been established, the element $u(E)$ of $H^t(G,k)$ restricts to zero on G', hence it suffices to show that the inclusion $G' \longrightarrow G$ induces isomorphisms on homology with coefficients in k in degrees $< d$. On the other hand, G is the semi-direct product of G' and the abelian normal subgroup N consisitng of elements inducing the identity on P_1 and P/P_1, hence it suffices to show that the projection $G \longrightarrow G'$ with kernel N induces isomorphisms on homology in degrees $< d$.

The action of D induces an action of the group of units D^* on P_1; extend the latter to P by making D^* act trivially on the given complement. This provides a homomorphism $h\colon D^* \longrightarrow G'$, and we make D^* act on G by conjugation. Then D^* acts trivially on G', and with respect to the canonical isomorphism

$$N = \mathrm{Hom}_{\underline{A}}(P/P_1, P_1)$$

the action of $u \in D^*$ on N corresponds to composition with the automorphism of P_1 associated to u. In particular, the D^*-action on N is multiplication with respect to a D-module structure.

This point established, we consider the induced D^*-action on the spectral sequence of the extension G of G' by N. By the lemma, $H_*(N,k)$ is a direct sum of eigenspaces associated to the characters of D^* over k. Using the fact that D^* acts trivially on G', it follows that $E^2_{st} = H_s(G', H_t(N,k))$, hence also E^r for $2 < r \leq \infty$, breaks up into eigenspaces preserved by the differentials. However, D^* acts trivially on the abutment, so the eigenspaces belonging to the trivial character form a spectral sequence

$$E^2_{st} = H_s(G',H_t(N,k)^{D^*}) \longrightarrow H_{s+t}(G,k).$$

By the lemma $H_t(N,k)$ has no non-trivial invariants for $0 < t < d$, so G and G' have the same homology in degrees $< d$. This concludes the proof of theorem 2 except for the lemma which will be demonstated in §9.

§8. The homology of an abelian group. In the proof of the lemma we will need a well-known formula for the homology $H_*(N,k)$ of an abelian group N with coefficients in a field k. Let $p = \mathrm{char}(k)$, and let ${}_pN$ be the subgroup of elements of N killed by p if $p > 0$, and ${}_pN = 0$ if $p = 0$. Let $\Lambda(V)$ and $\Gamma(V)$ be the exterior and divided-power algebra respectively of a k-vector space V.

Proposition 3. There exists an isomorphism of graded k-algebras

$$(*) \qquad \Lambda(N \otimes_{\mathbb{Z}} k) \otimes {}_k\Gamma({}_pN \otimes_{\mathbb{Z}} k) \xrightarrow{\sim} H_*(N,k)$$

with $N \otimes_{\mathbb{Z}} k$ of degree 1 and ${}_pN \otimes_{\mathbb{Z}} k$ of degree 2. There is a canonical such isomorphism, functorial in N, if either $p \neq 2$, or if $p = 2$ and N is restricted to the full subcategory of abelian groups such that ${}_2N \subset 2N$. When $p = 2$ and N is arbitrary, it is always possible to choose the isomorphism to be compatible with a given action of a finite group of odd order on N.

This result is contained for the most part in the Cartan seminar [2]. For the reader's convenience we shall now construct a homomorphism (*) with the required functorial properties. The fact that it is an isomorphism is proved by reducing to the case of cyclic groups and computing, and we refer to loc. cit. for the details. Recall that a "canonical" map or structure is always compatible with morphisms.

First of all, $H_*(N,K)$ has a canonical strictly anti-commutative algebra structure with divided prowers for elements of degree ≥ 2. The canonical isomorphism $N \otimes k = H_1(N,k)$, ($\otimes$ over $\mathbb{Z}$), thus extends to a canonical algebra homomorphism

$$\Lambda(N \otimes k) \longrightarrow H_*(N,k).$$

When $\mathrm{char}(k) = 0$, this is the map (*), so from now on we suppose $p > 0$.

In a moment, we shall describe a canonical exact sequence

$$(**) \qquad 0 \longrightarrow \Lambda^2(N \otimes k) \xrightarrow{\ i\ } H_2(N,k) \xrightarrow{\ j\ } {}_pN \otimes k \longrightarrow 0$$

and show that it splits canonically if either p is odd or if $p = 2$ and ${}_2N \subset 2N$, so that j has a canonical section s in these cases. On the other hand if $p = 2$ and N is endowed with an action of a group of odd order, the theorem of Maschke implies there exists a section s of j compatible with the given action. The section s extends uniquely to a homomorphism

$$\Gamma({}_pN \otimes k) \longrightarrow H_*(N,k)$$

compatible with divided powers. Combining this with the above homomorphism from $\Lambda(N \otimes k)$, we obtain a homomorphism (*) with the functorial properties described in the proposition.

It remains to describe (**) and show it splits canonically under the indicated conditions. Since for any k-module V, we have

$$\mathrm{Ext}^1_{\mathbb{Z}}(N,V) \cong \mathrm{Ext}^1_{\mathbb{Z}}({}_pN,V) \cong \mathrm{Hom}_k({}_pN \otimes k, V)$$

$$H^2(N,V) \cong \mathrm{Hom}_k(H_2(N,k),V)$$

it suffices to describe a canonical exact sequence

$$0 \longrightarrow \mathrm{Ext}^1_{\mathbb{Z}}(N,V) \xrightarrow{\ j^*\ } H^2(N,V) \xrightarrow{\ i^*\ } \mathrm{Hom}_k(\Lambda^2\bar{N},V) \longrightarrow 0.$$

($\bar{N} = N \otimes k$) splitting canonically under the indicated conditions. Elements of the Ext group (resp. $H^2(N,V)$) classify abelian group extensions (resp. central extensions) of N by V. Let j^* be the obvious inclusion and let i^* associate to a central extension its commutator pairing, i.e. the pairing obtained by lifting two elements of N into the extension and taking their commutator. As a central extension is abelian if and only if its commutator pairing is trivial, the sequence is exact except for the surjectivity of i^*.

To establish this point, suppose given a map $\Lambda^2 N \longrightarrow V$, that is an alternating $\mathbb{Z}$-bilinear map $h\colon N \times N \longrightarrow V$. Since k is a field, the map $\Lambda^2 N \longrightarrow N \otimes_k N$ sending $x \wedge y$ to $x \otimes y - y \otimes x$ is injective, hence there is a $\mathbb{Z}$-bilinear map $f\colon N \times N \longrightarrow V$ such that

$$h(n,n') = f(n,n') - f(n',n).$$

The f is a 2-cocycle, so the set $N \times V$ with the operation

$$(n,v)(n',v') = (nn', f(v,v') + v + v')$$

is a central extension of N by V. It is clear that the commutator pairing of this extension is h, so i^* is surjective.

When p is odd, there is a canonical choice for f:

$$f(n,n') = \tfrac{1}{2}h(n,n')$$

so i^* has a canonical section as claimed. (Note that i^* and this section are necessarily k-module homomorphisms, as they are morphisms of representable functors.) Suppose finally that $p = 2$ and $_2N$ $2N$. The commutator pairing of a central extension of N by V vanishes on $2N$, hence on $_2N$, so the restriction of the extension to $_2N$ is abelian. This gives us a commutative diagram

$$\begin{array}{ccc} \mathrm{Ext}^1_{\mathbb{Z}}(N,V) & \longrightarrow & H^2(N,V) \\ \cong\downarrow & \swarrow & \downarrow \\ \mathrm{Ext}^1_{\mathbb{Z}}({}_2N,V) & \longrightarrow & H^2({}_2N,V) \end{array}$$

where the vertical arrows are restriction from N to ${}_2N$, showing the exact sequence splits canonically in this case as claimed.

§9. Proof of the lemma. First suppose $\mathrm{char}(k) = 0$. We may assume that k is the algebraic closure of $\mathbb{Q}$ in $\mathbb{C}$. Let F be a totally real number field of degree d; it exists, for by Dirichlet's theorem there is an odd prime number ℓ such that d divides $\frac{1}{2}(\ell - 1)$, so one can take F to be a subfield of $\mathbb{Q}(\exp(2\pi i/\ell))$. We take D to be the ring of integers in F.

If N is a D-module, then $N \otimes_{\mathbb{Z}} k$, as a representation of D^*, is a direct sum of copies of $F \otimes_{\mathbb{Q}} k$ with D^* acting by multiplication on F. By Galois theory, there is a ring isomorphism: $F \otimes_{\mathbb{Q}} k \simeq k^d$ having for its components the homomorphisms $x \otimes y \mapsto \sigma(x)y$, where σ runs over the d distinct embeddings of F in k. Thus as a representation of D^*, $N \otimes_{\mathbb{Z}} k$ is a direct sum of one-dimensional representations with characters $\sigma\colon D^* \longrightarrow k^*$, hence

$$H_t(N,k) = \Lambda^t(N \otimes_{\mathbb{Z}} k)$$

is a direct sum of one-dimensional representations with characters $\prod \sigma^{n_\sigma}$ where the n_σ are integers ≥ 0 such that $\Sigma n_\sigma = t$. Assume the family $\{n_\sigma\}$ is such that this character is trivial. Then

$$\Sigma\, n_\sigma \log |\sigma(u)| = 0$$

for all u in D^*, where $|\ |$ is the absolute value in $\mathbb{C}$. By the Dirichlet unit theorem, this happens if and only if all the n_σ are equal. Thus if $t > 0$, all the n_σ are ≥ 1 and $t \geq d$, showing that $H_t(N,k)$ does not contain the trivial representation for $0 < t < d$. This proves the lemma when $\mathrm{char}(k) = 0$.

Suppose now that $p = \mathrm{char}(k) > 0$, and let k_d be the subfield of k with p^d elements. Since the norm $N\colon k_d^* \longrightarrow \mathbb{F}_p^*$ is surjective, its kernel is cyclic of order $(p^d - 1)/(p - 1)$; let x generate the kernel, and let $g(X)$ be the minimal polynomial of x over $\mathbb{F}_p$. Note that $k_d = \mathbb{F}_p(x)$, for if the latter field had degree $j < d$, then $(p^d - 1)/(p - 1)$ would divide $p^j - 1$, which is impossible as the former number is $> p^j$. Hence $g(X) = X^d + b_1X^{d-1} + .. + b_d$ has degree d, and $b_d = (-1)^d Nx = \pm 1$.

Let $f(X) = X^d + .. + a_d$, $a_i \in \mathbf{Z}$, $a_d = \pm 1$ reduce mod p to g(X), and let $D = \mathbf{Z}[X]/(f(X))$. Note that f is irreducible as g is, hence D is an order in a number field of degree d. The image of X in D is invertible as $a_d = \pm 1$, therefore we have an isomorphism $D/pD \simeq k_d$, and the image of the character ϕ: $D^* \longrightarrow k_d^*$ induced by this isomorphism contains the cyclic subgroup of order $(p^d - 1)/(p - 1)$.

Let N be a D-module. We claim there is an isomorphism

$$\Lambda(N\otimes_{\mathbf{Z}}k) \otimes_k \Gamma({}_pN\otimes_{\mathbf{Z}}k) \xrightarrow{\sim} H_*(N,k)$$

commuting with the action of D^*. This is clear from Proposition 3 if p is odd, or if $p = 2$ and ${}_2N \subset 2N$. On the other hand if $p = 2$ and $2N = 0$, then N is a module over $D/2D = k_d$, hence D^* acts on N through the group k_d^* of odd order, and the proposition furnishes the required isomorphism in this case. In general when $p = 2$, we choose a complement N" for the kernel (resp. Q for the image) of the homomorphism ${}_2N \longrightarrow N/2N$ of k_d-vector spaces, and let N' be the inverse image of Q in N. Then $N \simeq N' \oplus N''$ as D-modules, and ${}_2N' \subset 2N'$, $2N'' = 0$, so upon tensoring the isomorphisms obtained already for N' and N", we obtain the required isomorphism for N.

As D^*-modules, N/pN and ${}_pN$ are direct sums of copies of the one-dimensional representation over k_d with character ϕ. By Galois theory, there is a ring isomorphism: $k_d \otimes_{\mathbf{Z}} k \simeq k^d$ whose components are the homomorphisms $x \otimes y \mapsto x^{p^a}y$ for $0 \leq a < d$, hence as a representation of D^*, $k_d \otimes_{\mathbf{Z}} k$ is the direct sum of the one-dimensional representations with characters ϕ^{p^a} for $0 \leq a < d$. It follows that we have D^*-isomorphisms

$$N\otimes_{\mathbf{Z}}k \simeq \oplus V_a \otimes L^{\otimes p^a}, \quad {}_pN\otimes_{\mathbf{Z}}k \simeq \oplus W_a \otimes L^{\otimes p^a}$$

the sum being over $0 \leq a < d$, where L is the one-dimensional representation with character ϕ, and where D^* acts trivially on

V_a, W_a. Thus

$$\Lambda(N \otimes_{\mathbb{Z}} k) \otimes \Gamma({}_pN \otimes_{\mathbb{Z}} k) \cong \bigotimes_a [\Lambda(V_a \otimes L^{\otimes p^a}) \otimes \Gamma(W_a \otimes L^{\otimes p^a})]$$

$$\cong \bigoplus_{\{m_a, n_a\}} [\otimes_a (\Lambda^{m_a}(V_a) \otimes \Gamma^{n_a}(W_a))] \otimes L^{\otimes \Sigma(m_a + n_a)p^a}$$

where the sum is taken over the families of non-negative integers m_a, n_a for $0 \leq a < d$. This shows that $H_*(N,k)$ is a direct sum of one-dimensional representations of D^* as claimed.

It remains to show that the trivial representation does not occur in degrees t with $0 < t < d$. The direct summand corresponding to $\{m_a, n_a\}$ is homogeneous of degree $\Sigma(m_a + 2n_a)$, and if this summand contains the trivial representation, then

$$\sum_{0 \leq a < d} (m_a + n_a)p^a \equiv 0 \qquad \text{mod } (p^d - 1)/(p - 1)$$

because by construction $\phi(D^*)$ contains the cyclic subgroup of k^* of this order. Let $\{m_a, n_a\}$ be a non-zero solution of this congruence such that $t = \Sigma(m_a + 2n_a)$ is minimal. Clearly $n_a = 0$ for all a. If $m_b \geq p$ for some b, then by replacing m_b by $m_b - p$ and $m_b + 1$ by $m_{b+1} + 1$, (or m_o by $m_o + 1$ if $b = d - 1$), and keeping the others the same, we would get a new solution contradicting minimality. Thus $m_a < p$ for $0 \leq a < d$, so by the uniqueness of the p-adic expansion, we see that the minimal solution is $m_a = 1$, $n_a = 0$ for all a. The minimal degree t is d, so the proof of the lemma is complete.

References

[1] M. Artin and B. Mazur, Etale homotopy, Lecture Notes in Mathematics 100 (1969).

[2] H. Cartan, Séminaire Cartan 1954-55, Algèbres d'Eilenberg-Maclane et homotopie, Benjamin, New York, 1967.

[3] S. Gersten, K-theory of polynomial extensions (to appear).

[4] A. Grothendieck, Classes de Chern et representations linéaires des groupes discrets, Dix exposés sur la cohomologie des schémas, North-Holland Publ. Co., Amsterdam, 1968.

[5] R. G. Swan, The nontriviality of the restriction map in the cohomology of groups, Proc. Amer. Math. Soc. 11 (1960), p. 885-887.

Massachusetts Institute of Technology.

HIGHER ALGEBRAIC K-THEORY : II

[after Daniel Quillen]

by Daniel Grayson

The purpose of this paper is to prove three theorems announced in Higher Algebraic K-theory : I by Quillen.

The first theorem says that the two definitions of $K_i(R)$ offered by Quillen agree. The notion of monoidal category allows the construction of a fibration which shows that $K_0R \times BGl(R)^+$ is the loop space of $QP(R)$.

The localization theorem concerns the localization of a ring, $R \longrightarrow S^{-1}R$, and identifies the third term in the long exact sequence. It was first proved for K_0 and K_1 by Bass. Gersten gave a proof for the higher K_i and conjectured the non-affine version proved here.

The fundamental theorem describes extensions $R \longrightarrow R[t]$ and $R \longrightarrow R[t,t^{-1}]$, and was first proved by Bass. The proof here involves an application of the localization theorem to the projective line.

I thank Daniel Quillen for communicating to me his proofs of these theorems and for his helpful explanations of them.

Monoidal Categories and Localization

Suppose S is an abelian monoid which acts on a set X. S acts invertibly on X if each translation

$$X \longrightarrow X$$

$$x \longmapsto sx \quad ,$$

for $s \in S$, is a bijection. Define $S^{-1}X$ to be the quotient $S\times X/S$, where S acts on each factor of the product. Let S act on $S^{-1}X$ by $t\cdot(s,x) = (s,tx)$, and define $X \longrightarrow S^{-1}X$ to be $x \longmapsto (1,x)$. Then the translation $(s,x) \longmapsto (s,tx)$ has an inverse given by $(s,x) \longmapsto (ts,x)$, so S acts invertibly on $S^{-1}X$. The arrow $X \longrightarrow S^{-1}X$ respects the S-action, and is a universal arrow from X to a set upon which S acts invertibly.

If S is abelian, then the set $S^{-1}S$ is a group under the multiplication $(s,t)\cdot(u,v) = (su,tv)$. The map $S \longrightarrow S^{-1}S$ is a homomorphism, and is a universal arrow from S to a group.

This notion of localization is placed in the context of categories as follows.

Def: A monoidal category S is a category S with an operation $+ : S\times S \longrightarrow S$ and an object 0. There are natural isomorphisms $A+(B+C) \cong (A+B)+C$, $0+A \cong A$, $A+0 \cong A$. The following diagrams must commute :

$$\begin{array}{ccccc} A+(B+(C+D)) & \cong & (A+B)+(C+D) & \cong & ((A+B)+C)+D \\ \wr\| & & & & \wr\| \\ A+((B+C)+D) & & \cong & & (A+(B+C))+D \end{array} ,$$

$$\begin{array}{ccc} A+(0+C) & \cong & (A+0)+C \\ \wr\| & & \wr\| \\ A+C & = & A+C \end{array} . \quad \text{(see MacLane)}$$

Def: A left action of a monoidal category S on a category X is a functor $+ : S\times X \longrightarrow X$ with natural isomorphisms $A+(B+F) \cong (A+B)+F$ and $0+F \cong F$, where $A,B \in S$ and $F \in X$. Diagrams analogous to those above must commute.

Def: A monoidal functor is a functor $S \xrightarrow{f} T$ where S and T are monoidal categories, equipped with natural isomorphisms $f(A+B) \cong fA + fB$ and $f0 \cong 0$. The following diagrams must commute :

$$\begin{array}{ccccc} f((A+B)+C) & \cong & f(A+B)+fC & \cong & (fA+fB)+fC \\ \wr\| & & & & \wr\| \\ f(A+(B+C)) & \cong & fA+f(B+C) & \cong & fA+(fB+fC) \end{array}$$

$$\begin{array}{ccc} f(0+A) & \cong & f0 + fA \\ \wr\| & & \wr\| \\ fA & \cong & 0 + fA \end{array} \qquad \begin{array}{ccc} f(A+0) & \cong & fA + f0 \\ \wr\| & & \wr\| \\ fA & \cong & fA + 0 \end{array} .$$

Def: A functor $g : X \longrightarrow Y$ of categories with S-action, preserves the action if there is a natural isomorphism $A+gF \cong g(A+F)$, and

$$\begin{array}{ccccc} (A+B)+gF & \cong & g((A+B)+F) & \cong & g(A+(B+F)) \\ \wr\| & & & & \wr\| \\ A+(B+gF) & & \cong & & A+g(B+F) \end{array} \qquad \text{and}$$

$$\begin{array}{ccc} 0 + gF & \cong & g(0+F) \\ \wr\| & & \wr\| \\ gF & = & gF \end{array} \qquad \text{commute.}$$

The commutativity of these diagrams yields the commutativity of every diagram which should commute. For details, see (MacLane). This commutativity assures us that the constructions to be made will satisfy the axioms for category or functor.

For details and notation about topological notions applied to categories, and for the theorem about constructing fibrations, the reader should refer to (Quillen).

<u>Def</u>: If S is a monoidal category which acts on a category X, then S acts <u>invertibly</u> on X if each translation

$$\begin{array}{ccc} X & \longrightarrow & X \\ F & \longmapsto & A+F \end{array}$$

is a homotopy equivalence.

<u>Def</u>: If S acts on X, the category $\langle S,X\rangle$ has the same objects as X. An arrow is represented by an isomorphism class of tuples $(F,G,A,A+F \longrightarrow G)$ with $A \in S$ and F,G in X. This arrow is an arrow from F to G. An isomorphism of tuples is an isomorphism $A \cong A'$ which makes

$$\begin{array}{ccc} A+F & \cong & A'+F \\ & \searrow \quad \swarrow & \\ & G & \end{array} \qquad \text{commute.}$$

<u>Def</u>: The category $S^{-1}X$ is $\langle S, S\times X\rangle$, where S acts on both factors of the product. The action of S on $S^{-1}X$ is given by $A + (B,F) = (B,A+F)$, if S is commutative up to natural isomorphism.

Notice that S acts invertibly on $S^{-1}X$. The translation $(B,F) \mapsto (B,A+F)$ has homotopy inverse $(B,F) \mapsto (A+B,F)$, in light of the natural transformation $(B,F) \longrightarrow (A+B,A+F)$.

<u>Note</u>: If every arrow is an isomorphism in S, then $\langle S,S\rangle$ has initial object 0 and is contractible. We now make the blanket assumption for the rest of the paper that this condition holds. In practice, S is usually the groupoid of isomorphisms in an exact category.

We consider now the projection on the first factor $S^{-1}X \longrightarrow \langle S,S\rangle$. Call it ρ. The map ρ is given by

$$(B,F) \longmapsto (B) \qquad \text{on objects, and by}$$

$$\begin{pmatrix} A, & A+B, & A+F \\ & \downarrow & \downarrow \\ & B', & F' \end{pmatrix} \longmapsto \begin{pmatrix} A, & A+B \\ & \downarrow \\ & B' \end{pmatrix} \qquad \text{on arrows.}$$

Suppose we are given an arrow $B \longrightarrow B'$ in $\langle S,S\rangle$. It may be represented by some $(A, A+B \longrightarrow B')$, and the arrow determines A up to isomorphism, but not up to unique isomorphism. An automorphism of the data giving the arrow is an automorphism $a: A \cong A$ such that

$$\begin{array}{ccc} A+B & \overset{a+1}{\cong} & A+B \\ & \searrow \quad \swarrow & \\ & B' & \end{array} \qquad \text{commutes.}$$

We see that if $A+B \longrightarrow B'$ is monic and $\mathrm{Hom}(A,A) \longrightarrow \mathrm{Hom}(A+B,A+B)$ is injective, then the isomorphism a is necessarily the identity. So assume

1) every arrow of S is monic
2) translations $S \longrightarrow S$ are faithful.

Under these conditions, every arrow in $\langle S,S\rangle$ determines its A up to unique isomorphism, and ρ is cofibred. The cobase-change map for an arrow $(A, A+B \longrightarrow B')$ may be given by

$$\rho^{-1}B \longrightarrow \rho^{-1}B'$$

$$(B,F) \longmapsto (B',A+F).$$

If we identify the fibers with X via the second projection, then the cobase-change map is just translation by A on X. If every translation on X is a homotopy equivalence, then all the cobase-change maps are, so the square

$$\begin{array}{ccc} X & \longrightarrow & S^{-1}X \\ \downarrow & & \downarrow \\ pt & \longrightarrow & \langle S,S\rangle \end{array}$$

is homotopy cartesian. But $\langle S,S\rangle$ has initial object 0, so the map $X \longrightarrow S^{-1}X$ given by $(F) \longmapsto (B,F)$ is a homotopy equivalence.

On the other hand, suppose $X \longrightarrow S^{-1}X$ is a homotopy equivalence. This map is compatible with the action of S, and S acts invertibly on $S^{-1}X$. Therefore S acts invertibly on X.

We have shown :

<u>Th</u>: $X \longrightarrow S^{-1}X$ is a homotopy equivalence if and only if S acts invertibly on X.

Homology Computation

Now $\pi_0 S$ acts on $H_p X$, and acts invertibly on $H_p S^{-1}X$, so $X \longrightarrow S^{-1}X$, the map given by $(F) \overset{p}{\longmapsto} (0,F)$, induces a map

$$(\pi_0 S)^{-1} H_p X \longrightarrow H_p(S^{-1}X)$$

Th: This map is an isomorphism.

def: If M is a $\pi_0 S$-module, define a functor $\overline{M}$: $\langle S,S\rangle \longrightarrow$ (ab gps) which sends each object (B) to the abelian group M, and sends an arrow $(A, A+B \xrightarrow{\sim} B')$ to multiplication by the class of A on M.

Pf of Th: If $\pi_0 S$ acts invertibly on M, then $\overline{M}$ is a morphism-inverting functor, and the homology group $H_p(\langle S,S\rangle, \overline{M})$ reduces to singular homology on the classifying space $B\langle S,S\rangle$ with coefficients in the local coefficient system determined by $\overline{M}$. Since $\langle S,S\rangle$ is contractible, we know that

$$H_p(\langle S,S\rangle, \overline{M}) = \begin{cases} M & \text{if } p = 0, \text{ and} \\ 0 & \text{if } p > 0. \end{cases}$$

Every fiber of the cofibred map $S\ X \xrightarrow{f} \langle S,S\rangle$ is identified with X, and the cobase-change maps are given by the action of S on X (see p.4). The spectral sequence for the map is thus :

$$E^2_{pq} = H_p(\langle S,S\rangle, \overline{H_q X}) \Longrightarrow H_{p+q}(S^{-1}X).$$

This spectral sequence is obtained from the bicomplex :

$$E^0_{pq} = \coprod_{\substack{B_0 \to \cdots \to B_p \\ \in \\ N_p\langle S,S\rangle}} \quad \coprod_{N_q(\rho\backslash B_0)} \mathbb{Z}$$

An action of S on this bicomplex is determined by the action of S on $S^{-1}X$ (via the X-component) and the action of S on $\langle S,S\rangle$ (the trivial action). Taking homology first in the q direction yields :

$$E^1_{pq} = \coprod_{B_0 \to \cdots \to B_p} H_q(\rho\backslash B_0) = \coprod H_q(\rho^{-1}B_0) = \coprod H_q X,$$

$$E^2_{pq} = H_p(\langle S,S\rangle, \overline{H_q X})$$

The action of S on the abutment and the abutment itself are computed using the degeneracy of the opposite spectral sequence, which begins :

$$E^1_{pq} = \begin{cases} \coprod_{N_q S^{-1}X} \mathbb{Z} & \text{if } p = 0, \text{ and} \\ 0 & \text{if } p > 0. \end{cases}$$

The action on the abutment $H_{p+q}S^{-1}X$ is the one induced by the action of S on $S^{-1}X$.

Localization with respect to a multiplicative subset of a ring is exact, so it preserves our spectral sequence. We localize with respect to $\pi_0 S$ inside its own integral group ring, and obtain :

$$E^1_{pq} = (\pi_0 S)^{-1} \coprod_{N_p\langle S,S\rangle} H_q X \Longrightarrow (\pi_0 S)^{-1} H_{p+q}(S^{-1}X).$$

Now S acts componentwise on E^1_{pq}, and acts invertibly on $H_{p+q}(S^{-1}X)$, so we get:

$$E^1_{pq} = \coprod (\pi_0 S)^{-1} H_q X, \text{ and}$$

$$E^2_{pq} = H_p(\langle S,S\rangle, \overline{(\pi_0 S)^{-1} H_q X}) \Longrightarrow H_{p+q}(S^{-1}X).$$

By the remark at the beginning of the proof, we know that this localized sequence degenerates from E^2 on, and the edge map is an isomorphism:

$$(\pi_0 S)^{-1} H_q X \xrightarrow{\ \sim\ } H_q(S^{-1}X).$$

That the edge map is the map induced by $X \longrightarrow S^{-1}X$ can be seen by comparing the degenerate spectral sequences which result from the following map of fibrations :

$$\begin{array}{ccccc} X & \longrightarrow & X & \longrightarrow & pt \\ \| & & \downarrow & & \downarrow 0 \\ X & \longrightarrow & S^{-1}X & \longrightarrow & \langle S,S\rangle \end{array}$$

QED

Actions on fibers

Suppose $f : X \longrightarrow Y$ is a map of categories on which S acts, f is compatible with the actions, and S acts trivially on Y. Then the action of S on X is said to be fiberwise with respect to f, and S does act on the fibers f G. If f is fibered and the base change maps respect the action on the fibers, then the action is said to be cartesian. In this case, $S^{-1}X$ is fibred over Y, its fibers are of the form $S^{-1}f^{-1}G$, and the base change maps are induced by those of f.

We consider now the projection on the second factor $S^{-1}X \longrightarrow \langle S,X\rangle$. Call it q. Assume

1) every arrow in X is monic, and
2) for each F in X the map $S \longrightarrow X$ given by $B \longmapsto B+F$ is a faithful functor.

Reasoning as before, we see that q is cofibred, each fiber may be identified with S, and the cobase-change maps are translations.

Let S act on $S^{-1}X$ via the first factor. This action is cartesian with respect to q, so localization yields a cofibred map $S^{-1}S^{-1}X \longrightarrow \langle S,X\rangle$, each fiber of which may be identified with $S^{-1}S$. Since S acts invertibly on $S^{-1}S$, the cobase-change maps are homotopy equivalences, so

$$\begin{array}{ccc} S^{-1}S & \longrightarrow & S^{-1}S^{-1}X \\ \downarrow & & \downarrow \\ S^{-1}pt & \longrightarrow & \langle S,X\rangle \end{array} \quad \text{is homotopy cartesian.}$$

The map $S^{-1}S \longrightarrow S^{-1}S^{-1}X$ is given by $(A,B) \mapsto (A,(B,F))$ for some fixed F in X. Consider the following diagram :

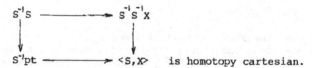

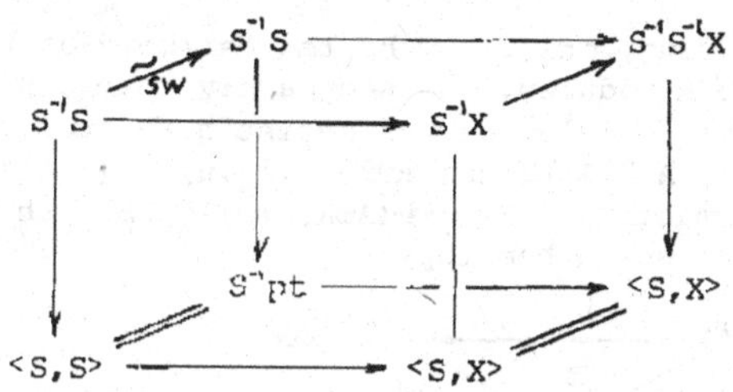

The back square is that which we just showed was homotopy cartesian. The map $S^{-1}S \xrightarrow{\sim} S^{-1}S$ is the switch isomorphism given by $(A,B) \mapsto (B,A)$. The map $S^{-1}X \longrightarrow S^{-1}S^{-1}X$ is the usual map given by $(A,F) \mapsto (0,(A,F))$. The map $pr_2 : S^{-1}S \longrightarrow \langle S,S\rangle$ is given by $(A,B) \longmapsto (B)$. Every square but the top is commutative, and the top square is homotopy commutative, as shown by the following natural transformations of functors $S^{-1}S \longrightarrow S^{-1}S^{-1}X$:

$$(0,(A,B+F)) \longrightarrow (B,(B+A,B+F)) \longleftarrow (B,(A,F)).$$

Notice that $S^{-1}X \longrightarrow S^{-1}S^{-1}X$ is a homotopy equivalence (see Th on p.4). Thus the front square is homotopy cartesian. Combine this with the fact that $\langle S,S\rangle$ is contractible, and we arrive at the following theorem :

Th: If $\langle S,X\rangle$ is contractible, then the map $S^{-1}S \longrightarrow S^{-1}X$ given by $(A,B) \longmapsto (A,B+F)$ for some fixed F in X is a homotopy equivalence.

The Plus Construction

The most important example of the previous constructions is the case where P is an exact category in which every exact sequence splits, and $S = \mathrm{Iso}(P)$ is the subcategory of P whose arrows are all isomorphisms of P. Direct sum is the operation which makes S into a monoidal category. $S^{-1}S$ becomes an H-space with multiplication $S^{-1}S \times S^{-1}S \longrightarrow S^{-1}S$ given by $((A,B),(C,D)) \longmapsto (A\oplus C, B\oplus D)$.

Suppose R is a ring, and let P be the category of finitely generated projective R-modules. We see easily that $\pi_0 S^{-1}S$ is K_0R. If A is a projective R-module, we can define a functor from Aut(A) to $S^{-1}S$ by sending $u : A \cong A$ to the arrow $(1_A, u) : (A,A) \cong (A,A)$. The natural transformation $(A,B) \longrightarrow (A\oplus R, B\oplus R)$ on $S^{-1}S$ shows that this diagram commutes up to homotopy :

$$\begin{array}{ccc} \mathrm{Aut}(A) & \longrightarrow & \mathrm{Aut}(A\oplus R) \\ & \searrow \quad \swarrow & \\ & S^{-1}S. & \end{array}$$

Thus we can define a map from $BGl(R) = \varinjlim \mathrm{Aut}(R^n)$ to $S^{-1}S$. In fact, this map lands in the connected component of the identity, $(S^{-1}S)_0$.

We can realize this map by using the telescope construction. If S_n is the component of S which contains R^n, then S_n is a groupoid equivalent to $\mathrm{Aut}(R^n) = Gl_n(R)$. Define $S_n \longrightarrow S_{n+1}$ by $(B) \longmapsto (R \oplus B)$, and $S_n \longrightarrow S_{n+m}$ to be the composite of m of these functors. If N is the ordered set of positive integers, we have defined a functor from N to the category of categories, and can construct the corresponding cofibred category L over N. The objects of L are pairs (n,B) with B in S_n, and an arrow from (n,B) to $(n+m,C)$ is an isomorphism $R^m \oplus B \cong C$. L is homotopy equivalent to $BGl(R)$.

Define $L \longrightarrow S^{-1}S$ by $(n,B) \longmapsto (R^n, B)$.

Let e in $\pi_0 S$ be the class of R. Since each projective module is a direct summand of a free one, the monoid generated by e is cofinal in $\pi_0 S$. Thus $H_p S^{-1}S \cong (\pi_0 S)^{-1} H_p S = H_p S[1/e]$. If $(R^n, B) \in (S^{-1}S)_0$, then for some m, $R^{n+m} \cong B \oplus R^m$. Thus any elememt of $H_p((S^{-1}S)_0)$ is of the form x/e^n for some n and some $x \in H_p S_n$. We see that $H_p((S^{-1}S)_0) \cong \varinjlim H_p S_n \cong H_p L$.

We can conclude that $L \longrightarrow (S^{-1}S)_0$ is an acyclic map. Since $(S^{-1}S)_0$ is an H-space as well, it must be $BGl(R)^+$.

The multiplication on the H-space $S^{-1}S$ has a homotopy inverse given by $(A,B) \longmapsto (B,A)$ so the components must all be homotopy equivalent. We have proved :

Th : $S^{-1}S$ is homotopy equivalent to $K_0R \times BGl(R)^+$.

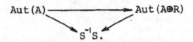

Cofinality

Suppose $M \subseteq P$ are exact categories in which every exact sequence splits. Then M is cofinal in P if given $A \in P$ there exist $B \in P$ and $C \in M$ so $A \oplus B \cong C$, and if M is a full subcategory of P.

Th : If M is cofinal in P, then $QM \longrightarrow QP$ is a covering space, and $K_qM \longrightarrow K_qP$ is an isomorphism for $q > 0$ and is injective for $q = 0$.

For a proof of this theorem, see (Gersten).

Suppose $f : S \longrightarrow T$ is a monoidal functor. Then f is cofinal if given $A \in T$ there exist $B \in T$ and $C \in S$ so that $A + B \cong fC$. Suppose T acts on X, and S acts on X through f.

Th : If $f : S \longrightarrow T$ is cofinal, then $S^{-1}X = T^{-1}X$.

Pf: The point is that S acts invertibly on X if and only if T acts invertibly on X. Thus

$$S^{-1}X = T^{-1}(S^{-1}X) = S^{-1}(T^{-1}X) = T^{-1}X.$$

The Extension Construction

Let P be an exact category in which every exact sequence splits. Then S = Iso(P). Given C in P let E_C be the category whose objects are all exact sequences $(0 \rightarrow A \rightarrow B \rightarrow C \rightarrow 0)$ from P, and whose arrows are all isomorphisms which are the identity on C :

$$\begin{array}{ccccccccc} 0 & \rightarrow & A & \rightarrow & B & \rightarrow & C & \rightarrow & 0 \\ & & \wr\uparrow & & \wr\uparrow & & \| & & \\ 0 & \rightarrow & A' & \rightarrow & B' & \rightarrow & C & \rightarrow & 0. \end{array}$$

We define a fibred category E over QP with fibres E_C. The base-change map $E_C \rightarrow E_{C'}$ for an arrow $C' \rightarrow C$ in QP can be described as follows :

a) for an injective arrow $C' \rightarrowtail C$, given $0 \rightarrow A \rightarrow B \rightarrow C \rightarrow 0$, construct the pullback $0 \rightarrow A \rightarrow B' \rightarrow C' \rightarrow 0$:

$$\begin{array}{ccccccccc} 0 & \rightarrow & A & \rightarrow & B' & \rightarrow & C' & \rightarrow & 0 \\ & & \| & & \downarrow & & \downarrow & & \\ 0 & \rightarrow & A & \rightarrow & B & \rightarrow & C & \rightarrow & 0. \end{array}$$

b) for a surjective arrow $C' \twoheadleftarrow C$, given $0 \rightarrow A \rightarrow B \rightarrow C \rightarrow 0$, compose the surjections to get a surjection $B \twoheadrightarrow C'$, and let A' be its kernel. We obtain $0 \rightarrow A' \rightarrow B \rightarrow C' \rightarrow 0$ in $E_{C'}$.

$$\begin{array}{ccccccccc} 0 & \rightarrow & A' & \rightarrow & B & \rightarrow & C' & \rightarrow & 0 \\ & & \uparrow & & \| & & \twoheaduparrow & & \\ 0 & \rightarrow & A & \rightarrow & B & \rightarrow & C & \rightarrow & 0. \end{array}$$

We see that E is the category whose objects are exact sequences $0 \rightarrow A \rightarrow B \rightarrow C \rightarrow 0$ from P, and whose arrows are represented by diagrams:

$$\begin{array}{ccccccccc} 0 & \rightarrow & A' & \rightarrow & B' & \rightarrow & C' & \rightarrow & 0 \\ & & \uparrow & & \| & & \twoheaduparrow & & \\ 0 & \rightarrow & A & \rightarrow & B' & \rightarrow & C_1 & \rightarrow & 0 \\ & & \| & & \downarrow & & \downarrow & & \\ 0 & \rightarrow & A & \rightarrow & B & \rightarrow & C & \rightarrow & 0, \end{array}$$

but that isomorphisms of such diagrams involving C_1 give rise to the same arrow in E. The fibred map $E \rightarrow QP$ is the projection $(0 \rightarrow A \rightarrow B \rightarrow C \rightarrow 0) \mapsto (C)$, and every arrow of E is cartesian.

We let S act on E by setting $(A') + (0 \rightarrow A \rightarrow B \rightarrow C \rightarrow 0) = (0 \rightarrow A' \oplus A \rightarrow A' \oplus B \rightarrow C \rightarrow 0)$ and observe that $E \rightarrow QP$ is fibrewise and cartesian with respect to this action.

Notice that the map $S \rightarrow E_0$ given by $(A) \mapsto (0 \rightarrow A \rightarrow A \rightarrow 0 \rightarrow 0)$ is an equivalence of categories.

<u>Th</u>: For any C in P, $\langle S,E_C\rangle$ is contractible

<u>Pf</u>: Let M denote $\langle S,E_C\rangle$. We show

i) M is connected,
ii) M is an H-space,
iii) the multiplication on M has a homotopy inverse,
iv) the endomorphism $x \longmapsto x^2$ on M is homotopic to the identity, and
v) M is contractible.

We define the product on E_C using pullback in P : given $F_i = (0 \to A_i \to B_i \to C \to 0)$, set

$$F_1 * F_2 = (0 \to A_1 \oplus A_2 \to B_1 \times_C B_2 \to C \to 0).$$

Projection on one factor gives :

$$\begin{array}{ccccccccc} (0 & \to & A_1 \oplus A_2 & \to & B_1 \times_C B_2 & \to & C & \to & 0) \\ & & \downarrow & & \downarrow & & \| & & \\ (0 & \to & A_1 & \to & B_1 & \to & C & \to & 0). \end{array} \tag{1}$$

We may choose a splitting for the surjections and obtain an isomorphism $A_2 + F_1 = F_1 * F_2$, and this determines an arrow $F_1 \longrightarrow F_1 * F_2$ in $\langle S,E_C\rangle$. Similarly, we may construct an arrow $F_2 \longrightarrow F_1 * F_2$, and we have connected F_1 and F_2 and proved i).

The constant functor to $(0 \to 0 \to C \to C \to 0)$ provides an identity for the operation just defined, so M is an H-space.

Any connected H-space has a homotopy inverse : consider

$$\begin{array}{ccccc} M & \xrightarrow{g} & M\times M & \xrightarrow{pr_2} & M \\ \| & & \downarrow f & & \| \\ M & \xrightarrow{g} & M\times M & \xrightarrow{pr_2} & M, \end{array}$$

where f is the map $(x,y) \longmapsto (xy,y)$, and g is $(x) \longmapsto (x,e)$, where e is the unit element. Since M is connected, the rows are fibrations, and the vertical maps on the fiber and on the base are homotopy equivalences, we know the total map f is a homotopy equivalence, with inverse h, say. One checks that the map $x \mapsto pr_1(h(e,x))$ is an inverse.

If $F_1=F_2$ then in the diagram above (1), the diagonal map provides a canonical splitting of the surjections, and yields a natural arrow $F_1 \longrightarrow F_1 * F_1$. This natural transformation gives the homotopy of iv).

Consider homotopy classes of maps $[M,M]$. By ii) this set is a monoid, by iii) this monoid is a group, by iv) the elements of this group satisfy the equation $x^2=x$, and is therefore trivial. Thus M is contractible.

QED

Th: The square

$$\begin{array}{ccc} S^{-1}S & \longrightarrow & S^{-1}E \\ \downarrow & & \downarrow \\ pt & \longrightarrow & QP \end{array}$$

is homotopy cartesian.

Pf: We must show that the base change maps for the fibred map $S^{-1}E \longrightarrow QP$ are homotopy equivalences. It is enough to consider those associated to injective and surjective arrows of QP of the form $0 \rightarrowtail C$ and $0 \overset{j}{\twoheadleftarrow} C$. We will treat only the surjective case since the injective case is similar.

Identifying E_0 and S, the base change map $j^* : E_C \longrightarrow E_0$ is $(0 \to A \to B \to C \to 0) \mapsto (B)$. Consider $f : E_0 \longrightarrow E_C$ given by $(A) \mapsto (0 \to A \to A \oplus C \to C \to 0)$. Since $\langle S, E_C \rangle$ is contractible, a previous theorem tells us that $S^{-1}f : S^{-1}E_0 \longrightarrow S^{-1}E_C$ is a homotopy equivalence. The composite $j^* \circ S^{-1}f : S^{-1}E_0 \longrightarrow S^{-1}E_0$ is given by $(A', A) \longmapsto (A', A \oplus C)$. This is a homotopy equivalence, as we have seen before, so j^* is a homotopy equivalence.

QED

Th: $S^{-1}E$ is contractible.

Pf: If X is a category, its <u>subdivision</u> Sub(X) is the category whose objects are the arrows of X, and where an arrow from f to g is a pair of arrows from X, h and k, such that $kfh = g$. One sees that the codomain map $\mathrm{Sub}(X) \longrightarrow X$ is a homotopy equivalence.

If X is the subcategory of QP of injective arrows, then E is equivalent to Sub(X). X has initial object 0, so E is contractible. Then S acts invertibly on E, so we know that E and $S^{-1}E$ are homotopy equivalent. The Theorem is proved.

QED

Th: $\Omega QP \sim K_0R \times BGl(R)^+$.

This Theorem is a corollary of previous theorems. Here P is the category of finitely generated projective R-modules.

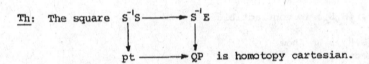

The Localization Theorem for projective modules

Suppose X is a quasi-compact scheme,
U is an affine open subscheme of X,
j is the inclusion $U \subseteq X$,
I is the sheaf of ideals defining the complement X-U in X,
I is locally principal and generated by a non-zero-divisor,
H is the category of quasi-coherent sheaves on X which are zero on U and admit a resolution of length 1 by vector bundles on X.

Th : There is an exact sequence

$$\cdots \longrightarrow K_{q+1}U \longrightarrow K_qH \longrightarrow K_qX \longrightarrow K_qU$$

for $q \geq 0$.

Pf: Let P be the category of vector bundles on X,
V the category of vector bundles on U which extend to vector bundles on X, and
P_1 the category of quasi-coherent sheaves on X which have a resolution of length 1 by vector bundles.

U is affine, so every exact sequence in V splits. Let E be the extension construction over QV.

Iso(P) is cofinal in Iso(V), so we may use it instead of Iso(V) ; let S = Iso (P).

We will construct a diagram of categories with S-action :

$$\begin{array}{ccccc} G & \xrightarrow{f} & F & \longrightarrow & E \\ {\scriptstyle h}\downarrow & \overset{g}{\searrow} & \downarrow{\scriptstyle p} & & \downarrow \\ QH & & QP & \longrightarrow & QV \end{array}$$

and show that f and h are homotopy equivalences. Localization will give :

$$\begin{array}{ccccc} S^{-1}G & \xrightarrow{\sim} & S^{-1}F & \longrightarrow & S^{-1}E \ \sim \ \text{pt} \\ \wr\downarrow & & \downarrow & \square & \downarrow \\ QH & & QP & \longrightarrow & QV, \end{array}$$

with the right-hand square homotopy cartesian. Combining this with the cofinality of V in all vector bundles on U gives the result. The map $K_qH \longrightarrow K_qX$ differs by a sign from the usual one.

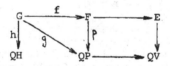

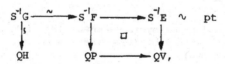

In order to simplify notation, it is convenient to replace E by the equivalent category whose objects are surjections $(B \twoheadrightarrow C)$ with $B, C \in V$. That this category is equivalent is clear because a surjection determines its kernel up to unique isomorphism. An arrow in E is now represented by :

$$\begin{array}{ccc} B' & \twoheadrightarrow & C' \\ \Vert & & \Uparrow \\ B' & \twoheadrightarrow & C_1 \\ \downarrow & \square & \downarrow \\ B & \twoheadrightarrow & C. \end{array}$$

F is defined as the pullback in
$$\begin{array}{ccc} F & \longrightarrow & E \\ \downarrow & & \downarrow \\ QP & \longrightarrow & QV. \end{array}$$

Its objects are pairs $(B, Z \twoheadrightarrow j^*B)$ with $B \in P$, $Z \in V$. An arrow may be represented by

$$\begin{array}{cccc} B' & Z' & \twoheadrightarrow & j^*B' \\ \Uparrow & \Vert & & \Uparrow \\ B_1 & Z' & \twoheadrightarrow & j^*B_1 \\ \downarrow & \downarrow & \square & \downarrow \\ B & Z & \twoheadrightarrow & j^*B. \end{array}$$

G is a sort of extension construction over $Q(H \times P)$. Its objects are surjections $(L \twoheadrightarrow M \oplus B)$ with $L, B \in P$ and $M \in H$. Its arrows are represented by diagrams :

$$\begin{array}{ccc} L' & \twoheadrightarrow & M' \oplus B' \\ \Vert & & \Uparrow \\ L' & \twoheadrightarrow & M_1 \oplus B_1 \\ \downarrow & \square & \downarrow \\ L & \twoheadrightarrow & M \oplus B, \end{array}$$

and isomorphic diagrams give the same arrow. The vertical arrows on the right are each direct sums of arrows from H and P.

$G \longrightarrow Q(H \times P)$ is defined by $(L \twoheadrightarrow M \oplus B) \mapsto (M, B)$. This map is fibred.

$g : G \longrightarrow QP$ is defined by $(L \twoheadrightarrow M \oplus B) \mapsto (B)$, and
$h : G \longrightarrow QH$ is defined by $(L \twoheadrightarrow M \oplus B) \mapsto (M)$. Both g and h are fibred.

$f : G \longrightarrow F$ is defined by $(L \twoheadrightarrow M \oplus B) \mapsto (B, j^*L \twoheadrightarrow j^*B)$.

S acts on G via $(A) + (L \overset{\varphi}{\twoheadrightarrow} M \oplus B) = (A \oplus L \overset{(0\ \varphi)}{\twoheadrightarrow} M \oplus B)$. The action on F is similar, and is that induced by the action on E. S acts trivially on QH, QP, and QV.

<u>Lemma 1</u> : $h : G \longrightarrow QH$ is a homotopy equivalence.

<u>Pf</u>: Let R be the category whose objects are surjections $(L \twoheadrightarrow M)$ with $L \in P$. M is a fixed object of H. The arrows of R are given by diagrams :

$$\begin{array}{ccc} L' & \twoheadrightarrow & M \\ \downarrow & & \Vert \\ L & \twoheadrightarrow & M, \end{array}$$

where $L' \rightarrowtail L$ is an admissible monomorphism of P, i.e. its cokernel is a vector bundle.

There are natural transformations

$$(L \twoheadrightarrow M) \longrightarrow (L \oplus L' \twoheadrightarrow M) \longleftarrow (L' \twoheadrightarrow M),$$

so R is contractible.

Its subdivision, Sub(R), is equivalent to the fiber $h^{-1}(M)$, which is therefore contractible.

Since each fiber of h is contractible, h is a homotopy equivalence.

QED

<u>Lemma 2</u> : If C is a vector bundle on X, then $C \subsetneq j_*j^*C$, and and

$$j_*j^*C = \bigcup_n I^{-n}C.$$

<u>Pf</u>: The question is local on X, so we may assume X is affine and I is generated by the function w on X. Let R be the ring of X, so $X = \mathrm{Spec}\ (R)$, $U = \mathrm{Spec}(R[1/w])$. C is a projective R-module, and w is a non-zero-divisor in R, so w is a non-zero-divisor on C. Thus $C \subsetneq C_w = C \otimes_R R[1/w]$, and $C_w = \bigcup_n w^{-n}C$.

QED

<u>Lemma 3</u> : $f : G \longrightarrow F$ is a homotopy equivalence.

<u>Pf</u>: Both g and p are fibred, so it is enough to show f is a homotopy equivalence on each fiber over QP. If $B \in QP$, consider the map $g^{-1}B \longrightarrow p^{-1}B$.

Let T be the category whose objects are surjections $(L \twoheadrightarrow B)$ with $L \in P$, and whose arrows are diagrams

$$\begin{array}{ccc} L' & \twoheadrightarrow & B \\ \downarrow & & \| \\ L & \twoheadrightarrow & B, \end{array}$$

where $L' \rightarrowtail L$ is an admissible mono from P_1 whose cokernel is in H, i.e. any injection which is an isomorphism on U.

Then the functor $\mathrm{Sub}(T) \longrightarrow g^{-1}B$ given by

$$\left(\begin{array}{ccc} L' & \twoheadrightarrow & B \\ \downarrow & & \| \\ L & \twoheadrightarrow & B \end{array}\right) \longmapsto (L \twoheadrightarrow (\mathrm{ckr}\ i) \oplus B)$$

is an equivalence of categories.

Let $W = j^*B$, so that $p^{-1}B = E_W$. We must show that Sub $(T) \longrightarrow E_W$ is a homotopy equivalence. This map is

$$\begin{pmatrix} L' \twoheadrightarrow B \\ \downarrow \quad \| \\ L \text{ --- } B \end{pmatrix} \longmapsto (j^*L \twoheadrightarrow j^*B = W).$$

It factors through the target map $\mathrm{Sub}(T) \longrightarrow T$, which is a homotopy equivalence, so it is enough to show that the map $T \xrightarrow{w} E_W$ given by $(L \twoheadrightarrow B) \mapsto (j^*L \twoheadrightarrow j^*B = W)$ is a homotopy equivalence. To do this we need only show that each fiber $w/(Z \twoheadrightarrow W)$ is contractible, where $(Z \twoheadrightarrow W)$ is an object of E_W.

An object of this fiber category is an object $(L \twoheadrightarrow B)$ of T with an isomorphism $w(L \twoheadrightarrow B) \cong (Z \twoheadrightarrow W)$ which is the identity on W, i.e.

$$\begin{pmatrix} L \twoheadrightarrow B \;, & j^*L \twoheadrightarrow j^*B \\ & \wr\| \qquad \| \\ & Z \twoheadrightarrow W \end{pmatrix}.$$

Define an ordered set <u>Lat</u> to be the set of vector bundles L on X such that

1) $L \subseteq j_*Z$,
2) $j^*L = Z$,
3) the image of the map $(L \subset j_*Z \twoheadrightarrow j_*W = j_*j^*B)$ is B.

Elements of Lat will be called <u>lattices</u>.
The obvious map from Lat to the fiber $w/(Z \twoheadrightarrow W)$ is an equivalence of categories, so we need only to show that Lat is contractible ; we show it is actually filtering.

We have an exact sequence $0 \longrightarrow Y \longrightarrow Z \longrightarrow W \longrightarrow 0$ in V which splits. Now $Y = j^*C$ for some $C \in P$, so $Z = j^*(C \oplus B)$. Consider the lattice $C \oplus B \subseteq j_*Z$. If $(L \subseteq j_*Z)$ is another lattice, then condition 3) insures that $L \subseteq j_*j^*C \oplus B$. Now by lemma 2 $j_*j^*C = \bigcup_n I^{-n}C$, and L is finitely generated locally on X, which is quasi-compact, so for large n, $L \subseteq I^{-n}C \supseteq C$.

Thus Lat is filtering, and Lemma 3 is proved.

QED

The end of the proof of the theorem is now near. S acts trivially on QH, so by lemmas 1 and 3 it acts invertibly on G and F. Thus G and F are homotopy equivalent to $S^{-1}G$ and $S^{-1}F$, respectively, and h and f remain homotopy equivalences after localization. We know that $S^{-1}E \longrightarrow QV$ is a fibration, so $S^{-1}F \longrightarrow QP$ is, too, since it has the same fibers. Since the homotopy fibers of these two maps are the same, the square

$$\begin{array}{ccc} S^{-1}F & \longrightarrow & S^{-1}E \\ \downarrow & & \downarrow \\ QP & \longrightarrow & QV \end{array}$$

is homotopy cartesian.

As indicated earlier, we now know that

$$BQH \longrightarrow BQP \longrightarrow BQV$$

has the homotopy type of a fibration, and the cofinality of V in the category of all vector bundles on U gives the long exact sequence we want.

It only remains to compute the map $K_q H \longrightarrow K_q P$. To do this we show that the square

$$\begin{array}{ccc} G & \longrightarrow & QP \\ \downarrow & & \downarrow \\ QH & \longrightarrow & QP_1 \end{array}$$ commutes up to sign.

The two functors $G \rightrightarrows QP_1$ are given by $(L \twoheadrightarrow M\oplus B) \mapsto (M)$ and by $(L \twoheadrightarrow M\oplus B) \mapsto (B)$. The map $(L \twoheadrightarrow M\oplus B) \mapsto (M\oplus B)$ is their sum, so we must show this map is homotopic to a constant map. The functor $(L \twoheadrightarrow M\oplus B) \mapsto (L)$ maps all arrows of G to injective arrows of QP_1, so $0 \rightarrowtail L \twoheadrightarrow M\oplus B$ exhibits two natural transformations which give the desired homotopy.

The theorem is proved.

Suppose R is a ring,
$S \subseteq R$ is a multiplicative set of central non-zero-divisors,
H is the category of finitely generated R-modules M of projective dimension ≤ 1 such that $M_S = 0$.

<u>Th</u> : There is an exact sequence

$$\cdots \longrightarrow K_{q+1}R_S \longrightarrow K_q H \longrightarrow K_q R \longrightarrow K_q R_S$$

for $q \geq 0$.

The proof is formally the same as the proof of the previous theorem, except that Lemma 2 is replaced by:

<u>Lemma 2'</u> : If C is a projective R-module, then $C \subseteq C_S$, and

$$C_S = \bigcup_{s \in S} s^{-1}C.$$

The Suspension of a ring

This section is included so we can complete the proof of the fundamental theorem in the next section. We must ensure that a certain computation of Loday's involving products using the +-construction is compatible with our use of the Q-construction.

Suppose A is a ring. Then the cone of A, CA, is the ring of infinite matrices with entries in A which have only a finite number of non-zero entries in any given row or column. The matrices which have only a finite number of non-zero entries form a two-sided ideal $I \subseteq CA$. The suspension of A, SA, is the quotient ring CA/I.

Let e be the element of CA whose only non-zero entry is a 1 in the corner :

$$\begin{pmatrix} 1 & & & \\ & 0 & & \\ & & 0 & \\ & & & \ddots \end{pmatrix}$$

It is idempotent, so $CA = eCA \oplus (1-e)CA$ is a decomposition of CA into A-CA-bimodules. We use this to define w as the composite

$$Gl_n A \longrightarrow Aut((eCA)^n) \longrightarrow Aut(CA^n) = Gl_n CA.$$

It sends a matrix (a_{ij}) to the matrix (b_{ij}) where $b_{ij} = a_{ij}e$ if $i \neq j$, and $b_{ii} = a_{ii}e + (1-e)$.

<u>Th</u>: [Gersten-Wagoner] (i) $K_0CA \times BGl(CA)^+$ is contractible.

(ii) $K_0A \times BGl(A)^+ \xrightarrow{w} K_0CA \times BGl(CA)^+ \longrightarrow K_0SA \times BGl(SA)^+$

is a fibration.

Let P(R) denote the category of finitely generated projective right R-modules.

<u>Def</u>: $v : P(A) \longrightarrow P(CA)$ is the exact functor $B \mapsto B \otimes_A (eCA)$.

<u>Th</u>: $QP(A) \xrightarrow{v} QP(CA) \longrightarrow QP(SA)$ is a fibration which is a delooping of the Gersten-Wagoner fibration.

<u>Pf</u>: We use the naturality of the extension construction to loop this sequence, yielding

$$S^{-1}S(A) \xrightarrow{v} S^{-1}S(CA) \longrightarrow S^{-1}S(SA).$$

Here $S^{-1}S(A)$ denotes $S^{-1}S$ where $S = Iso(P(A))$ (for any ring A).

The two functors $\quad Aut(A^n) \longrightarrow S^{-1}S(CA)$

$$V : \quad a \longmapsto (1_{eCA^n} , a \otimes 1_{eCA^n})$$

$$W : \quad a \longmapsto (1_{CA^n} , a \otimes 1_{eCA^n} \oplus 1_{(1-e)CA^n})$$

are homotopic, so we identify the looped sequence with the Gersten-Wagoner fibration. The sequence in the statement of the theorem consists of connected H-spaces, so it, too, must be a fibration.

QED

The Fundamental Theorem

Suppose A is a (not necessarily commutative) ring.

Def: $NK_qA = \mathrm{ckr}\,(K_qA \longrightarrow K_qA[t])$

Def: Nil(A) is the exact category whose objects are pairs (P,f), where P is a finitely generated projective A-module, and f is a nilpotent endomorphism of P.

Def: $Nil_q(A) = \ker\,(K_q\underline{Nil}(A) \longrightarrow K_qA)$

Th: 1) $NK_qA \cong Nil_{q-1}(A)$

2) $K_q(A[t,t^{-1}]) \cong K_qA \oplus K_{q-1}A \oplus NK_qA \oplus NK_qA$

Pf : Let X be the projective line over A. Then X has open subsets Spec(A[t]) and Spec(A[t⁻¹]) which satisfy the conditions of the localization theorem for projective modules. We get :

$$\begin{array}{ccccccccccc} \cdots \longrightarrow & K_qH & \longrightarrow & K_qX & \longrightarrow & K_qA[t^{-1}] & \longrightarrow & K_{q-1}H & \longrightarrow \cdots \\ & \| & & \downarrow & & \downarrow & & \| & \\ \cdots \longrightarrow & K_qH & \longrightarrow & K_qA[t] & \longrightarrow & K_qA[t,t^{-1}] & \longrightarrow & K_{q-1}H & \longrightarrow \cdots \end{array} \qquad (*)$$

The naturality of the long exact sequence with respect to flat maps is clear from the proof of the localization theorem. The vertical equalities involving H arise from the fact that the category of abelian sheaves on X which vanish on Spec(A[t⁻¹]) is equivalent to the category of abelian sheaves on Spec(A[t]) which vanish on Spec(A[t,t⁻¹]).

If (P,f) ∈ Nil(A), we have the characteristic sequence of f :

$$0 \longrightarrow P[t] \xrightarrow{t-f} P[t] \longrightarrow P_f \longrightarrow 0,$$

where P_f is the A[t]-module P with t acting as f. Since f is nilpotent, P_f is zero on Spec(A[t,t⁻¹]), so determines an object of H.

If M is an A[t]-module of projective dimension ≤ 1 killed by some power t^n of t, then M is a projective A-module. For, let $0 \longrightarrow P \longrightarrow Q \longrightarrow M \longrightarrow 0$ be a projective resolution of M by A[t]-modules. Then

$$0 \longrightarrow M \xrightarrow{t^n} P/t^nP \longrightarrow P/t^nQ \longrightarrow 0$$

is exact, and P/t^nP is a projective A-module, P/t^nQ is an A-module of projective dimension 1. Then M is projective.

Thus
$$\begin{array}{ccc} \underline{Nil}(A) & \longrightarrow & H \\ (P,f) & \longmapsto & P_f \end{array}$$

is an equivalence of categories.

The K-theory of the projective line was computed in (Quillen). We know

$$K_qX \cong K_qA \cdot 1 \oplus K_qA \cdot z,$$

where $1 = cl(O_X)$, $z = cl(O_X(-1))$ in K_0X. We alter this basis slightly :

$$K_qX = K_qA \cdot (1-z) \oplus K_qA \cdot 1.$$

Let $U = \text{Spec}\,(A[t])$, and $V = \text{Spec}\,(A[t^{-1}])$.

Now,

$$K_qA \xrightarrow{1-z} K_qX \longrightarrow K_qV$$ is zero, since $O_X|_V = O_X(-1)|_V$,

and

$$K_qA \xrightarrow{1} K_qX \longrightarrow K_qV$$ is the usual split injection

induced by $A \longrightarrow A[t^{-1}]$. Thus the top row above splits into pieces :

$$0 \longrightarrow K_qA \longrightarrow K_qV \longrightarrow K_{q-1}\underline{\text{Nil}}(A) \longrightarrow K_{q-1}A \longrightarrow 0. \qquad (\#)$$

If $(P,f) \in \underline{\text{Nil}}(A)$, then the characteristic sequence extends to all of X as :

$$0 \longrightarrow P_X(-1) \longrightarrow P_X \longrightarrow P_f \longrightarrow 0.$$

Thus, the square

$$\begin{array}{ccc} K_{q-1}\underline{\text{Nil}}(A) & \xrightarrow{\sim} & K_{q-1}H \\ {\scriptstyle pr_1}\downarrow & & \downarrow \\ K_{q-1}A & \xrightarrow{1-z} & K_{q-1}X \end{array}$$ commutes, and

the last map of (#) is the usual projection. Splitting off the first and last terms of (#) gives $NK_qA \cong \text{Nil}_{q-1}(A)$, proving 1).

From (*) we derive the Mayer-Vietoris sequence :

$$\cdots \longrightarrow K_qX \longrightarrow K_qA[t] \oplus K_qA[t^{-1}] \longrightarrow K_qA[t,t^{-1}] \longrightarrow K_{q-1}X \longrightarrow \cdots$$

and, as above, split it into shorter pieces :

$$0 \longrightarrow K_qA \longrightarrow K_qA[t] \oplus K_qA[t^{-1}] \longrightarrow K_qA[t,t^{-1}] \longrightarrow K_{q-1}A \longrightarrow 0.$$

According to (Loday, Coro 2.3.7), the map $K_qA[t,t^{-1}] \longrightarrow K_{q-1}A$ is split by the map induced by cup-product with t. All we need to do is verify that his definition of this map agrees with ours, so we must check that

$$\begin{array}{ccc} K_qA[t,t^{-1}] & \xrightarrow{t \mapsto T} & K_qSA \\ \downarrow & & \searrow{\scriptstyle \sim} \\ K_{q-1}H = K_{q-1}\underline{\text{Nil}}(A) & \xrightarrow{pr_1} & K_{q-1}A \end{array}$$ commutes.

Loday uses the +-construction for his definition of the isomorphism $K_q SA = K_{q-1} A$, but we saw in the previous section that we may as well use the Q-construction. Let

$$\tau = \begin{pmatrix} 0 & & & \\ 1 & 0 & & \\ & 1 & 0 & \\ & & 1 & \ddots \\ & & & \ddots \end{pmatrix} \quad \text{and } \sigma = \begin{pmatrix} 0 & 1 & & \\ & 0 & 1 & \\ & & 0 & 1 \\ & & & \ddots \\ & & & \ddots \end{pmatrix}$$ be elements of CA.

They satisfy $\tau\sigma=1$, $\sigma\tau=1-e$. We have the commutative diagram :

$$\begin{array}{ccc} A[t] & \longrightarrow & A[t,t^{-1}] \\ {\scriptstyle t \mapsto \tau}\downarrow & & \downarrow \\ CA & \longrightarrow & SA. \end{array}$$

We now refer to the proof of the localization theorem. If $P_1(R)$ denotes the exact category of finitely-generated R-modules of projective dimension $\leqslant 1$, we may conclude that

$$QH \longrightarrow QP_1(A[t]) \longrightarrow QP_1(A[t,t^{-1}])$$

is a fibration homotopy equivalent to the one produced by the theorem except for a change in sign of the left hand map. Notice that in the first part of the proof of the fundamental theorem we have implicitly used the maps with this more natural sign-sense.

Let H' be the exact category of right CA-modules B of projective dimension $\leqslant 1$ such that

$$B \otimes_{CA} SA = 0.$$

Since $eCA \in H'$, the map $v : QP(A) \longrightarrow QP(CA)$ yields a map $QP(A) \xrightarrow{v} QH'$.

Consider this diagram :

$$\begin{array}{ccccccc} Q\underline{Nil}(A) & \xleftarrow{\sim} & QH & \longrightarrow & QP_1'(A[t]) & \longrightarrow & QP_1'(A[t,t^{-1}]) \\ \downarrow & & \downarrow & & \downarrow & & \downarrow \\ QP(A) & \xrightarrow{v} & QH' & \longrightarrow & QP_1(CA) & \longrightarrow & QP_1(SA). \end{array}$$

A functor $H \longrightarrow H'$ is defined by $M \mapsto M \otimes_{A[t]} CA$, but we must check that this CA-module has the right projective dimension, and that this functor is exact. It is enough to see that the characteristic sequence of an element of $\underline{Nil}(A)$ remains exact under this tensor product. At issue is the injectivity of $\tau - f : P \otimes_A CA \longrightarrow P \otimes_A CA$, and this is true because the sum of an injective endomorphism and a nilpotent endomorphism which commute is injective.

We must also check that the left-hand square commutes. If (P,f) is in $\underline{Nil}(A)$ then there is a natural isomorphism

$$P_f \otimes_{A[t]} CA \cong P \otimes_A (eCA)$$

defined by the diagram

$$
\begin{array}{ccccccccc}
0 & \longrightarrow & P\otimes_A CA & \xrightarrow{\tau} & P\otimes_A CA & \longrightarrow & P\otimes_A (eCA) & \longrightarrow & 0 \\
 & & \| & & \uparrow \wr & & \vdots\uparrow & & \\
0 & \longrightarrow & P\otimes_A CA & \xrightarrow{\tau - f} & P\otimes_A CA & \longrightarrow & P_f\otimes_{A[t]} CA & \longrightarrow & 0
\end{array}
$$

where $1 + f\sigma + f^2\sigma^2 + f^3\sigma^3 + \ldots$ is the vertical isomorphism. This isomorphism yields the commutativity of the square in question.

Finally we define the categories $P'(A[t])$ and $P'(A[t,t^{-1}])$. $P'(A[t])$ is the full exact subcategory of $P_1(A[t])$ consisting of modules M satisfying

$$\mathrm{Tor}_1^{A[t]}(M,CA) = 0.$$

We saw above that this category contains H. $P'(A[t,t^{-1}])$ is defined in a similar fashion relative to SA. It is clear then that these categories fit into the diagram as indicated, and the resolution theorem says that the top row still contains a fibration equivalent to the original one.

We conclude that we have a map of fibrations, so the naturality of the boundary map in the long exact sequence of homotopy groups yields the commutativity of

$$
\begin{array}{ccc}
K_q A[t,t^{-1}] & \longrightarrow & K_{q-1}\underline{\mathrm{Nil}}(A) \\
\downarrow & & \downarrow \\
K_q SA & \xrightarrow{\sim} & K_{q-1}A
\end{array} .
$$

This concludes the proof of the fundamental theorem.

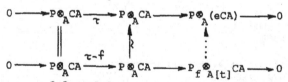

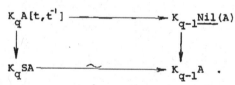

Bibliography

Bass Algebraic K-theory
Benjamin 1968

Gersten On the spectrum of algebraic K-theory
BAMS 78(1972) 216-220

Gersten The Localization Theorem for Projective Modules
Comm. in Alg., 2 (1974) 307-350

Loday K-théorie algébrique et représentations de groupes
Ann.Sc.Ec.Norm.Sup., 4ème série, 9, no.3 (1976)

MacLane Categories for the Working Mathematician
Springer-Verlag, GTM 5 (1971)

Quillen Higher Algebraic K-theory : I
Springer-Verlag Lecture Notes in Math., # 341

Wagoner Delooping classifying spaces in algebraic K-theory
Topology 11 (1972) 349-370

Continuous Cohomology and p-Adic K-Theory

J. B. Wagoner*

In [10] continuous algebraic K-theory groups K_i^{top} are defined for a complete discrete valuation ring $\mathcal{O}$ with finite residue field of positive characteristic $p > 0$ and also for the field of fractions F of $\mathcal{O}$. The group K_2^{top} agrees with the fundamental group of the special linear group as defined in [7] by means of universal topological central extensions. In [12] it is shown that for $i \geqslant 1$

(1) $$K_i^{top}(\mathcal{O}) \cong \varprojlim_n K_i(\mathcal{O}/\mathfrak{p}^n)$$

where $\mathfrak{p} \subset \mathcal{O}$ is the maximal ideal and K_i are the algebraic K-theory groups of Quillen. The present work is joint with R. J. Milgram. We use the continuous cohomology of the p-adic group $GL_n(\mathcal{O})$ to show

Theorem. *Let the field of fractions* F *of* $\mathcal{O}$ *be a finite extension of* Q_p. *Then for* $i \geqslant 1$

$$K_i^{top}(\mathcal{O}) \cong K_i(\mathcal{O}/\mathfrak{p}) \oplus V_i$$

where V_i *is a finitely generated module over* Z_p *such that*

$$\dim_{Q_p}(V_i \otimes_{Z_p} Q_p) = 0 \qquad \textit{for } i \textit{ even}$$

and

$$\dim_{Q_p}(V_i \otimes_{Z_p} Q_p) = \dim_{Q_p} F \qquad \textit{for } i \textit{ odd}.$$

*Supported in part by the National Science Foundation Grant MPS 74-03423 and the Alfred P. Sloan Foundation.

For example we see that $K_i^{top}(Z_p)$ is finite for i even and is Z_p modulo torsion for i odd. Recall from [2] that $K_i(Z)$ is finite if $i \not\equiv 1 \bmod 4$ and is Z modulo torsion if $i = 4k + 1$ and $k > 0$. The equation (1) shows the map

(2) $$K_i(Z) \to K_i^{top}(Z_p)$$

is induced by the ring homomorphisms $Z \to Z/p^s$. So it is natural to ask whether (2) is injective on the torsion free part when $i = 4k + 1 > 1$. More generally, if $\mathcal{O} \subset F$ is the ring of integers in a totally real number field F, p is a fixed rational prime, and S is the set of primes in $\mathcal{O}$ lying above p, is it true that

(3) $$K_{4k+1}(\mathcal{O}) \to \prod_{\mathfrak{p} \in S} K_{4k+1}^{top}(\mathcal{O}_{\mathfrak{p}})$$

is injective on the torsion free part? Here $\mathcal{O}_{\mathfrak{p}}$ is the completion of $\mathcal{O}$ at $\mathfrak{p}$. This question is related, at least by analogy and hopefully more precisely, to whether the higher p-adic regulators are non-zero. Compare [3].

The motivation and proof for the main theorem follows in broad outline Borel's computation for the rank of $K_*(A)$ of a ring of integers A in a number field. For simplicity let $A = Z$. The determination of

$$K_i(Z) \otimes_Z \mathbb{R} = \pi_i(BGL(Z)^+) \otimes_Z \mathbb{R}$$

uses the theorem of Milnor-Moore [6, Appendix] which gives an isomorphism via the Hurewicz map

$$\pi_i(BGL(Z)^+) \otimes_Z R \underset{\cong}{\to} \text{Prim } H_i(BGL(Z)^+;\mathbb{R})$$

for the H-space $BGL(Z)^+$. From the definition of the plus construction, computing the right hand side amounts to finding the dimension of the indecomposable elements in $H^i(BSL(Z);R)$ when $i > 1$. Borel [2] shows this cohomology group is isomorphic to the well-known Lie algebra cohomology $H^*(sl_n(\mathbb{R}), so_n(\mathbb{R})) \cong E(u_5, u_9, u_{13}, \ldots)$ for $i \ll n$. By the van-Est theorem this is the same as $H^i_{cont}(SL_n(\mathbb{R});\mathbb{R})$, the cohomology of the real Lie group $SL_n(\mathbb{R})$ with continuous cochains taking values in $\mathbb{R}$. As $K_i^{top}(\mathcal{O})$ is defined by the inverse limit (1) we must have a "p-adic" analogue of this program. The three main steps are computing $H^*_{cont}(GL_n(\mathcal{O});Q_p)$, the continuous cohomology of the p-adic group $GL_n(\mathcal{O})$ with coefficients in Q_p; a p-adic Milnor-Moore theorem; and a p-adic version of the Hurewicz-Serre theorem for classes of abelian groups. We wish to thank D. Rector for discussions concerning the last two steps. He has also given proofs of these results.

I. The continuous homology and cohomology of $GL_n(\mathcal{O})$.

For each $0 < r \leq s$ and $1 \leq n \leq \infty$ let $\Gamma_n(r,s)$ denote the congruence subgroup of $GL_n(\mathcal{O}/\mathfrak{p}^s)$ defined by the exact sequence

(4) $$1 \to \Gamma_n(r,s) \to GL_n(\mathcal{O}/\mathfrak{p}^s) \to GL_n(\mathcal{O}/\mathfrak{p}^r) \to 1$$

and let $\Gamma_n(r) \subset GL_n(\mathcal{O})$ be defined by

$$1 \to \Gamma_n(r) \to GL_n(\mathcal{O}) \to GL_n(\mathcal{O}/\mathfrak{p}^r) \to 1.$$

Then $GL_n(\mathcal{O}) = \varprojlim_s GL_n(\mathcal{O}/\mathfrak{p}^s)$ and $\Gamma_n(r) = \varprojlim_s \Gamma_n(r,s)$ provided $n < \infty$. For $1 \leq n \leq \infty$ define the p-adic homology of $GL_n(\mathcal{O})$ to be

$$H_*^{(p)}(GL_n(\mathcal{O})) = \varprojlim_s H_*(GL_n(\mathcal{O}/p^s);Z_p)$$

and the p-*adic* *cohomology* to be

$$H^*_{(p)}(GL_n(\mathcal{O})) = \varprojlim_t (\varinjlim_s H^*(GL_n(\mathcal{O}/p^s);Z/p^t))$$

Similarly we have the p-adic homology and cohomology of the congruence subgroup $\Gamma_n(r)$.

Let $\mathcal{O}$ and F be as in the main theorem.

Proposition I. $H_*^{(p)}(GL_\infty(\mathcal{O}))$ *and* $H^*_{(p)}(GL_\infty(\mathcal{O}))$ *are finitely generated* Z_p*-modules and upon tensoring over* Z_p *with* Q_p *they are dual Hopf algebras with the multiplication induced from direct sum of matrices. Moreover*

$$H_*^{(p)}(GL_\infty(\mathcal{O})) \otimes_{Z_p} Q_p \cong E(u_1,u_3,u_5,\ldots)$$

where u_{2k+1} *denotes a set of* $\dim_{Q_p} F$ *primitive generators in degree* $2k+1$.

There are two ways to obtain this calculation. The first is basically the work of Lazard on the continuous cohomology of p-adic Lie groups [5]. Let $i \geqslant 0$ be given. It follows from the stability results of [11] that for $n < \infty$ sufficiently large $H^i_{(p)}(GL'_\infty(\mathcal{O})) \cong H^i_{(p)}(GL_n(\mathcal{O}))$. From [5, Chap. V] we see for r sufficiently large (depending on $\mathcal{O}$) that $H^i_{(p)}(\Gamma_n(r)) \cong H^i_{cont}(\Gamma_n(r);Z_p)$ is a finitely generated Z_p-module and that $\varinjlim_s H^*(\Gamma_n(r,s);Z/p^t) \cong H^i_{cont}(\Gamma_n(r);Z/p^t)$ is finite. From the cohomology spectral sequence of Hochschild-Serre the exact sequences (4), We conclude from [1, §3] upon taking inverse limits that there is a spectral sequence

$$E_2^{k,\ell} = H^k(GL_n(\mathcal{O}/p^r);\ H^\ell_{(p)}(\Gamma_n(r))) \Rightarrow H^{k+\ell}_{(p)}(GL_n(\mathcal{O}))$$

Since $GL_n(\mathcal{O}/\mathfrak{p}^r)$ is finite, we have

$$H^i_{(p)}(GL_n(\mathcal{O})) \otimes_{Z_p} Q_p = (H^i_{(p)}(\Gamma_n(r) \otimes_{Z_p} Q_p)^{GL_n(\mathcal{O}/\mathfrak{p}^r)}$$

But again by [5, Chap. V] we see that

$$\begin{aligned} H^*_{(p)}(\Gamma_n(r)) \otimes_{Z_p} Q_p &\cong H^*_{cont}(\Gamma_n(r);Q_p) \\ &\cong H^*(gl_n(F);Q_p) \\ &\cong E(u_1,u_3,u_5,\ldots,u_{2n-1}) \end{aligned}$$

where $gl_n(F)$ is considered as a Lie algebra over Q_p and u_{2k+1} is a basis of $\dim_{Q_p} F$ indecomposable elements in degree $2k + 1$. One checks these generators are invariant under $GL_n(\mathcal{O}/\mathfrak{p}^r)$ to get the desired result.

The second approach is by a direct calculation due to Milgram. First one computes $\varprojlim_{s} H^*(\Gamma_\infty(r,s);Z/p)$ by starting with the abelian group $\Gamma_\infty(r,r+1)$ and proceeding to $\Gamma_\infty(r,s)$ for $s \geq r + 2$ using the standard Hochschild-Serre spectral sequence for $\Gamma_\infty(r,s) \to \Gamma_\infty(r,s-1)$. Then analysis of the Bockstein gives the inverse limit corresponding to the coefficient groups Z/p^t, and finally shows $H^*_{(p)}(\Gamma_\infty(r)) \otimes_{Z_p} Q_p$ is invariant under the action of $GL_\infty(\mathcal{O}/\mathfrak{p}^r)$.

II. The p-adic Hurewicz-Serre Theorem.

This is the analogue of the result in Serre's theory of classes of abelian groups that a simply connected space with finitely generated homology has finitely generated homotopy.

Proposition II. *Let* $X_1 \leftarrow X_2 \leftarrow \ldots \leftarrow X_k \leftarrow \ldots$ *be an inverse system of connected simple spaces such that* $H_i(X_k)$ *is a finite* p-*group for* $1 \leq i, k$

and such that $\lim\limits_{\leftarrow k} H_i(X_k)$ *is a finitely generated* Z_p*-module for* $1 \leq i$. *Then* $\lim\limits_{\leftarrow k} \pi_i(X_k)$ *is a finitely generated* Z_p*-module for* $1 \leq i$.

The proof is analogous to, say, Theorem 8.1 of [4] and uses the inverse system of connective towers of the X_k's. Actually, it probably suffices to assume only that each X_k is nilpotent.

III. The p-adic Milnor-Moore Theorem.

If $X = \{X_1 \leftarrow X_2 \leftarrow \ldots\}$ is an inverse system of spaces let the p-adic homology of X be defined as

$$H_i^{(p)}(X) = \lim_{\leftarrow k} H_i(X_k;Z_p)$$

If each $H_i(X_k)$ is a finite p-group for $1 \leq i, k$ then of course $H_i^{(p)}(X) = \lim\limits_{\leftarrow k} H_i(X_k)$ for $1 \leq i$.

Proposition III. *Let* $X = \{X_1 \leftarrow X_2 \leftarrow \ldots \leftarrow X_k \leftarrow \ldots\}$ *be an inverse system of connected* H*-spaces and* H*-maps such that* $H_i(X_k)$ *is a finite* p*-group for* $i, k \geq 1$ *and such that* $\lim\limits_{\leftarrow k} H_i(X_k)$ *is a finitely generated* Z_p*-module for* $i \geq 1$. *Then* $H_*^{(p)}(X) \otimes_{Z_p} Q_p$ *is a Hopf algebra and the Hurewicz map*

$$\left(\lim_{\leftarrow k} \pi_i(X_k)\right) \otimes_{Z_p} Q_p \to \mathrm{Prim}\left(H_*^{(p)}(X) \otimes_{Z_p} Q_p\right)$$

is an isomorphism for $i \geq 1$.

The proof is analogous to the proof of the usual Milnor-Moore Theorem [6, Appendix] and uses the inverse system of Postnikov towers of the H-spaces X_k.

Here is how I, II, and III go together to give the main theorem. For each $k \geq 1$ consider the fibration

$$X_k \to BGL_\infty(\mathcal{O}/\mathfrak{p}^k)^+ \to BGL_\infty(\mathcal{O}/\mathfrak{p})^+$$

where X_k is a connected H-space. By the stability results of [9] and [11] the homology, and therefore homotopy, of $BGL_\infty(\mathcal{O}/\mathfrak{p}^k)^+$ is finite in each dimension. Consequently each $\pi_i(X_k)$ is a finite p-group as it is well-known [8] that all the homotopy of $BGL_\infty(\mathcal{O}/\mathfrak{p})^+$ is prime to p and moreover that

$$H_i(X_k) = H_i(X_k;Z_p) = H_i(GL_\infty(\mathcal{O}/\mathfrak{p}^k);Z_p).$$

It follows from (I) that

$$\lim_{\overleftarrow{k}} H_i(X_k) = \lim_{\overleftarrow{k}} H_i(GL_\infty(\mathcal{O}/\mathfrak{p}^k);Z_p)$$

is a finitely generated Z_p-module. By (II) we get finite generation over Z_p for $V_i = \lim_{\overleftarrow{k}} \pi_i(X_k)$. Since the sequences

$$0 \to \pi_i(X_k) \to \pi_i BGL_\infty(\mathcal{O}/\mathfrak{p}^k)^+ \to \pi_i BGL_\infty(\mathcal{O}/\mathfrak{p})^+ \to 0$$

are exact and all the homotopy groups are finite, we obtain from [1, §3] upon taking inverse limits the exact sequence

$$0 \to V_i \to K_i^{top}(\mathcal{O}) \to K_i(\mathcal{O}/\mathfrak{p}) \to 0$$

for $i \geq 1$. Propositions I and III now given the rank of $V_i \otimes_{Z_p} Q_p$ as required.

References

1. M. F. Atiyah, Characters and cohomology of finite groups, Pub. Math. I.H.E.S., No. 9, 1961.

2. A. Borel, Stable real cohomology of arithmetic groups, Annales scientifiques de l'École Normale Supérieure 4e série, t. 7, fasc. 2, 1974.

3. J. Coates, On the values of the p-adic zeta function at the odd positive integers, to appear.

4. Sze-Tun Hu, Homotopy Theory, Academic Press, N. Y., 1959.

5. M. Lazard, Groupes analytiques p-adiques, Pub. Math. I.H.E.S. No. 26, 1965.

6. John W. Milnor and John C. Moore, On the structure of Hopf algebras, Ann. of Math. Ser. 2, Vol. 81, No. 2 (1965), pp. 211-264.

7. C. C. Moore, Group extensions of p-adic and adelic linear groups, Pub. Math. I.H.E.S., No. 35, 1968.

8. D. Quillen, On the cohomology and K-theory of the general linear groups over a finite field, Ann. of Math. 96(1972), pp. 552-586.

9. ________, to appear.

10. J. B. Wagoner, Homotopy theory for the p-adic special linear group, Comm. Math. Helv. 50(1975), pp. 535-559.

11. ________, Stability for homolopy of the general linear group of a local ring, to appear in TOPOLOGY.

12. ________, Delooping the continuous K-theory of a valuation ring, preprint, Univ. of Calif., Berkeley.

COHOMOLOGY OF GROUPS

This is a report on the activity of a seminar on group cohomology held during the conference. The speakers were R.C. Alperin, K.S. Brown, L. Evens, Z. Fiedorowicz, C. Soulé, and J.B. Wagoner. Summaries provided by the speakers appear below, following some introductory remarks.

Algebraic K-theory, as is well-known, is intimately related to the cohomology theory of groups. This relation is transparent in "classical" algebraic K-theory, where one has $K_1(R) = H_1(GL(R))$ and $K_2(R) = H_2(E(R))$, cf. [13]. A more subtle relation occurs in Quillen's higher K-theory (cf. [8]), where $K_i(R)$ for $i \geq 1$ is the i-th homotopy group of a certain space $BGL(R)^+$ whose cohomology (with constant coefficients) is the same as that of the group $GL(R)$. This relation was exploited by Quillen [14] in his calculation of the K-groups of a finite field and by Borel [1] in his calculation of the ranks of the K-groups of a ring of algebraic integers.

If one works instead with Quillen's second definition of the higher K-groups [15], then group cohomology arises again, in a somewhat different way, at least if R is a Dedekind ring (cf. [16]). This connection was exploited by Quillen [16] in his proof that the K-groups of a ring of algebraic integers are finitely generated and by Lee and Szczarba ([11], [12]) in their calculation of $K_3(\mathbb{Z})$.

The connections between K-theory and group cohomology motivated, at least in part, much of the research summarized below. The reader should be warned, however, that this research does not necessarily have any immediate application to K-theory. Indeed, in several of the cases it remains an important open problem to work out such applications.

K.S. Brown

Summaries of the talks

R.C. ALPERIN - Stability for $H_2(SU_n)$

(See Alperin's article elsewhere in these proceedings.)

* * *

K.S. BROWN - Tate cohomology of infinite groups

Tom Farrell [7] has recently discovered that Tate's cohomology theory for finite groups can be extended to a large class of infinite groups. I would like to briefly describe Farrell's theory and then mention some results I have obtained using it [4].

Let Γ be a group of finite virtual cohomological dimension n in the sense of Serre [19]. Farrell's theory associates cohomology groups $\hat{H}^i(\Gamma, M)$ to any integer i and any Γ-module M, and it has formal properties analogous to those of the Tate theory. One has $\hat{H}^i = H^i$ for $i > n$, and $\hat{H}^n(\Gamma, M)$ is the cokernel of the transfer map $H^n(\Gamma', M) \to H^n(\Gamma, M)$, where Γ' is a torsion-free subgroup of finite index. (This cokernel is independent of the choice of Γ'.) In case Γ is a virtual duality group (e.g., an arithmetic group, such as $GL_r(\mathbb{Z})$), one can also explicitly describe $\hat{H}^i$ for $i < 0$ in terms of the homology functor H_{n-1-i}. For $0 \leq i \leq n-1$ it is harder to describe $\hat{H}^i$, and one can only say (still assuming that Γ is a virtual duality group) that there is an exact sequence relating $\{H^i\}$, $\{\hat{H}^i\}$, and $\{H_{n-1-i}\}$. Of course, this problem does not arise in the classical theory (Γ finite), which is the case $n = 0$.

The Farrell groups can be non-trivial only if Γ has torsion. More precisely, the Farrell groups are always torsion-groups, and one can show fairly easily that they can have p-torsion only if Γ has p-torsion. My results arise from an attempt to explain exactly how the torsion in Γ produces cohomology.

Let S be the set of non-trivial finite subgroups of Γ. Then S is a partially ordered set with Γ-action (by conjugation). Introducing an equivariant version of Farrell's theory, one obtains groups $\hat{H}^i_\Gamma(S)$; one can allow here an arbitrary Γ-module of coefficients, which I have suppressed from the notation.

Theorem 1. There is an isomorphism $\hat{H}^*(\Gamma) \approx \hat{H}^*_\Gamma(S)$.

Roughly speaking, this result relates the Farrell cohomology of Γ to that of the normalizers of the non-trivial finite subgroups of Γ. The next result says that if one is interested for a fixed prime p in the p-primary component $\hat{H}^*(\Gamma)_{(p)}$ of the Farrell cohomology, then it is enough to look at the finite subgroups which are p-groups:

Theorem 2. Let S_p be the set of non-trivial finite p-subgroups of Γ. Then $\hat{H}^*(\Gamma)_{(p)} \approx \hat{H}^*_\Gamma(S_p)_{(p)}$.

The proofs of Theorems 1 and 2 are similar to the proofs of the analogous results on Euler characteristics in [2] and [3]. The essential feature of the Farrell theory that makes them go through is the fact that the equivariant Farrell cohomology $\hat{H}^*_\Gamma(X)$ is trivial if Γ acts freely on X.

For a simple example where Theorem 2 yields interesting results, one can take $\Gamma = SL_3(\mathbb{Z})$, $p = 3$. The ordered set S_3 is discrete and one can quite easily compute $\hat{H}^*_\Gamma(S_3)$. In particular, this leads to a determination of $H^*(\Gamma, \mathbb{Z})$, modulo 2-torsion. (See Soulé's summary below for more precise results.) One can also recover some of the Lee-Szczarba results [11] which they used to calculate the odd torsion in $K_3(\mathbb{Z})$.

Unfortunately, it does not seem possible at the moment to apply Farrell cohomology theory to the study of $K_i(\mathbb{Z})$ for $i > 3$. One would need for this a better understanding of $\hat{H}^i(GL_r(\mathbb{Z}))$ for $0 \leq i \leq n-1$, where $n = \frac{r(r-1)}{2}$ is the virtual cohomological dimension of $GL_r(\mathbb{Z})$.

* * *

L. EVENS - Chern classes of representations of finite groups

This is a report on joint work with Daniel S. Kahn.

For G a finite group, H a subgroup, ξ a (unitary) representation of H, one would like a formula for the Chern classes of the induced representation $\rho = \mathrm{ind}_{H\to G}\xi$. A partial answer to this question was given in [6]. Namely, if $\xi \in H^2(H,\mathbb{Z}) = \hat{H}$ is 1-dimensional,

(1) $c_i(\mathrm{ind}\ \xi) = \mathcal{N}_i(\xi)$ + additional terms.

$\mathcal{N}_i(\xi)$ is defined in terms of a multiplicative generalization of transfer studied in [5] while the additional terms involve the Chern classes $c_i(\pi_j)$ where π_j is the natural representation of $\mathcal{S}_j$ (symmetric group) by permutation matrices. Kahn and I can now give the orders of the $c_i(\pi_j)$, and also we have derived a formula analagous to (1) for ξ of $\deg > 1$ but where H is normal in G of index p. Our work was motivated by results of C.B. Thomas [21] applying bounds of Grothendieck [9] for rational representations to the $c_i(\pi_j)$. In many cases, we show that the Grothendieck bounds are attained exactly for these classes.

In more detail, let $\ell = p^m$, p prime. Then the p-primary component of $c_i(\pi_\ell)$ is of order

$$\begin{cases} 1 & i \not\equiv 0 \bmod p-1 \\ p^{1+\nu_p(i)} & i \equiv 0 \bmod p-1 \end{cases}$$

for p odd, and of order

$$\begin{cases} 2^{1+\nu_2(i)} & i \leq 2^m - 1 \\ 1 & i = 2^m \end{cases}$$

for $p=2$. (For p odd, this is what Grothendieck gives as upper bounds.)

In the process of deriving these results, we were able to prove theorems like the following which may be of independent interest.

Theorem. Let G be finite, H a normal subgroup of prime index p, and let ξ be a unitary representation of H of degree n. Write $\rho = \mathrm{ind}_{H\to G}\xi$. Then

$$c(\rho) = \mathcal{N}(c(\xi)) + \sum_{t=1}^{n-1} \mathcal{N}(c_t(\xi))[1-\gamma^{p-1})^{n-t} - 1]$$

$$+ [(1-\gamma^{p-1})^n - 1] .$$

Here $\gamma = \inf_{G/H\to G}\mu$, where μ is a generator of $H^2(G/H,\mathbb{Z})$.

It should be noted that to understand the structure of the cohomology ring $H^*(G,\mathbb{Z})$ one is forced to investigate "multiplicative" results of the type described above. However, there is a corresponding "additive" theory investigated by Knopfmacher [10], Thomas [21], and others. One may define integral classes $s_i(\rho) \in H^{2i}(G,\mathbb{Z})$ which are related to the components of the Chern character. Then it is reasonable to consider whether

$$s_k(\mathrm{ind}\ \xi) - tr_{H\to G}(s_k(\xi))$$

is zero or not. The answer is no, but Knopfmacher showed that for each k the difference is annihilated by an integer depending only on k. Knopfmacher also conjectured that the least integer with this property is

$$N_k = \frac{\Pi p^{[\frac{k}{p-1}]}}{k!} .$$

In 1971, in his (unpublished) thesis [17], Roush proved that the correct bound is in fact

$$\bar{N}_k = \prod_{p|N_k} p \ .$$

Being unaware of Roush's thesis, Kahn and I reproved this fact by substantially independent methods related to our arguments mentioned earlier in this report.

* * *

Z. FIEDOROWICZ - Homology of classical groups over a finite field

(See the article by Fiedorowicz and Priddy elsewhere in these proceedings.)

* * *

C. SOULÉ - The cohomology of $SL_3(\mathbb{Z})$

Let $p\mathbb{Z}_r$ denote the product of p copies of the group $\mathbb{Z}/r\mathbb{Z}$. The integral cohomology of $\Gamma = SL_3(\mathbb{Z})$ is the following:

n	$H^n(SL_3(\mathbb{Z}))$
12m+1	$(6m)\mathbb{Z}_2$
12m+2	$(6m)\mathbb{Z}_2$
12m+3	$(6m+2)\mathbb{Z}_2$
12m+4	$2\mathbb{Z}_3+2\mathbb{Z}_4+(6m)\mathbb{Z}_2$
12m+5	$(6m+1)\mathbb{Z}_2$
12m+6	$(6m+4)\mathbb{Z}_2$
12m+7	$(6m+3)\mathbb{Z}_2$
12m+8	$2\mathbb{Z}_3+2\mathbb{Z}_4+(6m+1)\mathbb{Z}_2$

(cont.)

n	$H^n(SL_3(\mathbb{Z}))$
12m+9	$(6m+5)\mathbb{Z}_2$
12m+10	$(6m+5)\mathbb{Z}_2$
12m+11	$(6m+4)\mathbb{Z}_2$
12m+12	$2\mathbb{Z}_3+2\mathbb{Z}_4+(6m+5)\mathbb{Z}_2$

(cont.)

To make this computation one constructs a "good" subspace Y of the symmetric space X associated to Γ (for $SL_2(\mathbb{Z})$, such a construction is due to J-P. Serre [18]). The space Y is a retract of X; its quotient by Γ is compact and homotopically trivial [20]. The classical spectral sequence given by the action of Γ on Y degenerates (but to prove this one must distinguish the 2-torsion from the 3-torsion) and it remains to compute the cohomology of some finite subgroups of $SL_3(\mathbb{Z})$.

By the same methods one can compute the Farrell cohomology $\hat{H}^i(\Gamma,M)$ (see [7], [4]). When $M=\mathbb{Z}$, $\hat{H}^i(\Gamma)=H^i(\Gamma)$ for $i>0$, and $\hat{H}^0(\Gamma) = 2\mathbb{Z}_3+\mathbb{Z}_8+\mathbb{Z}_4$. We deduce from this (cf. [4]) that $H_0(GL_3(\mathbb{Z}),St_3)=H_1(GL_3(\mathbb{Z}),St_3)=0$, $H_2(GL_3(\mathbb{Z}),St_3)=\mathbb{Z}/12\mathbb{Z}$, and $H_3(GL_3(\mathbb{Z}),St_3) = \mathbb{Z} + 2\mathbb{Z}_2$. The two subgroups $\mathbb{Z}_3$ of $H_3(GL_3(2))_{(3)} = H^4(SL_3(\mathbb{Z}))_{(3)}$, which come from the two conjugacy classes of subgroups of $SL_3(\mathbb{Z})$ of order 3, map isomorphically onto the 3-torsion of $K_3(\mathbb{Z})=\mathbb{Z}/48\mathbb{Z}$.

Furthermore, the action of $SL_3(\mathbb{Z})$ on Y implies that $SL_n(\mathbb{Z})$ $(n\geq 0)$ is the sum of a finite number of its finite subgroups, amalgamated over their intersections.

* * *

J.B. WAGONER - Homology stability

Let A be any associative ring with identity. In algebraic

K-theory it is useful to know whether the chain of homomorphisms

$$\to H_n(GL(k,A)) \to H_n(GL(k+1,A)) \to$$

eventually stabilizes to a sequence of isomorphisms. For homology with integer coefficients D. Quillen has recently proved a sharp stability theorem for fields with at least three elements and, more generally, has a stability result for any local ring A. However, in the local ring case the stability range seemed to depend on n and on A. We extend his methods to obtain a stability theorem depending only on n and not on A. For homology with rational coefficients A. Borel has proved stability when A is a ring of integers in a number field [1].

<u>Theorem. Let</u> A <u>be a local ring. For homology with integer coefficients</u>

$$H_n(GL(md,A)) \to H_n(GL((m+1)d,A))$$

<u>is an isomorphism provided</u> $d \geq n+3$ <u>and</u> $m > 0$ <u>satisfy</u> $\frac{m}{d} > 2n+3$.

For the details see [22]. The idea of the proof is that whenever a discrete group G acts on a highly connected complex there is a well-known spectral sequence converging to zero which gives information about $H_*(G)$ in a certain range. Applying this to $G_m = GL(md,A)$ acting on the <u>space of d-fold unimodular vectors</u> $Un_d(A^{md})$ and then comparing spectral sequences - or more precisely, constructing a relative spectral sequence - yields the vanishing of $H_*(G_{m+1},G_m)$ in a certain range by an inductive argument where homology is taken with coefficients in a fixed but arbitrary algebraically closed field. The theorem then follows from the Bockstein homology exact sequence for $0 \to \mathbb{Z} \to \mathbb{Q} \to \mathbb{Q}/\mathbb{Z} \to 0$.

The space of d-fold unimodular vectors $Un_d(A^{md})$ is defined as follows: Let P be a finitely generated projective module over A.

A sequence of elements $(p_1,\dots,p_k)$ in P is <u>unimodular</u> provided the map $A^k \to P$ given by $(a_1,\dots,a_k) \to \sum_i a_i p_i$ is an injection onto a direct summand. Let $d > 0$. Let $Un_d(P)$ be the simplicial complex whose k-simplices are ordered (k+1)-tuples $(v_0,\dots,v_k)$ where each v_i is itself an ordered sequence of exactly d elements in P such that the resulting sequence of $(k+1)\cdot d$ elements in P is unimodular. $Aut_A(P)$ acts on the left of $Un_d(P)$ by the formula

$$\alpha\cdot(v_0,\dots,v_k) = (\alpha(v_0),\dots,\alpha(v_k))$$

where if $v = (p_1,\dots,p_d)$ is a unimodular d-tuple then $\alpha(v) = (\alpha(p_1),\dots,\alpha(p_d))$.

<u>Proposition</u>. <u>Let</u> A <u>be a local ring and let rank</u> $P = md$. <u>Then</u> $Un_d(P)$ <u>is</u> $[\frac{m}{d} - 2]$ <u>- connected</u>.

When $d = 1$ the definition and proposition are due to Quillen. The reason for using d-fold unimodular vectors is that the proof of the main theorem requires the ring A to have "enough units" in the spirit of Quillen's appendix to [23]. To accomplish this A is replaced by the ring $M_d(A)$ of $d\times d$ matrices with entries in A, and $GL(m,A)$ is replaced by $GL(md,A) = GL(m,M_d(A))$. The key tool is

<u>Quillen's Lemma</u>. <u>Let</u> $d > 0$ <u>be given and let</u> k <u>be any algebraically closed field. Then there is an order</u> D <u>in a number field of degree</u> d <u>over</u> $\mathbb{Q}$ <u>with the following properties: Given any D-module</u> N, <u>let the group of units</u> D^* <u>act on it by multiplication and let the group homology</u> $H_*(N;k)$ <u>be endowed with the induced action of</u> D^*. <u>Then for each</u> t, $H_t(N;k)$ <u>is a direct sum of one dimensional representations of</u> D^* <u>over</u> k . <u>Moreover</u>, $H_t(N;k)$ <u>does not contain the trivial representation for</u> $0 < t < d$.

For the proof see the appendix to [23].

*　　　*　　　*

References

1. A. BOREL, Stable real cohomology of arithmetic groups, Ann. Sci. École Norm. Sup. (4) 7 (1974), 235-272.

2. K.S. BROWN, Euler characteristics of discrete groups and G-spaces, Invent. Math. 27 (1974), 229-264.

3. K.S. BROWN, Euler Characteristics of groups: The p-fractional part, Invent. Math. 29 (1975), 1-5.

4. K.S. BROWN, High dimensional cohomology of discrete groups, Proc. Nat. Acad. Sci. USA, to appear.

5. L. EVENS, A generalization of the transfer map in the cohomology of groups, Trans. Amer. Math. Soc. 108 (1963), 54-65.

6. L. EVENS, On the Chern classes of representations of finite groups, Trans. A.M.S. 115 (1965), 180-193.

7. F.T. FARRELL, An extension of Tate cohomology to a class of infinite groups, to appear.

8. S.M. GERSTEN, Higher K-theory of rings, Lecture Notes in Mathematics 341, pp. 3-42, Springer-Verlag, Berlin, 1973.

9. A. GROTHENDIECK, Classes de Chern et representations linéaires des groupes discrets, Dix exposés sur la cohomologie des schemas, North Holland, Amsterdam, 1968.

10. J. KNOPFMACHER, On Chern classes of representations of finite groups, J. London Math. Soc. 41 (1966), 535-541.

11. R. LEE and R. SZCZARBA, On the homology of congruence sugbroups and $K_3(\mathbb{Z})$, Proc. Nat. Acad. Sci. USA 72 (1975), 651-653.

12. R. LEE and R. SZCZARBA, On $K_3(\mathbb{Z})$, preprint, Yale University.

13. J. MILNOR, Introduction to algebraic K-theory, Ann. Math. Studies 72, Princeton University Press, 1971.

14. D.G. QUILLEN, On the cohomology and K-theory of the general linear groups over a finite field, Ann. of Math. 96 (1972), 552-586.

15. D.G. QUILLEN, Higher algebraic K-theory: I, Lecture Notes in Mathematics 341, pp. 85-147, Springer-Verlag, Berlin, 1973.

16. D.G. QUILLEN, Finite generation of the groups K_i of rings of algebraic integers, Lecture Notes in Mathematics 341, pp. 179-198, Springer-Verlag, Berlin, 1973.

17. F. ROUSH, Transfer in generalized cohomology theories, Princeton thesis, 1971.

18. J-P. SERRE, Arbres, amalgames, SL_2, mimeographed notes, Collège de France, 1968-69.

19. J-P. SERRE, Cohomologie des groupes discrets, Ann. Math. Studies 70, pp. 77-169, Princeton University Press, 1971.

20. C. SOULÉ, Cohomologie de $SL_3(\mathbb{Z})$, C.R. Acad. Sci. Paris sér. A 280 (1975), 251-254.

21. C.B. THOMAS, An integral Riemann-Roch formula for flat line bundles, Proc. London Math. Soc., to appear.

22. J.B. WAGONER, Stability for homology of the general linear group of a local ring, preprint, University of California, Berkeley.

23. J.B. WAGONER, Equivalence of algebraic K-theories, preprint, University of California, Berkeley.

On the homology and cohomology of the orthogonal and symplectic groups over a finite field of odd characteristic

by Jack M. Shapiro

In [5] a method is introduced to compute the homology and cohomology of $GL_n(k)$, the general linear group over a finite field, k. This is done by first computing the homology and cohomology of the space $F\Psi^q$. This space is related to the classifying space of the infinite unitary group BU, in fact it is the homotopy fibre of the map $\Psi^q - 1: BU \to BU$. Using these results we make the analogous computations for the orthogonal and symplectic groups. The coefficients are taken in the integers mod ℓ, where ℓ is an odd prime satisfying a special relationship with the order of k. This note can thus be seen as a compliment to [6] in which the results of [4] were extended to the other classical groups.

Throughout the exposition an attempt is made to strike a balance between recopying the many theorems and propositions of [5] and merely stating that proofs are analogous to those found in [5]. We have tried to include just enough information so that the exposition can be read, without full detail of the proofs, independent of [5].

At this conference a talk was delivered by Z. Fiedrowicz on the same topic. His talk was based on joint work with S. Priddy and represents another generalization of Quillen's results done from the vantage point of loop spaces.

§1

In this section we include the information regarding the space $F\Psi^q$ that will be needed later on. $F\Psi^q$ is a space defined in [5] as the homotopy theoretical fixed point set of the qth Adams operation, $\Psi^q: BU \to BU$. A map can be defined, $\phi: F\Psi^q \to BU$ which is essentially the inclusion of a fibre in a fibration. It is well known that any complex representation of a group G induces a map in the homotopy category $BG \to BU$. The essential fact about the map ϕ is that for a representation which is fixed by Ψ^q the induced map from BG to BU factors through ϕ (i.e., the representation induces a map $BG \to F\Psi^q$).

For q, a power of an odd prime p, we will describe the homology and cohomology of $F\Psi^q$ with coefficients in $Z/\ell Z$, ℓ an odd prime different from p (in fact cohomology and homology for all spaces discussed will be mod ℓ). Let us recall that

$$H^*(BU) \cong P[c_1, c_2, \ldots]$$

the polynomial ring over $Z/\ell Z$ in the universal Chern classes, c_i. It is shown in [5] that if $i = jr$, where r is the order of $q \bmod \ell$, then c_{jr} pulls back, via ϕ, to a non-zero class in $H^{2jr}(F\Psi^q)$ also denoted by c_{jr}. By using these classes, together with appropriate exact sequences, odd dimensional classes $e_{jr} \in H^{2jr-1}(F\Psi^q)$ are also constructed.

<u>Theorem 1</u> ([5], theorem 1)

$H^*(F\Psi^q) \cong P[c_r, c_{2r}, \ldots] \otimes \Lambda[e_r, e_{2r}, \ldots]$ the tensor product of a polynomial algebra and an exterior algebra.

For the integer r introduced above let C be the cyclic group of order $q^r - 1$, $Z/(q^r - 1)Z$. Let $v \in H^1(BC)$ be the class of the homomorphism $C \to Z/\ell Z$ sending $1 \to 1$ and $u = c_1(\xi) \in H^2(BC)$ the first Chern class of the complex representation $\xi : C \to \mathbb{C}$ defined by $\xi(1) = \exp(2\pi i/q^r - 1)$. Then ([5])

$$H^*(BC) \cong P[u] \otimes \Lambda[v].$$

Let $W + \overline{W} = \xi + \xi^q + \ldots + \xi^{q^{r-1}} + \xi^{-1} + \xi^{-q} + \ldots + \xi^{-q^{r-1}}$ be the $2r$-dimensional complex representation of C which extends the representation W of [5] to one which is fixed under the homomorphism $a \to a^{-1}$ of C. Since $W + \overline{W}$ is also fixed under Ψ^q it induces a map $BC \to F\Psi^q$ which will also be denoted by $W + \overline{W}$. In the following theorem $c_i(W + \overline{W})$ and $e_i(W + \overline{W})$ represent the images of the classes c_i and e_i in $H^*(BC)$ under the homomorphism $(W + \overline{W})^* : H^*(F\Psi^q) \to H^*(BC)$.

Theorem 2

(i) If the order of $q \bmod \ell$, r, is odd then:

$$c_i(W+\overline{W}) = \begin{cases} 1 & i=0 \\ -u^{2r} & i=2r \\ 0 & \text{otherwise} \end{cases}$$

$$e_{jr}(W+\overline{W}) = \begin{cases} -2u^{2r-1}v & j=2 \\ 0 & \text{otherwise} \end{cases}$$

(ii) If r is even then:

$$c_i(W+\overline{W}) = \begin{cases} 1 & i=0 \\ -2u^r & i=r \\ u^{2r} & i=2r \\ 0 & \text{otherwise} \end{cases}$$

$$e_{jr}(W+\overline{W}) = \begin{cases} -2u^{r-1}v & j=1 \\ 2u^{2r-1}v & j=2 \\ 0 & \text{otherwise} \end{cases}$$

Proof

This follows directly from proposition 1 and the formulas of section 5 in [5].

Finally, let us see how $W+\overline{W}$ relates $H_*(BC)$, the homology of BC, to $H_*(F\Psi^q)$. Let $x = (-1)^{r-1}u^r$ and $y = (-1)^{r-1}u^{r-1}v$ be the classes of $H^*(BC)$ introduced in [5]. By duality there exists ξ'_j and η'_j in $H_*(BC)$ dual to x^j and $x^{j-1}y$, respectively. Let us define classes in $H_*(F\Psi^q)$ by

$$\xi''_j = (W+\overline{W})_*(\xi'_j) \qquad \eta''_j = (W+\overline{W})_*(\eta'_j).$$

Note that while these are not the classes ξ_j, η_j introduced in [5] they are closely related (e.g., $\xi''_2 = \xi_1^2$, if r is odd).

Theorem 3

If r is odd then $P[\xi_2'', \xi_4'', \ldots] \otimes \Lambda[\eta_2'', \eta_4'', \ldots]$ is a subalgebra of $H_*(F\Psi^q)$ (i.e., the obvious monomials are linearlly independent in $H_*(F\Psi^q)$).

Proof

This follows by introducing the algebra homomorphism

$$h_N : H_*(F\Psi^q) \to P[t_1, t_2, \ldots] \otimes \Lambda[s_1, s_2, \ldots]$$

of [5] and showing that the image of the ξ_{2j}'' and η_{2j}'' are algebraically independent in the image algebra. The calculations are identical to those in [5] and yield $h_N(\xi_{2j}'') = j^{th}$ elementary symmetric function σ_j of $t_1^2, t_2^2, \ldots, t_N^2$ and $h_N(\eta_{2j}'') = d\sigma_j$.

Remark: A similar statement can be made for the case r even but this will not be needed for our exposition. The important point is that by the above calculations (Theorem 2) $0 \neq \xi_1'' \in H_{2jr}(F\Psi^q)$.

§2

Let k be the finite field with q elements and μ_ℓ a primitive ℓ^{th} root of unity in the algebraic closure of k. The simple extension of k, $k(\mu_\ell)$, is of degree r over k, where r is the order of $q \bmod \ell$. The multiplicative group of non-zero elements, $k(\mu_\ell)^*$, is a cyclic group of order q^r-1 and is therefore isomorphic to C.

If $e_1, \ldots, e_n$ is the standard basis for the vector space k^n over k we define a symmetric (skew-symmetric) nondegenerate bilinear form on k^n by $\langle e_i, e_{i+m} \rangle = 1$ $1 \le i \le m$, $n = 2m$ or $2m + 1$, and $\langle e_i, e_j \rangle = 0$ otherwise. If $n = 2m + 1$ we add that $\langle e_n, e_n \rangle = 1$ (for the skew-symmetric case $n = 2m$ and $\langle e_{i+m}, e_i \rangle = -1$). A non-singular linear transformation $T: k^n \to k^n$ will be an orthogonal (symplectic) transformation if $\langle Tv, Tw \rangle = \langle v, w \rangle$ for all $v, w \in k^n$. We denote the orthogonal (symplectic) transformations by $O_n(k)(Sp_{2m}(k))$. We point out that if the standard orthogonal inner product (i.e., $\langle e_i, e_j \rangle = \delta_{ij}$) is chosen and we take the group of transformations keeping that form invariant we get a group isomorphic to $O_n(k)$. (See e.g. [2].)

Let $k(\mu_\ell)^* \cong C$ act on $k(\mu_\ell)^2$ by $\alpha(v,w) = (\alpha v, \alpha^{-1} w)$, $\alpha \in k(\mu_\ell)^*$, $v,w \in k(\mu_\ell)$. Restriction of scalars induces a representation of $k(\mu_\ell)^*$ on k^{2r}. A basis for $k(\mu_\ell)$ over k can be found so that the matrix representation of multiplication by α, $A(\alpha)$, is a symmetric matrix ([1], corollary to lemma 3). Identifying this basis with $e_1,\ldots,e_r$ on the first factor of $k(\mu_\ell)^2$ and with $e_{r+1},\ldots,e_{2r}$ on the second factor gives us a matrix representation for the action of $k(\mu_\ell)^*$ on k^{2r} (described above) as

$$\begin{pmatrix} A(\alpha) & 0 \\ 0 & A(\alpha)^{-1} \end{pmatrix}$$

Since $A(\alpha)$ is symmetric this matrix is orthogonal (symplectic). If we embed the algebraic closure of k in the complex numbers (as roots of unity) we can use the embedding to induce a complex representation, called the Brauer lift, from any modular representation over k ([4]). The "Brauer lift" of the above representation is the complex representation $W + \overline{W}$ previously described ([5], proposition 3 (ii)).

§3

We will now restrict our attention to the group $O_n(k)$, the analogous results for $Sp_{2m}(k)$ being identical. Let us also point out that only in the case r odd do we get the full results on homology and cohomology. We therefore restrict ourselves to that case and will remark on the case r even only at the end.

As in [5] we consider the algebra

$$\bigoplus_{n \geq 0} H_*(O_n(k)).$$

The algebra structure is derived from the direct sum homomorphism

$0_m(k) \times 0_n(k) \to 0_{m+n}(k)$. The orthogonal action of C on k^{2r} previously described (§2) yields an embedding of C in $0_{2r}(k)$.

Let π be the cyclic group of order r and define an action of π on C by having the generator of π raise each element of C to the qth power. We can also define a Z_2 action on C by the homomorphism $a \to a^{-1}$, $a \in C$. The action of both these groups on C, embedded in $0_{2r}(k)$, can be realized by conjugation in $0_{2r}(k)$. This implies that we obtain a homomorphism

$$H_*(C)_{\pi \times Z_2} \to H_*(0_{2r}(k))$$

where the subscript denotes the coinvariants for the group action. $H^*(C)^{\pi \times Z_2}$ has a basis consisting of $\{x^{2j}, x^{2j-1}y\}$, therefore $H_*(C)_{\pi \times Z_2}$ has a dual basis consisting of elements $\xi'_{2j} \in H_{4jr}(C)$ and $\eta'_{2j} \in H_{4jr-1}(C)$. We will denote their images under the above homomorphism as

$$\xi''_{2j} \in H_{4jr}(0_{2r}(k)) \qquad j \geq 0$$

$$\eta''_{2j} \in H_{4jr-1}(0_{2r}(k)) \qquad j \geq 1$$

We now consider the homomorphism

$$\bigoplus_{n \geq 0} H_*(0_n(k)) \to H_*(F\Psi^q)$$

induced by the "Brauer lift" of the natural representation of $0_n(k)$ on k^n. If $e \in H_0(0_1(k))$ is the distinguished generator then $\xi_0 = e^{2r}$. By the remark at the end of section 2, the ξ''_{2j}, η''_{2j} lift to the corresponding elements of $H_*(F\Psi^q)$.

The following theorem and corollary should be thought of as an addendum to the theorems of section 8 and 9 in [5].

<u>Theorem 4</u> There is an algebra isomorphism

$$P[e,\xi''_2,\xi''_4,\ldots] \otimes \Lambda[\eta''_2,\eta''_4,\ldots] \to \bigoplus_{n \geq 0} H_*(0_n(k)).$$

<u>Proof</u>

The fact that the map is 1-1 follows from Theorem 3. For the surjectivity we must show that a "dual" map for cohomology is injective (see [5],§8).

This "dual" map is derived as follows.

Suppose $n = 2m$ or $2m+1$ and $m = tr + \delta$, $0 \leq \delta \leq r-1$. We get an embedding of C^t in $O_n(k)$ by taking the direct sum t-times of the action of C on k^{2t} plus a trivial action on k^{n-2tr}. Define an action of the semi-direct product $\Sigma_t \tilde{\times} (\pi \times Z_2)^t$ on C^t in the obvious way, where Σ_t (the symmetric group) permutes the factors. Once again this action on C^t, embedded in $O_n(k)$, can be realized by conjugation in $O_n(k)$. Hence we get a homomorphism

$$H^*(O_n(k)) \to H^*(C^t)^{\Sigma t \tilde{\times} (\pi \times Z_2)^t} .$$

It is for this "dual" map that we must prove injectivity to complete the theorem.

This can be done by stating and proving analogous lemmas to lemmas 12 and 13 of [5]. Translated to our case one states that the cohomology of $O_n(k)$ is detected by the family of abelian subgroups of exponent dividing ℓ^a, $a \geq 1$. The second then shows that any such subgroup is conjugate to a subgroup of C^t. In order to make this translation we need to make two observations.

(i) At a crucial point in the proof of the first lemma ([5] lemma 13) the index of the normalizer of C^t in $O_n(k)$ is shown to be an ℓ-adic unit. In our case this subgroup is

$$\Sigma_t \tilde{\times} (Z_2 \tilde{\times} C)^t$$

the semi-direct product, where Σ_t permutes the factors. The order of this subgroup is $t!\ 2^t(q^r-1)^t$ and will have as its index in $O_n(k)$ an ℓ-adic unit provided r is odd. We must use the fact that

$$\frac{q^{jr}-1}{j(q^r-1)}$$

is an ℓ-adic unit for $j \geq 1$ ([5]) together with the formulas for $|O_n(k)|$ ([4], §4).

(ii) For the second lemma (lemma 12) we begin where the proof in [5] ends. It is shown that any abelian subgroup of exponent ℓ^a is conjugate to a direct sum of r dimensional blocks, A_i, plus a trivial representation. Furthermore, each A_i comes from restriction of scalars of a 1-dimensional representation over $k(\mu_\ell)$. The rest follows by ([4], proof of theorem 4.4). Q.E.D.

Corollary

The homomorphism

$$H^*(O_n(k)) \to H^*(C^t)\Sigma t \tilde{\times}(\pi \times Z_2)^t$$

is an isomorphism. Furthermore, if $c_{jr}(k^n)$, $e_{jr}(k^n)$ denote the images of c_{jr}, $e_{jr} \in H^*(F\Psi^q)$ pulled back via the homomorphism induced by $BO_n(k) \to F\Psi^q$ (previously described), then

$$H^*(O_n(k)) \simeq P[c_{2r}(k^n),\ldots,c_{2tr}(k^n)] \otimes \Lambda[e_{2r}(k^n),\ldots,e_{2tr}(k^n)]$$

Proof See [5] theorem 4.

To conclude we make the following remarks.

(i) For the symplectic group the arguments are totally identical. All that has to be done is to keep a skew symmetric bilinear form in mind at the appropriate places. Instead of $\bigoplus_{n \geq 0} H_*(O_n(k))$ take $\bigoplus_{m \geq 0} H_*(Sp_{2m}(k))$ and theorem 4 plus the corollary can be rewritten replacing $Sp_{2m}(k)$ for $O_n(k)$.

(ii) The problem for r even is that we cannot be sure of injectivity for the crucial "dual" cohomology map. The analogous computations don't yield an ℓ-adic unit for the index of the normalizer. The surprising fact is that we do get a non-zero class in dimension 2r (see remark after theorem 3). This is especially surprising following the theorem in [3] which computes the homology of FO^q, the homotopy fibre of $\Psi^q-1:BO \to BO$.[1] It is shown there that even classes only appear in dimension 4jr. On the other hand if r is even then 2jr is divisible by 4 and in those dimensions we do get the universal Pontryagin classes in $H^*(BO)$.

(1) Priddy has informed me that Husemoller's result is not precisely correct as stated.

REFERENCES

1. E.A. Bender: Classes of matrices over an integral domain, Ill. J. Math. 11, 1967, 697-702.

2. R.W. Carter: Simple groups of Lie type, John Wiley and Sons, New York, 1972.

3. D. Husemoller: On the homology of the fibre of Ψ^q-1, Algebraic K-theory I, Battelle institute conference, 1972, Lecture Notes in Mathematics (341) Springer-Verlag, Berlin.

4. D. Quillen: The Adams conjecture, Topology 10, 1971, 67-80.

5. D. Quillen: On the cohomology and K-theory of the general linear group over a finite field, Anals of Mathematics, Vol. 96, 1972, 552-586.

6. J.M. Shapiro: On the cohomology of the classical linear groups, preprint.

Washington University
St. Louis, Missouri

Homology of Classical Groups over a Finite Field

by

Zbigniew Fiedorowicz and Stewart Priddy

Let $\mathbf{F}_q$ denote the field of q elements, $q = p^n$, p an odd prime. Let G_n, $0 \leq n \leq \infty$ denote one of the following classical groups over $\mathbf{F}_q$:

$GL_n(\mathbf{F}_q)$ — general linear group

$SL_n(\mathbf{F}_q)$ — special linear group: matrices with determinant = 1

$O_n^+(\mathbf{F}_q)$ — orthogonal group: matrices preserving the form

$$x_1^2 + x_2^2 + \cdots + x_n^2$$

$O_n^-(\mathbf{F}_q)$ — extraordinary orthogonal group: matrices preserving the form

$$\mu x_1^2 + x_2^2 + \cdots + x_n^2, \quad \mu \text{ a nonsquare}$$

$SO_n(\mathbf{F}_q)$ — special orthogonal group: orthogonal matrices with determinant = 1

$N_n(\mathbf{F}_q)$ — orthogonal matrices with spinor norm = 1 (see O'Meara [6])

$ND_n(\mathbf{F}_q)$ — orthogonal matrices with (spinor norm)×(determinant) = 1

$SN_n(\mathbf{F}_q) = SO_n(\mathbf{F}_q) \cap N_n(\mathbf{F}_q)$

$Spin_n(\mathbf{F}_q)$ — spinor group (see Dieudonné [2])

$Sp_{2n}(\mathbf{F}_q)$ — symplectic group: matrices preserving the alternating form

$$\langle x,y\rangle = \sum_{i=1}^{n}(x_{2i}y_{2i-1} - x_{2i-1}y_{2i})$$

$U_n(\mathbf{F}_{q^2})$ — unitary group: matrices over $\mathbf{F}_{q^2}$ preserving the Hermitian form

$$\langle x,y\rangle = \sum_{i=1}^{n} x_i y_i^q$$

$SU_n(\mathbf{F}_{q^2})$ — special unitary group: unitary matrices with determinant = 1.

We consider the following two problems:

(1) Identify the Quillen plus construction $(BG_\infty)^+$ as an infinite loop space.

(2) Compute the homology groups $H_*(BG_n; \Lambda)$ and cohomology rings $H^*(BG_n; \Lambda)$, $0 \leq n \leq \infty$.

Theorem 1. The following are fibration sequences of infinite loop spaces:

$$(BGL_\infty(\mathbf{F}_q))^+ \xrightarrow{\beta} BU \xrightarrow{\psi^q - 1} BU$$

$$(BSL_\infty(\mathbf{F}_q))^+ \xrightarrow{\beta} BSU \xrightarrow{\psi^q - 1} BSU$$

$$(BO_\infty^+(\mathbf{F}_q))^+ \coprod (BO_\infty^-(\mathbf{F}_q))^+ \xrightarrow{\beta} BO \xrightarrow{\psi^q - 1} BO$$

$$(BO_\infty^+(\mathbb{F}_q))^+ \xrightarrow{\beta} BO \xrightarrow{\psi^q - 1} BSO$$

$$(BSO_\infty(\mathbb{F}_q))^+ \xrightarrow{\beta} BSO \xrightarrow{\psi^q - 1} BSO$$

$$(BN_\infty(\mathbb{F}_q))^+ \approx (BND_\infty(\mathbb{F}_q))^+ \xrightarrow{\beta} BO \xrightarrow{\psi^q - 1} BSpin$$

$$(BSN_\infty(\mathbb{F}_q))^+ \xrightarrow{\beta} BSO \xrightarrow{\psi^q - 1} BSpin$$

$$(BSpin_\infty(\mathbb{F}_q))^+ \xrightarrow{\beta} BSpin \xrightarrow{\psi^q - 1} BSpin$$

$$(BSp_\infty(\mathbb{F}_q))^+ \xrightarrow{\beta} BSp \xrightarrow{\psi^q - 1} BSp$$

$$(BU_\infty(\mathbb{F}_{q^2}))^+ \xrightarrow{\beta} BU \xrightarrow{\psi^{-q} - 1} BU$$

$$(BSU_\infty(\mathbb{F}_{q^2}))^+ \xrightarrow{\beta} BSU \xrightarrow{\psi^{-q} - 1} BSU.$$

Here U, SU, Sp denote the unitary, special unitary and symplectic groups over the complex numbers; O, SO, Spin denote the orthogonal, special orthogonal and spinor groups over the real numbers. The map ψ^q is the Adams operation (see Adams [1]) and the map β is the Brauer lift (see Quillen [9]). All spaces are localized away from p.

We sketch the <u>proof</u>: Following Quillen [9] and May [7] we use Brauer lifting to obtain an infinite loop space map

$$\bar{\beta}: (BG_\infty)^+ \to F = \text{fibre } (\psi^q - 1).$$

For $\ell \neq p$, $H_*(F; \mathbb{Z}/\ell)$ may be computed using the Serre spectral sequence. Generators for $H_*(BG_\infty: \mathbb{Z}/\ell) = \varinjlim H_*(BG_n; \mathbb{Z}/\ell)$, described in Theorem 2, generate an algebra which is an upper bound for the homology algebra $H_*(BG_\infty^+; \mathbb{Z}/\ell)$. Using the Brauer character and Steenrod operations we show that these generators map under $\bar{\beta}_*$ to generators of the algebra $H_*(F; \mathbb{Z}/\ell)$, thus proving that $\bar{\beta}_*$ is an epimorphism of homology algebras. By dimension considerations this shows that our upper bound is also a lower bound and that $\bar{\beta}$ induces an isomorphism in homology.

This method was used by Quillen to obtain the first fibering. Our principal contribution is to the orthogonal case, where we deal with the fact that $\bar{\beta}$ is defined only up to an indeterminacy and that if chosen incorrectly will fail to be a homotopy equivalence. In fact, we show that if $\bar{\beta}$ is chosen to be an H-map then it will always be an equivalence.

We should point out that many of these fibrations (without the infinite loop structure) were first obtained by Friedlander [5].

Theorem 1 essentially solves (1). It identifies $(BG_\infty)^+$ as the homotopy fiber of a well known map between familiar spaces. Thus, for instance, the homotopy groups $\pi_i(BG_\infty)^+$ may be easily computed from the long exact sequence of homotopy groups of the fibration.

We have also obtained the following results on $H_*(BG_n,\Lambda)$.

Theorem 2. Let

$$\mathcal{Q}(\{G_n\};\ \Lambda) = \bigoplus_{n=0}^{\infty} H_*(BG_n;\ \Lambda)$$

be the homology algebra of the groups G_n under the operation of direct sum $G_n \times G_m \xrightarrow{\oplus} G_{n+m}$. Then

(a) $\mathcal{Q}(\{GL_n(\mathbb{F}_q)\};\ \mathbb{Z}/2) = \mathbb{Z}/2[\xi_i | i \geq 0] \otimes E[\eta_i | i \geq 1]$

where $\xi_i \in H_{2i}(BGL_1(\mathbb{F}_q);\ \mathbb{Z}/2)$, $\eta_i \in H_{2i-1}(BGL_1(\mathbb{F}_q);\ \mathbb{Z}/2)$

(b) $\mathcal{Q}(\{O_n^+(\mathbb{F}_q),\ O_n^-(\mathbb{F}_q)\};\ \mathbb{Z}/2) = \mathbb{Z}/2[v_i^+, v_i^- | i \geq 0]/(v_i^+)^2 = (v_i^-)^2$

where $v_i^\pm \in H_i(BO_1^\pm(\mathbb{F}_q);\ \mathbb{Z}/2)$

(c) $\mathcal{Q}(\{Sp_{2n}(\mathbb{F}_q)\};\ \mathbb{Z}/2) = \mathbb{Z}/2[\sigma_i | i \geq 0] \otimes E[\tau_i | i \geq 1]$

where $\sigma_i \in H_{4i}(BSp_2(\mathbb{F}_q);\ \mathbb{Z}/2)$, $\tau_i \in H_{4i-1}(BSp_2(\mathbb{F}_q);\ \mathbb{Z}/2)$

(d) $\mathcal{Q}(\{U_n(\mathbb{F}_{q^2})\};\ \mathbb{Z}/2) = \mathbb{Z}/2[\bar{\xi}_i | i \geq 0] \otimes E[\bar{\eta}_i | i \geq 1]$

where $\bar{\xi}_i \in H_{2i}(BU_1(\mathbb{F}_{q^2});\ \mathbb{Z}/2)$, $\bar{\eta}_i \in H_{2i-1}(BU_1(\mathbb{F}_{q^2});\ \mathbb{Z}/2)$.

In parts (e) – (h) let ℓ be an odd prime, $\ell \neq p$.

(e) $\mathcal{Q}(\{GL_n(\mathbb{F}_q)\};\ \mathbb{Z}/\ell) = \mathbb{Z}/\ell[[1], \alpha_i | i \geq 1] \otimes E[\beta_i | i \geq 1]$

where $[1] \in H_0(BGL_1(\mathbb{F}_q);\ \mathbb{Z}/\ell)$, $\alpha_i \in H_{2ir}(BGL_r(\mathbb{F}_q);\ \mathbb{Z}/\ell)$,

$\beta_i \in H_{2ir-1}(BGL_r(\mathbb{F}_q);\ \mathbb{Z}/\ell)$ and $r = \min\{m | q^m \equiv 1 (\mathrm{mod}\ \ell)\}$

(f) $\mathcal{Q}(\{O_n^+(\mathbb{F}_q),\ O_n^-(\mathbb{F}_q)\};\ \mathbb{Z}/\ell) = \mathbb{Z}/\ell[[1],\ [\bar{1}],\ \gamma_i | i \geq 1]/([1]^2 = [\bar{1}]^2)$

$\otimes E\{w_i | i \geq 1\}$, where $[1] \in H_0(BO_1^+(\mathbb{F}_q);\ \mathbb{Z}/\ell)$, $[\bar{1}] \in H_0(BO_1^-(\mathbb{F}_q);\ \mathbb{Z}/\ell)$,

$\gamma_i \in H_{4id}(BO_{2d}^\delta(\mathbb{F}_q);\ \mathbb{Z}/\ell)$, $w_i \in H_{4id-1}(BO_{2d}^\delta(\mathbb{F}_q);\ \mathbb{Z}/\ell)$ and

$d = \min\{m | q^{2m} \equiv 1 (\mathrm{mod}\ \ell)\}$, $\varepsilon = \pm 1$ according as $q \equiv \pm 1 (\mathrm{mod}\ 4)$ and

$\delta = \pm$ according as $(q/\varepsilon)^d \equiv \pm 1 (\mathrm{mod}\ \ell)$

(g) $\mathcal{Q}(\{Sp_{2n}(\mathbb{F}_q)\}; \mathbb{Z}/\ell) = \mathbb{Z}/\ell[[1], \lambda_i | i \geq 1] \otimes E[\mu_i | i \geq 1]$

where $[1] \in H_0(BSp_2(\mathbb{F}_q); \mathbb{Z}/\ell)$, $\lambda_i \in H_{4id}(BSp_{2d}(\mathbb{F}_q), \mathbb{Z}/\ell)$,

$\mu_i \in H_{4id-1}(BSp_{2d}(\mathbb{F}_q); \mathbb{Z}/\ell)$ and $d = \min\{m | q^{2m} \equiv 1 (\mathrm{mod}\ \ell)\}$

(h) $\mathcal{Q}(\{U_n(\mathbb{F}_{q^2}); \mathbb{Z}/\ell) = \mathbb{Z}/\ell[[1], \bar{\alpha}_i | i \geq 1] \otimes E[\bar{\beta}_i | i \geq 1]$

where $[1] \in H_0(BU_1(\mathbb{F}_{q^2}); \mathbb{Z}/\ell)$, $\bar{\alpha}_i \in H_{2is}(BU_s(\mathbb{F}_{q^2}); \mathbb{Z}/\ell)$,

$\beta_i \in H_{2is-1}(BU_s(\mathbb{F}_{q^2}); \mathbb{Z}/\ell)$ and $s = \min\{m | (-q)^m \equiv 1 (\mathrm{mod}\ \ell)\}$.

(It is useful to note in connection with parts (b) and (f) that the direct sum $O_n^{\varepsilon_1}(\mathbb{F}_q) \oplus O_m^{\varepsilon_2}(\mathbb{F}_q)$ lies in $O_{m+n}^{\varepsilon_1\varepsilon_2}(\mathbb{F}_q)$, $\varepsilon_i = \pm$.)

The <u>proof</u> may be outlined as follows: Except in the case $G = Sp(\mathbb{F}_q)$ and $\ell = 2$, the generators of $\mathcal{Q}(\{G_n\}; \mathbb{Z}/\ell)$ are obtained from a G-representation of a cyclic group. In the case $\mathcal{Q}(\{Sp_{2n}(\mathbb{F}_q)\}; \mathbb{Z}/2)$, the generators come from a 2-dimensional symplectic representation of a quaternionic group.

Let G_a denote the group in which the generators first appear (e.g. in (c),(g), $G_a = Sp_2(\mathbb{F}_q)$, $Sp_{2d}(\mathbb{F}_q)$ resp.) then we must show that the inclusion $(G_a)^m \to G_{am}$ induces an epimorphism in homology. Since $\mathcal{S}_m \wr G_a$ contains a Sylow ℓ-subgroup we are reduced to analyzing the wreath product $\mathbb{Z}/\ell \wr G_a$. According to Quillen [10], $H^*(B\,\mathbb{Z}/\ell \wr G_a)$ is detected by $\mathbb{Z}/\ell \times G_a$ and $(G_a)^\ell$. In this case we produce explicit conjugations which show that $\mathbb{Z}/\ell \times G_a$ is conjugate to a subgroup of $(G_a)^\ell$. Thus $H^*(B\mathbb{Z}/\ell \wr G_a; \mathbb{Z}/\ell)$ and hence $H^*(BG_{a\ell}; \mathbb{Z}/\ell)$ is detected by $(G_a)^\ell$. The method of Theorem 1 is then used to show that there are no relations among these generators other than those listed in (a) - (h).

Homology operations in $H_*(BG_\infty; \mathbb{Z}/\ell)$ can be computed by the method employed in Priddy [8]. We have derived explicit formulas for homology operations if $\ell = 2$ (cf. [3]).

We now turn to computing the cohomology rings $H^*(BG_n; \Lambda)$ by dualizing the results of Theorem 2. Because of their complicated structure we first consider $H^*(BO_n^{\pm}(\mathbb{F}_q); \mathbb{Z}/2)$. According to Theorem 2 these rings are detected by products of $O_1^{\pm}(\mathbb{F}_q) = \mathbb{Z}/2$. We will need the following notation:

(a) T_m and $\bar{T}_m$ will denote two sets of degree one indeterminates $\{t_i\}_{i=1}^m$ and $\{\bar{t}_i\}_{i=1}^m$.

(b) $\sigma_i(T_m)$ (respectively $\sigma_i(\bar{T}_m)$) will denote the i-th elementary symmetric polynomial in the variables $\{t_i\}_{i=1}^m$ (respectively in $\{\bar{t}_i\}_{i=1}^m$) if $1 \leq i \leq m$. We define further

$$\sigma_0(T_m) = \sigma_0(\bar{T}_m) = 1, \quad \sigma_i(T_m) = \sigma_i(\bar{T}_m) = 0 \text{ if } i < 0 \text{ or } i > m.$$

(c) We define elements $x_k(m,n)$, $\bar{x}_{2k-1}(m,n) \in \mathbb{Z}/2[T_m] \otimes \mathbb{Z}/2[\bar{T}_n]$ by

$$x_{2k-1}(m,n) = \sum_{p+q=k} \sigma_{2p-1}(T_m)\sigma_{2q}(\bar{T}_n)$$

$$\bar{x}_{2k-1}(m,n) = \sum_{p+q=k} \sigma_{2p}(T_m)\,\sigma_{2q-1}(\bar{T}_n)$$

$$x_{2k}(m,n) = \sum_{p+q=k} \sigma_{2p}(T_m)\sigma_{2q}(\bar{T}_n).$$

(d) $w_k \in H^k(BO_n^{\pm}(\mathbb{F}_q); \mathbb{Z}/2)$ will denote the image of the k-th Stiefel Whitney class under Brauer lift

$$BO_n^{\pm}(\mathbb{F}_q) \to BO_\infty(\mathbb{F}_q) \xrightarrow{\beta} BO$$

<u>Theorem 3.</u> (a) $H^*(BO_{2n-1}(\mathbb{F}_q); \mathbb{Z}/2)$ is the subalgebra of the ring direct product

$$\prod_{i=1}^n \mathbb{Z}/2[T_{2i-1}] \otimes \mathbb{Z}/2[\bar{T}_{2n-2i}]$$

generated by elements $\{x_k\}_{k=1}^{2n-1}$ and elements $\{\bar{x}_{2k-1}\}_{k=1}^{n-1}$ where $x_k = (x_k(2i-1,2n-2i))_{i=1}^n$ and $\bar{x}_{2k-1} = (\bar{x}_{2k-1}(2i-1,2n-2i))_{i=1}^n$.

(b) $H^*(BO_{2n-1}^-(\mathbb{F}_q); \mathbb{Z}/2)$ is the subalgebra of the ring direct product

$$\prod_{i=0}^{n-1} \mathbb{Z}/2[T_{2i}] \otimes \mathbb{Z}/2[\bar{T}_{2n-2i-1}]$$

generated by elements $\{x_k\}_{k=1}^{2n-2}$ and elements $\{\bar{x}_{2k-1}\}_{k=1}^n$ where $x_k = (x_k(2i,2n-2i-1))_{i=0}^{n-1}$ and $\bar{x}_{2k-1} = (\bar{x}_{2k-1}(2i,2n-2i-1))_{i=0}^{n-1}$.

(c) $H^*(BO_{2n}(\mathbb{F}_q); \mathbb{Z}/2)$ is the subalgebra of the ring direct product

$$\prod_{i=0}^n \mathbb{Z}/2[T_{2i}] \otimes \mathbb{Z}/2[\bar{T}_{2n-2i}]$$

generated by elements $\{x_k\}_{k=1}^{2n}$ and elements $\{\bar{x}_{2k-1}\}_{k=1}^n$ where $x_k = (x_k(2i,2n-2i))_{i=0}^n$ and $\bar{x}_{2k-1} = (\bar{x}_{2k-1}(2i,2n-2i))_{i=0}^n$.

(d) $H^*(BO_{2n}^-(\mathbb{F}_q); \mathbb{Z}/2)$ is the subalgebra of the ring direct product

$$\prod_{i=1}^n \mathbb{Z}/2[T_{2i-1}] \otimes \mathbb{Z}/2[\bar{T}_{2n-2i+1}]$$

generated by elements $\{x_k\}_{k=1}^{2n-1}$ and elements $\{\bar{x}_{2k-1}\}_{k=1}^{n}$ where $x_k = (x_k(2i-1,2n-2i+1))_{i=1}^{n}$ and $\bar{x}_{2k-1} = (\bar{x}_{2k-1}(2i-1,2n-2i+1))_{i=1}^{n}$.

(e) The Stiefel Whitney classes $w_i \in H^i(BO_n^{\pm}(\mathbb{F}_q); \mathbb{Z}/2)$, $i = 1,2,\dots,n$ are given by the formula

$$w_i = x_i + \bar{x}_i \quad \text{if } i \text{ is odd}$$

and by the following recursion formulas if $i = 2k$ is even:

$$R_{2k}(x,\bar{x},w) = \sum_{p=1}^{k} x_{2p-1}\bar{x}_{2k-2p+1} + \sum_{q=1}^{k} (x_{2q} + w_{2q})x_{2k-2q} = 0, \; k = 1,2,\dots,[n/2].$$

(f) For $k > [n/2]$, $w_{2k} = 0$ and hence the formulas

$$R_{2k}(x,\bar{x},w) = 0$$

reduce to relations between the generators $\{x_k\}$, $\{\bar{x}_{2k-1}\}$.

(g) The relations (f) are the only relations in $H^*(BO_n^{\pm}(\mathbb{F}_q); \mathbb{Z}/2)$ with the single exception of $H^*(BO_n(\mathbb{F}_q); \mathbb{Z}/2)$, n even where there is an additional relation

$$x_n + w_n = 0.$$

In all cases the Poincaré series of $H^*(BO_n^{\pm}(\mathbb{F}_q); \mathbb{Z}/2)$ is

$$P(H^*(BO_n^{\pm}(\mathbb{F}_q); \mathbb{Z}/2),t) = \frac{\prod_{i=1}^{n-1}(1+t^i)}{\prod_{i=1}^{n}(1-t^i)}.$$

<u>Remark</u>. It should be noted that the groups $O_{2n-1}^{\varepsilon}(\mathbb{F}_q)$, $\varepsilon = \pm$, are isomorphic and hence have isomorphic cohomology rings. Although the groups $O_{2n}^{\varepsilon}(\mathbb{F}_q)$, $\varepsilon = \pm$, are not isomorphic, $H^*(BO_{2n}^{\varepsilon}(\mathbb{F}_q); \mathbb{Z}/2)$, $\varepsilon = \pm$, are isomorphic as $\mathbb{Z}/2$ modules. However they have different ring structures. For instance

$$H^*(BO_2(\mathbb{F}_q); \mathbb{Z}/2) = \mathbb{Z}/2[x_1,\bar{x}_1,x_2]/x_1\bar{x}_1 = 0$$

$$H^*(BO_2^-(\mathbb{F}_q); \mathbb{Z}/2) = \mathbb{Z}/2[x_1,\bar{x}_1].$$

<u>Corollary</u>. $H^*(BO_\infty(\mathbb{F}_q); \mathbb{Z}/2) = H^*(BO_\infty^-(\mathbb{F}_q); \mathbb{Z}/2)$ is a polynomial algebra on generators $\{x_k, \bar{x}_{2k-1}\}_{k=1}^{\infty}$.

We now proceed to compute the other cohomology rings $H^*(BG_n; \Lambda)$. In the case of $GL_n(\mathbb{F}_q)$ these rings were computed by Quillen [9]. We include the general linear groups for completeness and because we have new results for $q \equiv 3 \pmod 4$.

According to Theorem 2 the rings $H^*(BGL_n(\mathbb{F}_q); \mathbb{Z}/2)$ and $H^*(BU_n(\mathbb{F}_{q^2}); \mathbb{Z}/2)$ are detected by products of $GL_1(\mathbb{F}_q)$ and $U_1(\mathbb{F}_{q^2})$ respectively. If $q \equiv 3 \pmod 4$ then $H^*(BGL_1(\mathbb{F}_q); \mathbb{Z}/2) = \mathbb{Z}/2[t]$ where t has degree 1. Similarly if $q \equiv 1 \pmod 4$ then $H^*(BU_1(\mathbb{F}_{q^2}); \mathbb{Z}/2) = \mathbb{Z}/2[t]$.

Theorem 4. Let $\mathcal{A}$ denote one of the cohomology rings

$$H^*(BGL_n(\mathbb{F}_q); \mathbb{Z}/2) \qquad q \equiv 3 \pmod 4$$

$$H^*(BU_n(\mathbb{F}_{q^2}); \mathbb{Z}/2) \qquad q \equiv 1 \pmod 4.$$

Then $\mathcal{A}$ is the subalgebra of $\mathbb{Z}/2\,[t_1,\ldots,t_n]$ with generators $\{s_{4k} = \sigma_{2k}(t_1,t_2,\ldots,t_n)^2\}_{k=1}^{[n/2]}$ and $\{x_{2k-1} = \sum_{i=1}^{n} t_i^{2k-1}\}_{k=1}^{n}$. The generators $\{s_{4k}\}_{k=1}^{[n/2]}$ and $\{x_{2k-1}\}_{k=1}^{[\frac{1}{2}(n+1)]}$ are algebraically independent. The only relations are

$$(\star) \qquad x_{2k-1}^2 = p_{2k-1}(x_1^2, x_3^2, \ldots, x^2_{2[\frac{n+1}{2}]-1}, s_4, s_8, \ldots, s_{4[n/2]}),$$

for $[\frac{n+1}{2}] < k \leq n$, which are obtained as follows: The polynomials $\{x_1, x_3, \ldots, x_{2[\frac{n+1}{2}]-1}, \sigma_2, \sigma_4, \ldots, \sigma_{2[n/2]}\}$ generate the symmetric subalgebra of $\mathbb{Z}/2[t_1,t_2,\ldots,t_n]$. Hence

$$x_{2k-1} = p_{2k-1}(x_1, x_3, \ldots, x_{2[\frac{n+1}{2}]-1}, \sigma_2, \sigma_4, \ldots, \sigma_{2[n/2]})$$

for some polynomial p_{2k-1}. This of course is not a relation in $\mathcal{A}$ since $\sigma_{2i} \notin \mathcal{A}$. However squaring this relation gives $(\star)$ which is a relation in $\mathcal{A}$.

Corollary. The algebras

$$H^*(BGL_\infty(\mathbb{F}_q); \mathbb{Z}/2) \qquad q \equiv 3 \pmod 4$$

$$H^*(BU_\infty(\mathbb{F}_{q^2}); \mathbb{Z}/2) \qquad q \equiv 1 \pmod 4$$

are polynomial algebras on generators $\{x_{2k-1}, s_{4k}\}_{k=1}^{\infty}$.

In the case of $H^*(BGL_n(\mathbb{F}_q); \mathbb{Z}/2)$, $q \equiv 3 \pmod 4$, Quillen [9] chose different generators and relations. His presentation has the advantage of providing explicit formulas for the relations. However it has the serious disadvantage of obscuring the fact that $H^*(BGL_\infty(\mathbb{F}_q); \mathbb{Z}/2)$ $q \equiv 3 \pmod 4$ is a polynomial algebra.

Let $\mathcal{A}$ denote one of the remaining cohomology rings for $n \leq \infty$.

(1) $\quad H^*(BGL_n(\mathbb{F}_q); \mathbb{Z}/2) \qquad q \equiv 1 \pmod 4$

(2) $\quad H^*(BU_n(\mathbb{F}_{q^2}); \mathbb{Z}/2) \qquad q \equiv 3 \pmod 4$

(3) $H^*(BSp_{2n}(\mathbb{F}_q);\ \mathbb{Z}/2)$

(4) $H^*(BO^{\pm}_n(\mathbb{F}_q);\ \mathbb{Z}/\ell)$ ℓ odd prime, $\ell \neq p$

(5) $H^*(BGL_n(\mathbb{F}_q);\ \mathbb{Z}/\ell)$ " " "

(6) $H^*(BU_n(\mathbb{F}_{q^2});\ \mathbb{Z}/\ell)$ " " "

(7) $H^*(BSp_{2n}(\mathbb{F}_q);\ \mathbb{Z}/\ell)$ " " "

Let $c = \min\{k \mid \tilde{H}_*(BG_k;\ \mathbb{Z}/\ell) \neq 0\}$ then for cases (1),(2), $c = 1$; for case (3), $c = 2$; for cases (4)-(7), $c = 2d, r, s, 2d$, respectively (see Theorem 2 for d, r, s). Let m denote the maximal $(G_c)^m$ contained in G_n, then $H^*((BG_c)^m;\ \mathbb{Z}/\ell) = \mathbb{Z}/\ell[v_1, v_2, \ldots, v_m] \otimes E[u_i \mid 1 \leq i \leq m]$ where $\deg v_i = 2c$, $\deg u_i = 2c - 1$.

Theorem 5. $\mathcal{Q} = \mathbb{Z}/\ell[s_{1c}, s_{2c}, \ldots, s_{mc}] \otimes E[e_{ic} \mid 1 \leq i \leq m]$ where

$$s_{jc} = \sigma_j(v_1, v_2, \ldots, v_m)$$

$$e_{jc} = \sum_{\substack{i_1 < \cdots < i_j \\ 1 \leq k \leq j}} v_{i_1} \cdots \hat{v}_{i_k} \cdots v_{i_j} u_{i_k}.$$

We have also obtained the following results about $H^*(BG_\infty;\ \mathbb{Z})$ using the fibrations of Theorem 1.

Theorem 6. (a) $H^*(BGL_\infty(\mathbb{F}_q);\ \mathbb{Z}) = \mathbb{Z}[c_i \mid i \geq 1]/\{(q^i - 1)c_i = 0\}$

where $c_i \in H^{2i}(BGL_\infty(\mathbb{F}_q);\ \mathbb{Z})$

(b) $H^*(BO^+_\infty(\mathbb{F}_q);\ \mathbb{Z}) = \mathbb{Z}[p_i \mid i \geq 1]/\{\frac{1}{2}(q^{2i} - 1)p_i = 0\} \oplus T$

where $p_i \in H^{4i}(BO^+_\infty(\mathbb{F}_q);\ \mathbb{Z})$ and T denotes the $\mathbb{Z}/2$-module of 2-torsion in $H^*(BO \times SO;\ \mathbb{Z})$.

(c) $H^*(BSp_\infty(\mathbb{F}_q);\ \mathbb{Z}) = \mathbb{Z}[u_i \mid i \geq 1]/\{(q^{2i} - 1)u_i = 0\}$

where $u_i \in H^{4i}(BSp_\infty(\mathbb{F}_q);\ \mathbb{Z})$.

(d) $H^*(BU_\infty(\mathbb{F}_{q^2});\ \mathbb{Z}) = \mathbb{Z}[\bar{c}_i \mid i \geq 1]/\{((-q)^i - 1)\bar{c}_i = 0\}$

where $\bar{c}_i \in H^{2i}(BU_\infty(\mathbb{F}_{q^2});\ \mathbb{Z})$.

This theorem follows from a careful analysis of the Serre spectral sequence.

Complete details including corresponding results for finite fields of characteristic 2 will appear in our forthcoming paper [4].

References

[1] J. F. Adams, Vector fields on spheres, Ann. of Math. (2) 75(1962), 603-632.

[2] J. Dieudonné, Le Geometrie des Groupes Classiques, Ergebnisse der Mathematik und ihrer Grenzgebiete, 5(1955), Springer-Verlag.

[3] Z. Fiedorowicz and S. B. Priddy, Loop spaces and finite orthogonal groups, Bull. A. M. S. 81(1975), 700-702.

[4] ———, Homology of classical groups over finite fields and their associated infinite loop spaces (to appear).

[5] E. M. Friedlander, Computations of K-theories of finite fields, Topology, 15(1976), 87-109.

[6] O. F. O'Meara, Introduction to Quadratic Forms, Springer-Verlag, 1971.

[7] J. P. May, E_∞ ring spaces and E_∞ ring spectra (to appear).

[8] S. B. Priddy, Dyer-Lashof operations for the classifying spaces of certain matrix groups, Quart. J. of Math. 26(1975), 179-194.

[9] D. G. Quillen, On the cohomology and K-theory of the general linear group over a finite field, Ann. of Math. 96(1972), 552-586.

[10] ———, The Adams conjecture, Topology 10(1971), 67-80.

University of Michigan
Ann Arbor, Michigan 48104

Northwestern University
Evanston, Illinois 60201

GROUP COHOMOLOGY CLASSES WITH DIFFERENTIAL FORM COEFFICIENTS

by Bruno Harris
Brown University

We construct certain cohomology classes of linear groups over a commutative ring R ; the cohomology classes have coefficients in the R-modules of differentials $\Omega^i(R)$ $(i = 1,2, \ldots)$ and are analogues of classes of a vector bundle over a complex manifold constructed by Atiyah in [1]. An extended discussion of classes of this type was given by Grothendieck [2]. Our development (which also dates back a number of years) is very elementary and makes it easy to exhibit non-zero cohomology classes if the module Ω^1_R of Kahler differentials is large enough.

Let P be a finitely generated projective module over the commutative ring R . A <u>semilinear automorphism</u> (T,τ) of P consists of a ring automorphism τ of R and an abelian group automorphism T of P such that

$$T(rp) = \tau(r)T(p)$$

if $r \in R$, $p \in P$. These compose as: $(S,\sigma)(T,\tau) = (ST,\sigma\tau)$. Let now G be a group of semilinear automorphisms of P : extreme cases are 1) The elements of G are all R-linear or 2) the homomorphism $(T,\tau) \to \tau$ is injective on G ; we will eventually specialize to case (1), but (2) has geometric interest.

The elements of G extend to semilinear automorphisms of the R-module $\Omega^1_R \otimes_R P$, (T,τ) acting on this last module as $\tau \otimes T$. Now since P is projective it is easy to show the existence of <u>connections</u> D for P , i.e. abelian group homomorphisms $D : P \to \Omega^1_R \otimes_R P$ satisfying

$$D(rp) = rD(p) + d\,r \otimes p \tag{1}$$

for $r \in R$, $p \in P$. Connections are R-module splittings of

$$0 \to \Omega^1 \otimes P \to (\Omega^1 \otimes P) \times P \to P \to 0$$

where the R-module structure of $(\Omega^1 \otimes P) \times P$ is given by $r(\omega \otimes p, p^1) = (dr \otimes p^1 + r\omega \otimes p, rp^1)$. If D, D^1 are connections then, by (1), $D - D^1 \in \mathrm{Hom}_R(P, \Omega^1 \otimes P)$ which, since P is finitely-generated projective, can be rewritten as $\Omega^1 \otimes \mathrm{End}_R(P)$: thus the set of all connections for P is a principal homogeneous space over the R-module $\Omega^1 \otimes \mathrm{End}_R(P)$. (T,τ) acts on $\Omega^1_R \otimes_R \mathrm{End}_R(P)$ as $\tau \otimes \mathrm{Ad}\, T$, making $\Omega^1 \otimes \mathrm{End}_R P$ a G-module.

We now define an element α of $H^1(G, \Omega^1 \otimes \mathrm{End}_R(P))$: an Eilenberg-MacLane 1-cocycle (crossed homomorphism) representing α is given by choosing a connection D for P ; then for any $\gamma \in G$, $\gamma D \gamma^{-1}$ is again a connection and so $D - \gamma D \gamma^{-1} \in \Omega^1 \otimes \mathrm{End}_R(P)$, and the cocycle is $\gamma \to (D - \gamma D \gamma^{-1})$.

Let $\Omega^i = \Lambda^i(\Omega^1_R)$ (i^{th} exterior power of this R-module), and $\overbrace{\Omega^1 \otimes \cdots \otimes \Omega^1}^{i} \to \Lambda^i \Omega^1$, $\overbrace{\mathrm{End}(P) \otimes \cdots \otimes \mathrm{End}\, P}^{i} \to \mathrm{End}\, P$ be the maps given by multiplication.

These products on coefficients allow us to take the cup product of α with itself i times:

$$\alpha^i = \alpha \cup \cdots \cup \alpha \in H^i(G, \Omega^i \otimes \mathrm{End}_R P)$$

Finally, the <u>trace</u> : $\mathrm{End}_R(P) \to R$ gives

$$A_i = \mathrm{tr}(\alpha^i) \in H^i(G; \Omega^i_R)$$

(G acts on Ω^i_R with (T,τ) acting as τ ; the action is trivial if the elements of G are R-linear). The A_i are the cohomology classes we want.

If G is an algebraic subgroup of $GL(n,R)$ with lie algebra $\mathfrak{g} \subset \mathrm{End}(R^n)$ then we may replace the multilinear polynomial $\mathrm{tr}(X_1 \ldots X_i)$ by other $\mathrm{Ad}\, G$ invariant polynomials on $\mathfrak{g}$. However we will look from now on only at the case $P = R^n$, $G \subseteq GL(n,R)$, $\Omega^1 \otimes \mathrm{End}\, P = \Omega^1 \otimes M_n(R) = M_n(\Omega^1)$: matrices with coefficients in Ω^1 . A connection D is defined by specifying that $D(e_i) = 0$ where $e_1, \ldots, e_n$ is the standard basis of R^n . The cocycle then takes the matrix $g \in GL(n,R)$ to $dg \cdot g^{-1}$ where $d[g_{i,j}] = [dg_{i,j}]$. To take cup product, use the cochain formula:

if f_p is a p-cochain (function of p group elements) and f'_q is a q-cochain then

$$f_p \cup f'_q(g_1, \ldots, g_{p+q}) = f_p(g_1, \ldots, g_p) \cdot g_1 \ldots g_p f'_q(g_{p+1}, \ldots, g_{p+q})$$

Then $A_i \in H^i(GL(n,R);\Omega^i_R)$ is represented by the cocycle: $(g_1, \ldots, g_i) \to \mathrm{trace}(dg_1 \ldots dg_i g_i^{-1} \ldots g_1^{-1})$. If we restrict the cocycle to the abelian subgroup of diagonal matrices in G , then the cocycle above is skew-symmetric in its variables; if we take R as the group ring of a free abelian group of rank i then the cocycle restricted to the diagonal matrices is not cohomologous to zero. In dimension $i = 2$, $G = E(n,R)$, we get the "differential symbol". More precisely the classes A_i are additive with respect to block direct sum of matrices and are defined on the infinite general linear group $GL(R)$, i.e.

$$A_i \in H^i(BGL(R);\Omega^i) \approx H^i(BGL(R)^+;\Omega^i)$$

hence A_i define homomorphisms

$$K_i(R) \to H_i(BGL(R)^+) \to \Omega^i_R$$

for $i \geq 1$, which coincide for $i = 1$ with $g \to d \log \det g$ and for $i = 2$ with the differential symbol $(x,y) \to \frac{dx}{x} \wedge \frac{dy}{y}$.

Now let $R = \mathbb{C}$, $G = SU_n$, $(n \geq 2)$, T = diagonal matrices in SU_n .

If $n = 2$ then A_2 restricted to T is represented by the cocycle taking the diagonal matrices $\mathrm{diag}(\lambda,\lambda^{-1})$, $\mathrm{diag}(\mu,\mu^{-1})$ ($\lambda,\mu \in S^1$, circle group) to $2\frac{d\lambda}{\lambda} \wedge \frac{d\mu}{\mu} \in \Omega^2_{\mathbb{C}}$. In particular we get homomorphisms

$$H_2(S^1) \to H_2(SU_2) \to \Omega^2_{\mathbb{C}}$$

To describe this in a little more detail, note that
$H_1(S^1) = S^1 \approx \mathbb{R}/\mathbb{Q} \oplus \mathbb{Q}/\mathbb{Z}$

$H_2(S^1) \approx H_2(\mathbb{R}/\mathbb{Q}) \approx \Lambda^2(S^1)$ (second exterior power over $\mathbb{Z}$)

$H_n(S^1) \approx H_n(\mathbb{R}/\mathbb{Q}) \oplus H_n(\mathbb{Q}/\mathbb{Z}) \approx \Lambda^n(S^1) \oplus H_n(\mathbb{Q}/\mathbb{Z})$ where $H_n(\mathbb{Q}/\mathbb{Z}) = \mathbb{Q}/\mathbb{Z}$ for n odd, 0 for n even. The kernel of $d\,\ell og : S^1 \to \Omega^1_{\mathbb{C}}$ is the subgroup of algebraic numbers in S^1. For $n > 1$ we took $\Lambda^n(S^1)$ over $\mathbb{Z}$ and $\Lambda^n(\Omega^1_{\mathbb{C}})$ over $\mathbb{C}$; we have to find a torsion free divisible subgroup T of S^1 such that under $\Lambda^n(S^1) \to \Omega^n_{\mathbb{C}}$, $\Lambda^n(T)$ maps injectively. To do this choose a maximal subset of S^1 of elements algebraically independent over the field of all algebraic numbers, and let T be a divisible torsion free subgroup of S^1 generated by it; if this subset of S^1 is extended to a maximal subset of $\mathbb{C}^*$ of algebraically independent elements, then the differentials of these elements are a $\mathbb{C}$-basis for $\Omega^1_{\mathbb{C}}$. It is now clear that $\Lambda^q T$ maps isomorphically to a direct summand of $\Omega^q_{\mathbb{C}}$ which is a $\mathbb{Q}$-vector space of uncountable $\mathbb{Q}$-dimension $(q \geq 1)$.

Next we want to factor a suitable multiple of the above homomorphism $H_q(S^1) \to \Omega^q_{\mathbb{C}}$ through $H_q(SU_n)$. If q is even and $n \geq 2$ then the map $\Lambda^q(S^1) \to \Omega^q_{\mathbb{C}}$ taking $(\lambda_1, \ldots, \lambda_q)$ to $2\frac{d\lambda_1}{\lambda_1} \wedge \ldots \wedge \frac{d\lambda_q}{\lambda_q}$ is the restriction of the cocycle on SU_n previously described to the copy of S^1 embedded in SU_n as the set of diagonal matrices of the form $(\lambda, \lambda^{-1}, 1, \ldots, 1)$; in this case $\Lambda^q(T)$ maps isomorphically onto a direct summand of $H_q(SU(n))$. If q is odd this cocycle of SU_n is zero on S^1 embedded as just described; however if $n \geq 3$ and we embed S^1 in SU_n by sending λ to $\mathrm{diag}(\lambda, \lambda, \lambda^{-2}, 1, \ldots, 1)$ then the cocycle on this S^1 takes $(\lambda_1, \ldots, \lambda_q)$ to $(2 + (-2)^q)\frac{d\lambda_1}{\lambda_1} \wedge \ldots \wedge \frac{d\lambda_n}{\lambda_n}$ and we obtain again a splitting. To summarize

Theorem Let S^1 denote the circle group, T a torsion free divisible subgroup generated by a maximal subset of S^1 algebraically independent over the algebraic numbers. If $n \geq 2$ and q is even ≥ 2, or if $n \geq 3$ and $q \geq 2$ then $H_q(SU_n; \mathbb{Z})$ has a direct summand isomorphic to $\Lambda^q(T)$, a Q-vector space of dimension equal to the cardinality of the continuum. These subspaces of $H_q(SU_n)$ are in the image of $H_q(S^1)$ under embeddings of S^1 in SU_2, respectively SU_3, and map isomorphically under inclusions $SU_n \to SU_{n+1}$.

In closing, we note that the construction for cohomology classes we used works just as well with SU_n replaced by any subgroup of $GL_n(\mathbb{C})$ containing the copy of S^1 described, or with S^1 replaced by $\mathbb{C}^*$ and SU_n by a subgroup of GL_n containing $\mathbb{C}^*$. With regard to SU_n we may ask whether the image in $H_2(SU_n)$ of the homology of the subgroup of S^1 of algebraic numbers is zero (as is the case with $SL_n(\mathbb{C})$). Finally we remark that the construction of cohomology classes of a group described above can be summarized as follows: if G is a group, M a G-module and $\alpha \in H^1(G,M)$ is given then $\alpha \cup \ldots \cup \alpha \in H^i(G,M^{\otimes i})$. To obtain a simpler coefficient module we may take a quotient of $M^{\otimes i}$, such as $\mathbb{Z} \otimes_{\mathbb{Z}G} M^{\otimes i} = M_i$, and obtain $A_i \in H^i(G,M_i)$. As an example, let $M = IG$ (contained in $\mathbb{Z}G$) and α given by $g \to g - 1$.

Bibliography

[1] M. F. Atiyah, "Complex Analytic Connections in Fibre Bundles", Transactions Amer. Math. Soc., vol. 85, no. 1 (1957), p. 181-207.

[2] A. Grothendieck, "Classes de Chern et Representations Lineaires des Groupes Discrets" in "Dix exposés sur la cohomologie des schemas", Amsterdam: North-Holland, 1968.

Brown University

STABILITY FOR $H_2(SU_n)$

by Roger Alperin
Brown University

The methods of algebraic K-theory have been applied to the study of stability for $H_2(SL_n(k))$, k a field, and more generally for $H_2(E_n(R))$ where R is any commutative ring and $E_n(R)$ is the group of elementary matrices. In these investigations the elementary transvections play a major role; in fact, one can give a nice presentation for the universal central extension of $E_n(R)$. Recently, Quillen has shown how to handle stability for all of the homology groups of $GL_n(k)$, k a field $\neq \mathbb{Z}_2$, [Q1]. More precisely he shows that the natural maps

$$H_i(GL_n(k)) \to H_i(GL_{n+1}(k))$$

are isomorphisms for $n > i$. The sharpness of this stability theorem is facilitated by the use of elementary transvections in $GL_n(k)$. Our situation is concerned with the group SU_n of $n \times n$ complex unitary matrices of determinant one which contains no non-trivial transvections. However, we show that the natural maps

$$H_2(SU_n) \to H_2(SU_{n+1}) \qquad \text{are}$$

surjections for $n \geq 3$ and isomorphisms for $n \geq 6$. It will become evident that there is also a stability theorem for the higher homology of SU_n . B. Harris has shown that these homology groups are non-trivial, and in fact rather large [H].

§1: Let V be a vector space of dimension $n \geq 2$ over $\mathbb{C}$. The Tits building T(V) is the simplicial complex associated to the poset of proper subspaces of V . It is well known that T(V) has the homotopy type of a bouquet of spheres of dimension $n-2$ [Q2]. Denote by St(V) the GL(V) module $\tilde{H}_{n-2}(T(V))$; this is the Steinberg module. If dim V is 0 or 1 put $St(V) = \mathbb{Z}$. Observe that if $\dim V = 2$ then St(V) is the kernel of the augmentation map $\mathbb{Z}P^1 \to \mathbb{Z}$ where $\mathbb{Z}P^1$ is the free abelian group on the lines of V .

Theorem [L] There is an exact sequence of Steinberg modules for V of dimension n

$$0 \to St(V) \to \bigoplus_{\substack{W\subset V\\ \dim W=n-1}} St(W) \to \ldots \to \bigoplus_{\substack{W\subset V\\ \dim W=1}} St(W) \to \mathbb{Z} \to 0 \quad .$$

§2: If H is a subgroup of a group G, $\mathbb{Z}(H\backslash G)$ is the free abelian group on the left cosets of H in G and is a module over ZG. It is the induced module from H of the trivial H module $\mathbb{Z}$. Let $I(H\backslash G)$ be the kernel of the augmentation $\mathbb{Z}(H\backslash G) \to \mathbb{Z}$. If M is a G module, then put

$$H_i(G,H;M) = H_{i-1}(G;I(H\backslash G) \otimes M) \quad .$$

It is immediate then that there is a long exact sequence of homology groups

$$\ldots \to H_i(H;M) \to H_i(G,M) \xrightarrow{\partial} H_i(G,H;M) \to \ldots$$

Proposition 0: $H_0(SU_2;St(\mathbb{C}^2)) = H_0(U_2;St(\mathbb{C}^2)) = 0$

proof: The standard inclusion of the circle group S^1 into SU_2 with quotient space $\mathbb{P}^1(\mathbb{C})$ yields the identification $\mathbb{Z}P^1(\mathbb{C}) = ZSU_2 \otimes_{\mathbb{Z}S^1} \mathbb{Z}$. The homology exact sequence corresponding to the exact sequence of SU_2 modules

$$0 \to St(\mathbb{C}^2) \to \mathbb{Z}P^1 \to \mathbb{Z} \to 0 \qquad \text{is}$$

$$\ldots \to H_1(SU_2) \to H_0(SU_2;St(\mathbb{C}^2)) \to H_0(S^1) \to H_0(SU_2)$$

As SU_2 is a perfect group, and the last map in the sequence above is an isomorphism $H_0(SU_2;(\mathbb{C}^2)) = 0$. Since $H_0(U_2;St(\mathbb{C}^2))$ is a quotient of $H_0(SU_2;St(\mathbb{C}^2))$, it also is trivial.

It should be mentioned that recently J. Mather [M] has shown that the natural map $H_2(S^1) \to H_2(SU_2)$ is surjective. It would be of considerable interest to know the kernel of this homomorphism or equivalently the image of $H_3(SU_2,S^1) = H_2(SU_2,St(\mathbb{C}^2))$ in $H_2(S^1)$.

§3: In the context of the relative homology groups introduced above we shall require two spectral sequences. For the first, given a pair of group extensions, i.e. a commutative diagram,

$$\begin{array}{ccccccccc} 1 & \to & H & \to & G & \to & K & \to & 1 \\ & & \downarrow & & \downarrow & & \| & & \\ 1 & \to & \tilde{H} & \to & \tilde{G} & \to & K & \to & 1 \end{array}$$

with exact rows and G a subgroup of $\tilde{G}$, H a subgroup of $\tilde{H}$, then there is a relative Serre-Hochschild-Lyndon spectral sequence.

Proposition 1: With the data as above and M a $\tilde{G}$ module, there is first quadrant spectral sequence

$$E^2_{a,b} = H_a(K,H_b(\tilde{H},H;M)) \Rightarrow H_{a+b}(\tilde{G},G;M)$$

For the second spectral sequence, suppose C_* is an exact complex of G modules.

Proposition 2: There is a spectral sequence

$$E^1_{p,q} = H_q(G,C_p) \Rightarrow 0$$

proof: Apply the two hyperhomology spectral sequences ([W]),

$$E^1_{p,q} = H_q(G,C_p) \Rightarrow H_{p+q}(G,C_*) \quad \text{and}$$

$$E^2_{p,q} = H_p(G,H_q(C_*)) \Rightarrow H_{p+q}(G,C_*)$$

to the exact complex C_* ; so that $E^2_{p,q} = 0$ and hence the abutment of E^1 is 0 .

§4: Consider now the Lusztig exact complex

$$C_*(n): \quad 0 \to St(V) \to \bigoplus_{\substack{W\subset V \\ \dim W=n-1}} St(W) \to \dots \to \bigoplus_{\substack{W\subset V \\ \dim W=1}} St(W) \to \mathbb{Z} \to 0$$

of SU_n modules $(\dim V = n)$. From the hyperhomology spectral sequence above we have

$$E^1_{p,q}(n) = H_q(SU_n, \bigoplus_{\dim W=p} St(W))$$

$$\cong H_q(SU_{n,p}; St(\mathbb{C}^p)) \qquad \text{by dint}$$

of $\mathbb{Z}SU_n \underset{\mathbb{Z}SU_{n,p}}{\otimes} St(\mathbb{C}^p) \cong \bigoplus_{\dim W=p} St(W)$ where $SU_{n,p}$ is the subgroup of SU_n stabilizing the standard p-space in $\mathbb{C}^n$. Considering SU_n as a subgroup of SU_{n+1} in the standard way and the corresponding E^1 term

$$E^1_{p,q}(n+1) \cong H_q(SU_{n+1,p}; St(\mathbb{C}^p))$$

we obtain the following.

Proposition 3: There are first quadrant spectral sequences with

$$E^1_{p,q}(n) = \begin{cases} H_q(SU_{n+1,p}, SU_{n,p}; St(\mathbb{C}^p)) & p \leq n \\ H_q(SU_{n+1}, St(\mathbb{C}^{n+1})) & p = n+1 \\ 0 & p > n+1 \end{cases}$$

$$\Rightarrow 0$$

For the construction of the spectral sequence one uses the mapping cone construction on the double complex level together with Proposition 2 ([W]).

<u>Theorem</u>: $H_2(SU_{n+1}, SU_n) = 0 \quad n > 2$

proof: Observe first that since $H_1(SU_n) = 0$ for $n \geq 0$, then $H_1(SU_{n+1}, SU_n) = 0 \quad n \geq 0$. Consider the extension of groups

$$\begin{array}{ccccccccc} 1 & \to & SU_{n-p} & \longrightarrow & SU_{n,p} & \longrightarrow & U_p & \to & 1 \\ & & \downarrow & & \downarrow & & \| & & \\ 1 & \to & SU_{n+1-p} & \to & SU_{n+1,p} & \to & U_p & \to & 1 \end{array} \qquad p < n$$

and the associated spectral sequences

$$E^2_{a,b}(p,n) = H_a(U_p, H_b(SU_{n+1-p}, SU_{n-p}) \otimes St(\mathbb{C}^p))$$

$$\Rightarrow H_{a+b}(SU_{n+1,p}, SU_{n,p}; St(\mathbb{C}^p))$$

where SU_{n-p}, SU_{n+1-p} act trivially on $St(\mathbb{C}^p)$. This abutment is seen to be $E^1_{p,a+b}(n)$ as in Prop. 3. Since $E^2_{a,b}(p,n) = 0$ when $b = 0,1$ and $p < n$ then $E^1_{p,q}(n) = 0$ for $p < n$, $q = 0,1$; also $E^1_{n,0}(n) = 0$. It follows that $E^2_{0,2} = E^\infty_{0,2}(p,n) \cong E^1_{p,2}(n)$ for $p < n$. The terms

$$E^1_{p,2}(n) \cong H_0(U_p, H_2(SU_{n+1-p}, SU_{n-p}) \otimes St(\mathbb{C}^p))$$

can now be identified for $p \leq 2$: for $p = 0$, $E^1_{0,2} \cong H_2(SU_{n+1}, SU_n)$; for $p = 1$, observe that by extraction of roots the action of U_1 on SU_n, SU_{n-1} can be chosen as conjugation by scalar matrices and thus trivial, so that $E^1_{1,2} \cong H_2(SU_n, SU_{n-1})$; for $p = 2$, $E^1_{2,2} = 0$ by Proposition 0. For $n \geq 4$ no higher differentials will affect $E^r_{0,2}(n)$ or $E^r_{1,2}(n)$ and thus

$$d_1 : H_2(SU_n, SU_{n-1}) \to H_2(SU_{n+1}, SU_n)$$

is an isomorphism. By a chase of the commutative diagram (exact rows) below one finds that $d_1 d_1 = 0$ and the theorem results.

$$\begin{array}{ccccccc}
 & & H_2(SU_n) & \to & H_2(SU_n, SU_{n-1}) & \to & 0 \\
 & & \downarrow & & \downarrow d_1 & & \\
 & & H_2(SU_{n+1}) & \to & H_2(SU_{n+1}, SU_n) & \to & 0 \\
 & \diagup\!\!\diagup & \downarrow & & \downarrow d_1 & & \\
H_2(SU_{n+1}) & \to & H_2(SU_{n+2}) & \to & H_2(SU_{n+2}, SU_{n+1}) & \to & 0
\end{array}$$

Corollary: $H_3(SU_{n+1}, SU_n) = 0$ for $n > 5$

proof: Arguing as above we find that
$E^1_{p,q}(n) = 0$ for $q = 0,1$, $p < n$; $q = 0$, $p = n$ and $q = 2$, $p \leq n - 3$.

And thus if $n \geq 5$ $H_3(SU_n, SU_{n-1}) \xrightarrow{d_1} H_3(SU_{n+1}, SU_n)$ is surjective. With a diagram chase similar to that above $d_1 d_1 = 0$ and we obtain

$$H_3(SU_{n+2}, SU_{n+1}) = 0 \qquad \text{for } n \geq 5$$

Corollary: The natural maps

$$H_2(SU_n) \to H_2(SU_{n+1}) \qquad \text{are surjective}$$

for $n \geq 3$ and injective for $n \geq 6$.

§5. A well known theorem of Hopf gives an expression for the second homology in terms of relations which occur among the commutators of a given group G. One consequence of this formula is that every element of $H_2(G)$ is carried by a compact oriented surface of some genus. One can bound the genus of surfaces needed to generate $H_2(G)$ given information about the lengths of commutators needed to generate $G' = [G,G]$.

Lemma: Suppose G is a group in which every element of G' is a product of at most t commutators of elements of G, then $H_2(G)$ is generated by commutator relations $\prod_{i=1}^{s} [a_i, b_i]$ for $s \leq 2t + 1$.

Corollary: $H_2(SU_n)$ is generated by commutator relations of length $s \leq 3$.

proof: Every element of SU_n can be expressed as a single commutator [G].

Corollary: $H_2(SU)$ is generated by commutator relations of length $s \leq 3$ and these relations need only involve elements from SU_3.

Bibliography

[L] Lusztig, G., The Discrete Series of GL_n Over A Finite Field, Annals of Mathematics Studies, Number 81, 1974, p. II.

[Q1] Quillen, D., Finite generation of the groups K_i of rings of algebraic integers, Lecture Notes in Mathematics, vol. 341, Springer-Verlag, 1973.

[Q2] Quillen, D., MIT Lectures 1974-75.

[M] Mather, J., Letter to Sah, July 1975.

[W] Wagoner, J., Stability for homology of the general linear group of a local ring (to appear).

[G] Goto, M., A theorem on compact semi simple groups, J. Math. Soc. Japan, 1, 270-272 (1949).

[H] Harris, B., Group cohomology classes with differential form coefficients (to appear).

Brown University

HOMOLOGICAL STABILITY FOR CLASSICAL GROUPS OVER FINITE FIELDS

Eric M. Friedlander

Dan Quillen has proved (but not yet published) that

$$H_i(GL_n(k),Z) \xrightarrow{\sim} H_i(GL_{n+1}(k),Z)$$

if $i < n$ and if k is any field with more than two elements. As discussed at this conference, no such stability theorem for other classical groups is known. In this paper, we give a prime-to-residue-characteristic homological stability theorem for classical groups over finite fields (Theorem 2), a stability theorem for the integral homology of classical groups over the algebraic closure of a finite field (Theorem 3), and a stability theorem for the integral homology of the special linear groups for finite fields with more than two elements (Theorem 6).

We begin with the following well known stability results for the classical Lie groups. The proof is readily obtained by using the fibrations $S^{2n+1} \to BU_n \to BU_{n+1}$, $S^{4n+3} \to BSp_n \to BSp_{n+1}$, and $S^n \to BO_n \to BO_{n+1}$.

Lemma 1. a.) Let U_n be the Lie group of unitary $n \times n$ complex matrices. Then the natural inclusion $U_n \to U_{n+1}$ induces isomorphisms

$$H_i(BU_n,Z) \xrightarrow{\sim} H_i(BU_{n+1},Z), \qquad \pi_j(BU_n) \xrightarrow{\sim} \pi_j(BU_{n+1})$$

for all $i \leq 2n + 1$, $j \leq 2n$.

b.) Let Sp_n be the Lie group of $n \times n$ quaternionic matrices which preserve the quaternionic inner product. Then the natural inclusion $Sp_n \to Sp_{n+1}$ induces isomorphisms

$$H_i(BSp_n,Z) \xrightarrow{\sim} H_i(BSp_{n+1},Z), \qquad \pi_j(BSp_n) \xrightarrow{\sim} \pi_j(BSp_{n+1})$$

for all $i \leq 4n+3$, $j \leq 4n+2$.

c.) Let O_n be the Lie group of n x n real orthogonal matrices. Then the natural inclusion $O_n \to O_{n+1}$ induces isomorphisms

$$H_i(BO_n,Z) \xrightarrow{\sim} H_i(BO_{n+1},Z), \qquad \pi_j(BO_n) \xrightarrow{\sim} \pi_j(BO_{n+1})$$

for all $i \leq n - 1$, $j \leq n - 1$.

Granted Lemma 1, the following theorem is essentially implicit in [2]. Recent work of Z. Fiedorowicz and S. Priddy provides a completely different method of obtaining various special cases of Theorem 2.

<u>Theorem 2.</u> Let k be a finite field and let q be a prime different from p, the residue characteristic of k.

a.) The natural inclusion $GL_n(k) \to GL_{n+1}(k)$ induces isomorphisms

$$H_i(GL_n(k),Z/q) \xrightarrow{\sim} H_i(GL_{n+1}(k),Z/q) \text{ for } i \leq 2n.$$

b.) The natural inclusion $SL_n(k) \to SL_{n+1}(k)$ induces isomorphisms

$$H_i(SL_n(k),Z/q) \xrightarrow{\sim} H_i(SL_{n+1}(k),Z/q) \text{ for } i \leq 2n.$$

c.) The natural inclusion $U_n(k) \to U_{n+1}(k)$ induces isomorphisms

$$H_i(U_n(k),Z/q) \xrightarrow{\sim} H_i(U_{n+1}(k),Z/q) \text{ for } i \leq 2n.$$

d.) The natural inclusion $Sp_{2n}(k) \to Sp_{2n+2}(k)$ induces isomorphisms

$$H_i(Sp_{2n}(k), Z/q) \xrightarrow{\sim} H_i(Sp_{2n+2}(k), Z/q) \text{ for } i \leq 4n+2.$$

e.) The natural inclusion $O_n(k) \to O_{n+1}(k)$ induces isomorphisms

$$H_i(O_n(k), Z/q) \xrightarrow{\sim} H_i(O_{n+1}(k), Z/q) \text{ for } i \leq n-2,\ p \neq 2.$$

f.) The natural inclusion $SO_n(k) \to SO_{n+1}(k)$ induces isomorphisms

$$H_i(SO_n(k), Z/q) \xrightarrow{\sim} H_i(SO_{n+1}(k), Z/q) \text{ for } i \leq n-2.$$

<u>Proof</u>. Let $\varprojlim(\ )^{\wedge}: \mathscr{H} \to \mathscr{H}$ be the Sullivan prime-to-p profinite completion functor defined on the homotopy category of connected C. W. complexes [3]. Let $G_n(\ )$ denote $GL_n(\ )$ (respectively, $SL_n(\ )$; resp., $Sp_{2n}(\ )$; resp., $SO_n(\ )$). We recall that the frobenius self-map φ of algebraic varieties defined over k determines maps

$$\varphi: \varprojlim(BG_n(\mathbb{C}))^{\wedge} \to \varprojlim(BG_n(\mathbb{C}))^{\wedge}$$

where the complex Lie group $G_n(\mathbb{C})$ is viewed as a topological group. (The map $\varphi: G_{n,\bar{k}} \to G_{n,\bar{k}}$ of algebraic groups over the algebraic closure of k is the $\bar{k}$-linear map induced by the p^d-th power map $G_{n,k} \to G_{n,k}$, where p^d is the order of k).

By Theorem 2.9 of [2], there is a homotopy commutative square

$$\text{(2.1)} \qquad \begin{array}{ccc} BG_n(k) & \longrightarrow & \varprojlim(BG_n(\mathbb{C}))^{\wedge} \\ \Big\downarrow \alpha_n & & \Big\downarrow \Delta_n \\ \varprojlim(BG_n(\mathbb{C}))^{\wedge} & \xrightarrow{1\times\varphi} & (\varprojlim(BG_n(\mathbb{C}))^{\wedge})^{\times 2} \end{array}$$

with the property that any associated map of homotopy theoretic fibres $\mathrm{fib}(\alpha_n) \to \mathrm{fib}(\Delta_n)$ induces isomorphisms

$$H_*(\mathrm{fib}(\alpha_n), Z/q) \xrightarrow{\sim} H_*(\mathrm{fib}(\Delta_n), Z/q).$$

Moreover, diagram (2.1) is naturally determined by a commutative diagram of simplicial schemes (together with a choice of embedding into $\mathbb{C}$ of the Witt vectors of the algebraic closure $\bar{k}$ of k). Thus, $G_n(\ ) \to G_{n+1}(\ )$ determines a homotopy commutative square.

(2.2)

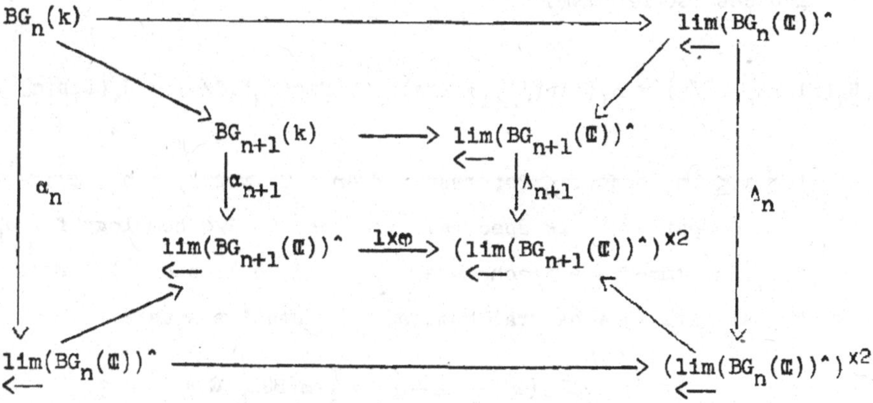

Using Lemma 1 together with the fact that $U_n \to GL_{2n}(\mathbb{C})$, $Sp_n \to Sp_{2n}(\mathbb{C})$, and $O_n \to O_n(\mathbb{C})$ are homotopy equivalneces, we conclude that $\underleftarrow{\lim}(BG_n(\mathbb{C}))^\wedge \to \underleftarrow{\lim}(BG_{n+1}(\mathbb{C}))^\wedge$ induces isomorphisms

$$H_i(\underleftarrow{\lim}(BG_n(\mathbb{C}))^\wedge, Z/q) \xrightarrow{\sim} H_i(BG_n(\mathbb{C}), Z/q)$$

$$\xrightarrow{\sim} H_i(BG_{n+1}(\mathbb{C}), Z/q) \xrightarrow{\sim} H_i(\underleftarrow{\lim}(BG_{n+1}(\mathbb{C}))^\wedge, Z/q)$$

for $i \leq 2n + 1$ (respectively, $\leq 2n + 1$; resp., $\leq 4n + 3$; resp., $\leq n - 1$). Using a spectral sequence comparison theorem comparing the Serre spectral sequences in Z/q homology for the maps Δ_n and Δ_{n+1}, we conclude that any maps on homotopy theoretic fibres

$$\mathrm{fib}(\Delta_n) \to \mathrm{fib}(\Delta_{n+1}),\ \mathrm{fib}(\alpha_n) \to \mathrm{fib}(\alpha_{n+1})$$

fitting in a homotopy commutative square of homotopy fibres associate to (2.2)

$$(2.3)\qquad \begin{array}{ccc} \mathrm{fib}(\alpha_n) & \to & \mathrm{fib}(\Delta_n) \\ \downarrow & & \downarrow \\ \mathrm{fib}(\alpha_{n+1}) & \to & \mathrm{fib}(\Delta_{n+1}) \end{array}$$

induces isomorphisms

$$H_i(\mathrm{fib}(\Delta_n),Z/q) \xrightarrow{\sim} H_i(\mathrm{fib}(\Delta_{n+1}),Z/q),\ H_i(\mathrm{fib}(\alpha_n),Z/q) \xrightarrow{\sim} H_i(\mathrm{fib}(\alpha_{n+1}),Z/q)$$

for $i \leq 2n$ (resp., $\leq 2n$; resp., $\leq 4n + 2$; resp., $\leq n - 2$).

Comparing Serre spectral sequences in Z/q homology for α_n and α_{n+1}, we immediately conclude a.), b.), d.), and f.). Moreover, $BU_n(k)$ fits in a natural homotopy commutative square

$$(2.4)\qquad \begin{array}{ccc} BU_n(k) & \longrightarrow & \underleftarrow{\lim}(BGL_n(\mathbb{C}))^\wedge \\ \downarrow \alpha_n & & \downarrow \Delta_n \\ \underleftarrow{\lim}(BGL_n(\mathbb{C}))^\wedge & \xrightarrow{1\times\psi} & (\underleftarrow{\lim}(BGL_n(\mathbb{C}))^\wedge)^{\times 2} \end{array}$$

where ψ is the composition of the map ϖ induced by frobenius and the transpose-inverse map. As for (2.1), $\mathrm{fib}(\alpha_n) \to \mathrm{fib}(\Delta_n)$ induces isomorphisms $H_*(\mathrm{fib}(\alpha_n),Z/q) \xrightarrow{\sim} H_*(\mathrm{fib}(\Delta_n),Z/q)$ and fits in a square of the form (2.3). We thus prove c.) exactly as we proved a.). Finally, e.) follows directly from f.) and the map of Hochschild-Serre spectral sequences in Z/q cohomology associated to the natural map of extensions:

$$\begin{array}{ccccccccc} 1 & \to & SO_n(k) & \to & O_n(k) & \to & Z/2 & \to & 1 \\ & & \downarrow & & \downarrow & & \downarrow & & \\ 1 & \to & SO_{n+1}(k) & \to & O_{n+1}(k) & \to & Z/2 & \to & 1 \end{array}$$

The following integral stability theorem is essentially implicit in [1], Theorem 2.2. The essential ingredient in the proof of that theorem is the "homologically cartesian square" (2.1) in the proof of Theorem 2 above. One could also prove Theorem 3 directly from Theorem 2 arguing as in the proof of Theorem 6 below.

<u>Theorem 3</u>. Let F be the algebraic closure of the prime field F_p.

a.) The natural inclusion $GL_n(F) \to GL_{n+1}(F)$ induces isomorphisms

$$H_i(GL_n(F),Z) \xrightarrow{\sim} H_i(GL_{n+1}(F),Z) \text{ for } i \leq 2n - 1.$$

b.) The natural inclusion $SL_n(F) \to SL_{n+1}(F)$ induces isomorphisms

$$H_i(SL_n(F),Z) \xrightarrow{\sim} H_i(SL_{n+1}(F),Z) \text{ for } i \leq 2n - 1.$$

c.) The natural inclusion $Sp_{2n}(F) \to Sp_{2n+2}(F)$ induces isomorphisms

$$H_i(Sp_{2n}(F),Z) \xrightarrow{\sim} H_i(Sp_{2n+2}(F),Z) \text{ for } i \leq 4n + 1.$$

d.) The natural inclusion $O_n(F) \to O_{n+1}(F)$ induces isomorphisms

$$H_i(O_n(F),Z) \xrightarrow{\sim} H_i(O_{n+1}(F),Z) \text{ for } i \leq n - 2,\ p \neq 2.$$

e.) The natural inclusion $SO_n(F) \to SO_{n+1}(F)$ induces isomorphisms

$$H_i(SO_n(F),Z) \xrightarrow{\sim} H_i(SO_{n+1}(F),Z) \text{ for } i \leq n - 2.$$

<u>Proof</u>. By [1], Theorem 2.2, there is a natural fibre triple

$$BG_n(F)^+ \to BG_n(\mathbb{C}) \to (BG_n(\mathbb{C}))_{(p)}$$

where $G_n(\ ) = GL_n(\ )$ (respectively, $SL_n(\ )$; resp., $Sp_{2n}(\ )$; resp., $O_n(\ )$ with $p \neq 2$; resp., $SO_n(\ )$), where $BG_n(\mathbb{C})$ is the classifying space of $G_n(\mathbb{C})$ as a topological group, and where $BG_n(\mathbb{C}) \to (BG_n(\mathbb{C}))_{(p)}$ is localization at p. Since $BG_n(F) \to BG_n(F)^+$ induces isomorphisms in integral homology, the theorem follows from Lemma 1 together with the long exact sequence in homotopy. (The reader should recall that a map $f: X \to Y$ of connected topological spaces which induces isomorphisms $\pi_i(X) \xrightarrow{\sim} \pi_i(Y)$ for $i \leq n$ must also induce isomorphisms $H_i(X,Z) \xrightarrow{\sim} H_i(Y,Z)$ for $i \leq n$.)

To complement Theorem 2, one would like to prove that

$$H_i(G_n(k),Z/p) \xrightarrow{\sim} H_i(G_{n+1}(k),Z/p)$$

for $i \ll n$. Unfortunately, such a result seems difficult. What we proceed to prove in Proposition 5 below is the existence of isomorphisms

$$H^i(SL_n(k),Z/p) \xleftarrow{\sim} H^i(GL_n(k),Z/p)$$

for $i < n/2$ (in fact, $i \leq n/2$ if k has more than 3 elements). Quillen's stability theorem for $GL_n(k)$ then implies that

$$H_i(SL_n(k),Z/p) = H_i(GL_\infty(k),Z/p) = 0$$

for any finite field k with more than 2 elements and any $i < n/2$.

Proposition 4. Let k be a finite field. Let $T \subset GL_n(k)$ be the group of diagonal matrices, let $U^1 \subset GL_n(k)$ be the group of "strictly upper triangular matrices" (upper triangular with diagonal entries equal to 1). Let $U^m \subset U^1$ be the group of $n \times n$ matrices (a_{ij}) satisfying $a_{ij} = 0$ if $i + m > j$ and $i \neq j$. For any positive integer r, $H^r(U^1, Z/p)$ has a filtration stable under the action of T whose associated graded module is isomorphic as a $Z/p[T]$ module to a subquotient of

$$M^r = \bigoplus_{s_1+\cdots+s_{n-1}=r} \left(\bigotimes_{j=1}^{n-1} H^{s_j}(U^j/U^{j+1}, Z/p) \right)$$

where T acts on U^m by conjugation in $GL_n(k)$.

Proof. Define $M^{r,m} = \bigoplus_{s_1+s_2+\cdots+s_m=r} \left(\bigotimes_{j=1}^{m} H^{s_j}(U^j/U^{j+1}, Z/p) \right)$. We inductively prove for any r that $H^r(U^1/U^{m+1}, Z/p)$ admits a filtration whose associated graded module is isomorphic as a $Z/p[t]$ module to subquotient of $M^{r,m}$. Assume that this is known for $m - 1$ (for $m - 1 \leq 1$, this is trivial). Since

$$(4.1) \qquad 0 \to U^m/U^{m+1} \to U^1/U^{m+1} \to U^1/U^m \to 1$$

is a central extension, the Hochshild-Serre spectral sequence of this extension is of the form

$$(4.2) \quad E_2^{s,t} = H^s(U^1/U^m, Z/p) \otimes H^t(U^m/U^{m+1}, Z/p) \Longrightarrow H^N(U^1/U^{m+1}, Z/p)$$

Moreover, conjugation by any element $\tau \epsilon T$ determines a self-map of extension (4.1) and thus a self-map of spectral sequence (4.2) whose effect on $E_2^{s,t}$ is obtained by conjugating U^1/U^m and U^m/U^{m+1} by $\tau \epsilon T$. Therefore, the self-map of (4.1) induced by $\tau \epsilon T$ induces an action of τ on the associated graded module $\bigoplus_{s+t=r} E_\infty^{s,t}$ of

$H^r(U^1/U^{m+1},Z/p)$; moreover, $\bigoplus_{s+t=r} E_\infty^{s,t}$ is a subquotient of $\bigoplus_{s+t=r} E_2^{s,t}$ as a $Z/p[T]$ module. Applying induction and the fact that a subquotient of a subquotient of a module is itself a subquotient of that module, we conclude the proposition.

Proposition 4 provides us with a sufficient computational tool to apply to the Z/p cohomology of U^1 (which is a p-Sylow subgroup of $GL_n(k)$) that we can now conclude the following proposition.

<u>Proposition 5</u>. Let k be a finite field of characteristic p. If $i < n/2$, then $SL_n(k) \to GL_n(k)$ induces an isomorphism

$$H^i(GL_n(k),Z/p) \xrightarrow{\sim} H^i(SL_n(k),Z/p).$$

Moreover, if $k \neq F_2, F_3$, then

$$H^i(GL_{2i}(k),Z/p) \xrightarrow{\sim} H^i(SL_{2i}(k),Z/p).$$

<u>Proof</u>. Let B_n (respectively, sB_n) be the subgroup of upper triangular matrices of $GL_n(k)$ (resp., $SL_n(k)$); let T (resp., sT) be the subgroup of diagonal matrices of $GL_n(k)$ (resp., $SL_n(k)$). Then

$$H^*(GL_n(k),Z/p) \xrightarrow{\sim} H^0(T/sT,H^*(SL_n(k),Z/p))$$

$$H^*(B_n,Z/p) \xrightarrow{\sim} H^0(T/sT,H^*(sB_n,Z/p))$$

Since sB_n contains a p-Sylow subgroup of $SL_n(k)$ (namely, the group U^1 of strictly upper triangular matrices),

$$H^*(SL_n(k),Z/p) \to H^*(sB_n,Z/p)$$

is injective. Thus, it suffices to prove that T/sT acts trivially on $H^i(sB_n, Z/p)$ for $i < n/2$ or $i \leq n/2$ as asserted. Moreover,

$$H^*(B_n, Z/p) \xrightarrow{\sim} H^0(T, H^*(U^1, Z/p))$$

$$H^*(sB_n, Z/p) \xrightarrow{\sim} H^0(sT, H^*(U^1, Z/p))$$

so that it suffices to prove that T acts trivially on the elements of $H^i(U^1, Z/p)$ invariant under sT for $i < n/2$ or $i \leq n/2$ as asserted.

Since the order of T (respectively, sT) is prime to the order of the Z/p vector space $H^i(U^1, Z/p)$, Proposition 4 implies that the invariant submodule of M^i with respect to the action of T (resp., sT) has as subquotient the associated graded module of the invariant submodule of $H^i(U^1, Z/p)$ with respect to the action of T (resp., sT). Therefore, it suffices to prove that T acts trivially on the elements of M^i invariant under sT for $i < n/2$ or $i \leq n/2$ as asserted.

Since U^j/U^{j+1} is a k-vector space,

$$H^{s_j}(U^j/U^{j+1}, Z/p) \xrightarrow{\sim} \bigoplus_{a_j+2b_j=s_j} (\wedge^{a_j}(U^j/U^{j+1}) \otimes S^{b_j}(U^j/U^{j+1}))$$

as a Z/p[T] module, where $\wedge^{\cdot}(V)$ (respectively, $S^{\cdot}(V)$) is the free exterior (resp., symmetric) algebra associated to V viewed as a Z/p vector space. Therefore, it suffices to prove that T acts trivially on the elements of each $M^i(a_1, b_1, \ldots, a_{n-1}, b_{n-1})$ invariant under sT, where

$$M^i(a_1, b_1, \ldots, a_{n-1}, b_{n-1}) = \wedge^{a_1}(U^1/U^2) \otimes S^{b_1}(U^1/U^2) \cdots \otimes S^{b_{n-1}}(U^{n-1})$$

with $a_1 + 2b_1 + \cdots + a_{n-1} + 2b_{n-1} = i$, $a_j \geq 0$, $b_j \geq 0$.

Let F be the algebraic closure of k. Then

$$(U^j/U^{j+1}) \underset{Z/p}{\otimes} F = \mathrm{Span}\{x_{c,c+j,e} \mid 1 \leq c \leq n-j,\ 0 \leq e < [k: F_p]\}$$

where the diagonal matrix $\tau = \text{diag}(t_1,\dots,t_n)$ conjugates $x_{c,c+j,e}$ to $(t_c t_{c+j}^{-1})^{p^e} x_{c,c+j,e}$. Therefore, $M^i(a_1,b_1,\dots,a_{n-1},b_{n-1}) \otimes_{Z/p} F$ is spanned by elements which transform as a tensor product of at most i such $x_{c,c+j,e}$'s. If $x \in M^i(a_1,\dots,b_{m-1}) \otimes_{Z/p} F$ transforms as such a tensor product and if $i < n/2$, then some integer d with $1 \leq d \leq n$ does not occur as a first or second subscript of any $x_{c,c+j,e}$ in the tensor product; for such an x and any $\tau = \text{diag}(t_1,\dots,t_n) \in T$, x transforms under conjugation by τ and conjugation by $\tau' = \text{diag}(t_1,\dots,t_{d-1},\det(\tau^{-1})\cdot t_d,t_{d+1},\dots,t_n) \in sT$ in the same way. Consequently, any element of $M^i(a_1,\dots,b_{n-1}) \otimes_{Z/P} F$ invariant under sT is also invariant under T, whenever $i < n/2$. Therefore, T acts trivially on the elements of each $M^i(a_1,\dots,b_{n-1})$ invariant under sT for $i < n/2$.

If $i = n/2$ is an integer and if every integer d between 1 and n occurs among the first and second subscripts of the tensor factors of some $x \in M^i(a_1,\dots,b_{n-1}) \otimes_{Z/p} F$ transforming as a tensor product of $x_{c,c+j,e}$'s, then $x = x_{j_1,j_2,e_2} \wedge \dots \wedge x_{j_{n-1},j_n,e_n}$ in $M^i(a_1,0,\dots,a_{n-1},0) \otimes_{Z/p} F$. Conjugation by $\tau = \text{diag}(t_1,\dots,t_n) \in T$ sends x to $(\prod_{k=1}^{n} t_{j_k}^{(-1)^{k+1}p^{e_k}})\cdot x$ where $e_{2m} = e_{2m-1}$. If $\tau \in sT$ so that $t_{j_n} = (t_{j_1} \cdots t_{j_{n-1}})^{-1}$, then x transforms to

$$(\prod_{k=1}^{n-1} t_{j_k}^{(-1)^{k+1}p^{e_k}+p^{e_n}})\cdot x$$

This implies that $(-1)^{k+1}p^{e_k}+p^{e_n}$ is congruent to 0 mod $p^{[k:F_p]} - 1$ for every k between 1 and $n - 1$ with $e_{2m} = e_{2m-1}$, whenever such an x is invariant under sT. Since this is possible only if $p = 2$ or 3 and if $[k: F_p] = 1$, the proposition is proved.

The interested reader should be able to improve the range of equality in Proposition 5 if he eliminates a few troublesome examples. For example, $H^1(SL_2(F_2),Z/2) \neq 0$, $H^1(SL_2(F_3),Z/3) \neq 0$, and $H^2(SL_3(F_4),Z/2) \neq 0$.

By combining Theorem 2, Proposition 5, and Quillen's stability theorem for $GL_n(k)$, we conclude the following stability theorem for $SL_n(k)$.

Theorem 6. Let k be a finite field with more than 2 elements. The natural inclusion $SL_n(k) \to SL_{n+1}(k)$ induces isomorphisms

$$H_i(SL_n(k),Z) \xrightarrow{\sim} H_i(SL_{n+1}(k),Z)$$

for $i < n/2$. Moreover, $H_i(SL_{2i}(k),Z) \xrightarrow{\sim} H_i(SL_{2i+1}(k),Z)$ provided that k has more than 3 elements.

Proof. Since $SL_n(k)$ and $SL_{n+1}(k)$ are finite, $H_i(SL_n(k),Z)$ and $H_i(SL_{n+1}(k),Z)$ are finite abelian groups for any positive integer i. Moreover, Theorem 2 and Proposition 5 imply that

$$(6.1) \qquad H_i(SL_n(k),Z/N) \xrightarrow{\sim} H_i(SL_{n+1}(k),Z/N)$$

for $i < n/2$ or $i \leq n/2$ as asserted, and for any positive integer N (to obtain (6.1) use the duality of $H_i(SL_n(k),Z/q)$ and $H^i(SL_n(k),Z/q)$ for any prime q, and the long exact sequence in homology for a short exact sequence of coefficient groups). We inductively assume

$$H_{i-1}(SL_n(k),Z) \xrightarrow{\sim} H_{i-1}(SL_{n+1}(k),Z)$$

for some $i < n/2$ or $i \leq n/2$ as asserted, and let N be the product of the orders of $H_i(SL_n(k),Z)$ and $H_i(SL_{n+1}(k),Z)$. Then the followin commutative diagram

$$\begin{array}{ccccccccc}
0 \to & H_i(SL_n(k),Z) & \to & H_i(SL_n(k),Z/N) & \to & H_{i-1}(SL_n(k),Z) & \xrightarrow{N} & H_{i-1}(SL_n(k) \\
 & \downarrow & & \downarrow & & \downarrow & & \downarrow \\
0 \to & H_i(SL_{n+1}(k),Z) & \to & H_i(SL_{n+1}(k),Z/N) & \to & H_{i-1}(SL_{n+1}(k),Z) & \xrightarrow{N} & H_{i-1}(SL_{n+1}(
\end{array}$$

has exact rows. Since the second, third, and fourth vertical arrows are isomorphisms, $H_i(SL_n(k),Z) \to H_i(SL_{n+1}(k),Z)$ is an isomorphism as well.

References

[1] E. Friedlander, Unstable K-theories of the algebraic closure of a finite field, Comment. Math. Helvetici 50(1975), 145-154.

[2] E. Friedlander, Computations of K-theories of finite fields, Topology Vol. 15(1976), 87-109.

[3] D. Sullivan, Genetics of homotopy theory and the Adams conjecture, Annals of Math. 100(1974), 1-79.

HERMITIAN K-THEORY IN TOPOLOGY: A SURVEY OF SOME RECENT RESULTS

William Pardon

Columbia University

In this part of the survey, we discuss only developments subsequent to Julius Shaneson's survey [SS], which relate geometry (so far this means manifold theory) to Hermitian K-theory. Many of the results require considerable preparation even to state, so we will not always be precise. It goes without saying that the resultant simplification restricts us to giving vague and incomplete descriptions.

I. W.-C. Hsiang and R. Sharpe [HS] have introduced a new surgery obstruction group $L_n^{st}(\pi)$ in their study of the homotopy groups of the group $\mathrm{Diff}(M^n)$ of diffeomorphisms of a closed smooth n-manifolds M^n (Charles Giffen[G] has independently defined and studied the L_n^{st}-groups from an algebraic point of view).

Let $n \geq 5$, M^n as above, let $\mathrm{Aut}(M)$ be the H-space of simple homotopy equivalences of M and let $\mathfrak{J}(M)$ be the homotopy fibre of the map $\mathrm{Diff}(M) \to \mathrm{Aut}(M)$. $\mathfrak{J}(M)$ has points $\{\psi, \psi_t\}$ $\psi \in \mathrm{Diff}(M)$, ψ_t a path in $\mathrm{Aut}(M)$ connecting ψ to the identity. Using the mapping torus of ψ_t, there is a natural map $\mathfrak{J}(M) \xrightarrow{\tau} \mathcal{S}_0(M \times (S^1, 1))$, (see [SS, §1]) for $\mathcal{S}_0$) the components of whose homotopy fibre can be calculated using [HW]. We also have $\eta: \mathcal{S}_0(M \times (S^1, 1) \to (G/O)^{\Sigma M_+}$ ([SS, 1.3.3]). Letting $\mathfrak{C}(M)$ be the fibre of $\eta\tau$ it is shown that $\pi_i\, \mathfrak{C}(M)$ is represented by cobordism classes of normal maps $W \to M \times D^{i+2}$, satisfying certain extra conditions. Most of [HS] is then devoted to a geometric analysis of this kind of surgery problem, using the algebraic and geometric discussion of pseudo-isotopies in [HW]. The result is this. In [Sh], Sharpe defines groups $St(\pi)$, $E(\pi)$, and $\widetilde{SU}^{\varepsilon}(\pi)$, analogous to $St(\mathbb{Z}\pi)$, $E(\mathbb{Z}\pi)$, and $SU^{\varepsilon}(\mathbb{Z}\pi)$, the Steinberg group, elementary group and special unitary group of the stable ε-hermitian hyperbolic form. He also defines a "split elementary unitary group" $\widetilde{EU}(\pi)$ and its universal central extension $\widetilde{StU}(\pi)$. Let P^ε be the pullback in

$$\begin{array}{ccc} P^\varepsilon & \longrightarrow & St(\pi) \\ \downarrow & & \downarrow \\ \widetilde{SU}^\varepsilon(\pi) & \longrightarrow & E(\pi) \end{array}$$

Then $L^{st}_{2k+1}(\pi) = P^\varepsilon/([P^\varepsilon, P^\varepsilon], \sigma_\varepsilon)$, $\sigma_\varepsilon = \binom{0\ 1}{\varepsilon\ 0}$,

$\varepsilon = (-1)^k$, and $L^{st}_{2k}(\pi) = \ker\ (\widetilde{StU}^\varepsilon(\pi) \to \widetilde{EU}^\varepsilon(\pi))$.

There is an exact sequence

$[\Sigma^2 M_+, G/O] \to \pi_0 \mathcal{C}(M) \to \pi_0\ \mathcal{OJ}(M) \overset{\eta\tau^\#}{\to} [\Sigma M_+, G/O]$ relating L^{st}-groups to $\mathcal{OJ}(M)$; another exact sequence and results from [HW] show for example

<u>THEOREM</u>: If $n \geq 5$ and $i < n$, then π_i $(\mathrm{Diff}(T^n))$ contains an infinite direct sum of copies of $\mathbb{Z}/2\mathbb{Z}$.

II. Sylvain Cappell has refined his splitting theorems [C1] by studying the groups $UNil_n$ introduced in [C1]. Here are some results. Let H, G_1, G_2 be finitely-presented groups $H \subset G_1, G_2$, let R be a subring of $\mathbb{Q}$ and let $R[\hat{G_i}] \subset R[G_i]$ denote the augumentation ideal, $i = 1, 2$.

<u>THEOREM [C2]</u>. There is for each $n \geq 0$ a 2-primary abelian group $UNil^h_n$ $(R[H]; R[\hat{G_i}]$), $i = 1, 2, 3)$ such that

1) It is zero if $\frac{1}{2} \in R$ or if H is square-root closed in G_1 and G_2 ([C2, Def, p. 5])

2) If Φ is the quadrad of rings $\left(\begin{array}{ccc} R[H] & \longrightarrow & R[G_1] \\ \downarrow & & \downarrow \\ R[G_2] & \longrightarrow & R[G_1 *_H G_2] \end{array}\right)$

$UNil^h_n$ is a summand of L^h_n (Φ) ([W, §3])

3) There is a split monomorphism

$UNil^h_n(R[H]; R[\hat{G_i}]) \longrightarrow L^h_n\ (R[G_1 *_H G_2])$

whose cokernel $\hat{L}^h_n$ $(R[G_1 *_H G_2])$ appears (under some K_0-conditions) in the long exact sequence

$$\cdots \to L^h_n(R[H]) \to L^h_n(R[G_1]) \oplus L^h_n(R[G_2]) \to \hat{L}^h_n(R[G_1 *_H G_2]) \to L^h_{n-1}(R[H] \to \cdots$$

THEOREM [C3]: There is a closed differentiable manifold W, simple homotopy equivalent to $RP^{4k+1} \# RP^{4k+1}$, $k \geq 2$, which is not differentiably, p.ℓ, or topologically a non-trivial connected sum.

The proof of the second theorem uses the existence of a non-trivial element of order two in $L_{4k+2}(\mathbb{Z}_2 * \mathbb{Z}_2)/L_{4k+2}(\mathbb{Z}_2) \oplus L_{4k+2}(\mathbb{Z}_2)$, which can be detected by UNiℓ methods. Thus $UNil_n^h$ seems to appear as the obstruction to good Mayer-Vietoris sequences and to connected-sum decompositions of manifolds.

III. Let R be a subring of $\mathbb{Q}$ and π a group. In [M1] Mishchenko introduced a notion of chain complexes over $R[\pi]$ with Poincaré duality built in, and the relation of cobordism of such chain complexes. If $\frac{1}{2} \in R$, then Mishchenko shows "cobordim" gives an equivalence relation, whose equivalence classes form a group, $\Omega_n(R\pi) \otimes \mathbb{Q} \cong L_n^h(R\pi) \otimes \mathbb{Q}$. Further, if $(f, b) : M^n \to N^n$ is a normal map of Poincaré complexes, then the difference of the chain complexes $[C_*(M)] - [C_*(N)]$ in $\Omega_n(R\pi) \otimes \mathbb{Q}$ is the "rational part" of surgery obstruction for (f, b) in $L_n^h(R\pi) \otimes \mathbb{Q}$. By adding to his definition of chain complexes the structure of the higher chain approximation to the diagonal (reduced mod 2 they give part of the Steenrod operations [Sp, 5.9]), Mishchenko [M2] succeeded in removing the restriction $1/2 \in R$, without, however, being able to extract the entire surgery obstruction from chain information in the normal map; for example, if $\pi = 1$, the Kervaire-Arf invariant was not detectable. Recently Andrew Rainicki [R] has completed Mishchenko's program by giving a definition of chain complexes having even more structure than Mishchenko's and capturing the surgery obstruction from chain information in the normal map, $(f, b) : M^n \to N^n$. There are many applications of this to Hermitian K-theory, although the identification of the surgery obstruction for a specific normal map by chain methods seems quite difficult. To mention a few applications the product pairing $\Omega_m \times L_n(\pi) \to L_{m+n}(\pi)$ recently determined geometrically by John Morgan can easily be analyzed by Rainicki's methods; many "higher signature" maps can be defined; and "L-theory" can be relativized in many ways leading to localization sequences, Mayer-Vietoris sequences, and the Cappell-Shaneson Γ-groups.

IV. Motivated by the localization sequence of [Sw, 8.2], the reviewer proved the existence of a long exact localization sequence for Wall groups [Pa 1]

$$\cdots \to L^h_{n+1}(\mathbb{Q}\pi) \to L^h_n(\mathbb{Q}\pi/\mathbb{Z}\pi) \to L^h_n(\mathbb{Z}\pi) \to L^h_n(\mathbb{Q}\pi) \to L^h_{n-1}(\mathbb{Q}\pi/\mathbb{Z}\pi) \to \cdots$$

where π is finite, and if $n \geq 5$, $L^h_n(\mathbb{Q}\pi/\mathbb{Z}\pi)$ is represented by cobordism classes of normal maps (in the style of [W, §9]) $(f, b) : M^n \to X^n$ such that f induces a rational homology equivalence and such that (f, b) and (f', b') are cobordant if there is a normal cobordism between them which is a rational homology h-cobordism. The notion of surgery on framed embedded $\mathbb{Z}_n$-manifolds used in place of surgery on spheres, and $L^h_n(\mathbb{Q}\pi/\mathbb{Z}\pi)$ becomes a "surgery theory" in the sense that it satisfies a "$\pi - \pi$" theorem [W, §5] and can be described algebraically as forms (n odd) or formations (n even). The geometry gives the method of proof of the localization sequence for Witt groups in these Proceedings. Karoubi proved part of the localization sequence in [K] and Rainicki [R'] has announced that it is a consequence of the algebraic theory of surgery on chain complexes.

V. Various subsets of the set S = {P. Conner, F. Raymond, J.W. Alexander, G. Hamrick, J. Vick} have done work in Hermitian K-theory algebraically distinguishable from the above in [AH], [ACHV], [C'1], [C'2], [C R]. Let π be a finite group. The Grothendieck group $GU_0^{\varepsilon}(\mathbb{Z}\pi)$ of ε-hermitian forms $\psi : M \times M \to \mathbb{Z}\pi$ ($\varepsilon = \pm 1$) they consider requires only that M be a $\mathbb{Z}$-free $\mathbb{Z}\pi$-module, <u>not</u> necessarily $\mathbb{Z}\pi$-projective or -free (there is also no "self-intersection" or quadratic form associated to ψ). Let $WU_0^{\varepsilon}(\mathbb{Z}\pi)$ denote $GU_0^{\varepsilon}(\mathbb{Z}\pi)$ modulo ε-hermitian forms $h : M \times M \to \mathbb{Z}\pi$ for which there is a $\mathbb{Z}\pi$-submodule $N \subseteq M$ such that $\psi | N \times N \equiv 0$ and the induced form

$N \times (M/N) \to \mathbb{Z}\pi$ is nonsingular. We summarize what seems to be the most complete result of S([AH]).

Suppose given an orientation - preserving action of the group π on a closed smooth 2n-manifold M^{2n}. The G-signature theorem of Atiyah - Segal - Singer determines, in terms of orbit data, the class $w_{\mathbb{R}}(\lambda) \in WU_0^{\varepsilon}(\mathbb{R}\pi)$ of the intersection pairing $\lambda : (H_n(M^{2n})/\text{tor}) \times (H_n(M^{2n})/\text{tor}) \to \mathbb{Z}\pi, \varepsilon = (-1)^n$, a theorem of Fröhlich states that $WU_0^{\varepsilon}(\mathbb{Z}\pi) \to (WU_0^{\varepsilon}(\mathbb{R}\pi)$ is injective with kernel equal to the torsion subgroup of $WU_0^{\varepsilon}(\mathbb{Z}\pi)$, and so it is natural to ask whether one can determine from orbit data the class $w_{\mathbb{Z}}(\lambda)$ in $WU_0^{\varepsilon}(\mathbb{Z}\pi)$ of λ itself. To answer this question [HV] defines for each $T \in \pi$ of order n and each prime p, $p|n$, a <u>torsion signature</u> $f_{\lambda}(T, p) \in \mathbb{Z}_2$.

<u>THEOREM [AH]</u>: For a group π of odd order, $w_{\mathbb{Z}}(\lambda)$ is completely determined by the multisignature of Atiyah - Segal - Singer and the torsion signatures $f_{\lambda}(T, P) \equiv \chi(M^T \cap (M^{T^{n_p}})_2) \bmod 2$ where M^T is the fixed point set of T, $n_p \cdot p^r = n$, $(n_p, p) = 1$ and $(M^{T^{n_p}})_2$ is the union of the components of the fixed point set of codimension $\equiv 2 \pmod 4$.

There is an example with $\pi = \mathbb{Z}/15\,\mathbb{Z}$ where the multisignature vanishes, but $f_{\lambda}(T, p)$ does not, for some T, p.

VI. Given a locally flat, codimension - two piecewise - linear (p.l.) embedding of p.l. manifolds $i : M^n \to W^{n+2}$, it is well known to topologists that one may make the embedding i non-locally-flat by replacing a disc pair $(i(D^n), D^{n+2}) \subseteq (i(M^n), M^{n+2})$ by the cone on a non-trivial knot $j : S^{n-1} \to S^{n+1}$. Here "non-trivial knot" means that j is not concordant to the standard imbedding. One very special case of the theorems of [C S 1] is that the knot group C_{n-1} (concordance classes of imbeddings $S^{n-1} \to S^{n+1}$) is isomorphic to a group of "singular hermitian forms", a Γ-group [C S 1],

$$\Gamma_{n+2}\begin{pmatrix} \mathbb{Z}[\mathbb{Z}] & \to & \mathbb{Z}[\mathbb{Z}] \\ \downarrow & & \downarrow \text{aug} \\ \mathbb{Z}[\mathbb{Z}] & \xrightarrow{\text{aug}} & \mathbb{Z} \end{pmatrix}$$

(See [C S 1, 6.4])

In [C S 2] Sylvain Cappell and Julius Shaneson have undertaken a systematic study of non-locally-flat p.l. embeddings $i : M^n \to W^{n+2}$ using the techniques and results of [C S 1]. For example,

THEOREM [C S 2, 6.1]: Let $f : M^n \to W^{n+2}$ be a homotopy equivalence of compact, oriented, p.l. manifolds, M closed, $n \geq 3$. If n is even, assume $\pi_1 W = 0$. Then f is homotopic to a p.l. embedding.

THEOREM [C S 2, 6.4]: Let M^n, $n \geq 3$ be a closed, orientable, p.l. manifold. Suppose that the Hurewicz homophormism $\pi_{n+1}(\Sigma M) \to H_{n+1}(\Sigma M)$ is surjective, ΣM the suspension of M. Then there is a p.l. embedding of M in S^{n+2}.

Sometimes the embeddings produced by these theorems have to be non-locally-flat. In fact, Cappell and Shaneson show that (in the case of the second theorem above) the Poincaré duals of the rational Pontriagin classes of M^n are carried by various levels of a natural stratification of the non-locally-flat part of the embedding [C S 2, 6.8]. Thus, if the rational Pontriagin classes of M are non-zero, the embedding cannot be locally flat; and one even gets an estimate of the dimension of the set of bad points by knowing which Pontriagin classes are non-zero !

VII. Jean-Claude Hausmann [H] has used the Cappell-Shaneson Γ-groups to study a cobordism group of homology spheres.

THEOREM 1: Let $N \rightarrowtail G \twoheadrightarrow Q$ be a short exact sequence of groups where N is normally generated by a finitely-generated perfect group (in particular, N is perfect). Then there is an isomorphism

$$\Gamma_k^e(\mathbb{Z}G \to \mathbb{Z}Q) \cong L_k^e(\mathbb{Z}Q),\ e = s \text{ or } h.$$

Using the theorem when $Q = \{e\}$ Hausmann obtains the following description of a group of homology spheres. Let G be a perfect group and let $\Omega_n^H(G)$ be homology h-cobordism classes of homology n-spheres X with $\pi_1 X = G$. Let $c : X \to B_G$ (= K(G, 1)) be the natural map; applying the "plus-construction" of Quillen we obtain

$$d : X^+ \to B_G^+.$$

Since the plus-construction preserves homology ($H_*(X;\mathbb{Z}) \overset{\cong}{\to} H_*(X^+;\mathbb{Z})$) and kills $\pi_1 X$, X^+ is a (homotopy) sphere. This gives an element $[d] \in \pi_n(B_G^+)$.

THEOREM 2: The correspondence $X \mapsto [d]$ induces an isomorphism

$$\Omega_n^H(G) \overset{\cong}{\to} \pi_n(B_G^+).$$

Given a Poincaré complex Y, recall that Sullivan defined the notion of homotopy triangulations (smoothings, topological structures) on Y, $\mathcal{S}_{PL}(Y)$ ($\mathcal{S}_O(Y)$, $\mathcal{S}_{TOP}(Y)$) [SS]. Hausmann defines an analogous notion of homology triangulation (smoothing, etc.) of a homology n-manifold. He then (using Theorem 1) shows that this set of structures on X is in one-to-one correspondence with $\mathcal{S}_{CAT}(X^+)$, where CAT = O, PL, ot TOP, provided X^+ is Poincaré.

BIBLIOGRAPHY

[AH] J.P. Alexander and G.C. Hamrick, "Torsion G-signature theorem for groups of odd order," preprint.

[ACHV] J.P. Alexander, P.D. Conner, G.C. Hamrick, and J.W. Vick, "Witt Classes of integral representations of an abelian p-group," Bull. A.M.S., 80 (1974), 1179-1182.

[C 1] S. Cappell, "Mayer-Vietoris sequences in Hermitian K-theory," Springer Lecture Notes, No. 343, pp. 478-512

[C 2] __________, "Unitary nilpotent groups and Hermitian K-theory," preprint.

[C 3] __________, "Splitting obstructions for Hermitian forms and manifolds with $\mathbb{Z}_2 \subseteq \pi_1$."

[C'1] P. Conner, "Metabolic, Hyperbolic, Split," preprint.

[C'2] __________, "Witt ring invariants for periodic maps," preprint.

[CR] P. Conner and F. Raymond, "The quadratic form on the quotient of a periodic map," Semigroup Forum.

[C S1] S.E. Cappell and Julius L. Shaneson, "The codimension two placement problem and homology equivalent manifolds," Ann. of Math, 99 (1974), 277-348.

[C S2] ____________________, "Piecewise linear embeddings and their singularities," Ann. of Math., January, 1976.

[HS] W.-C. Hsiang and R. Sharpe, "Parameterized surgery and isotopy preprint.

[HW] A. Hatcher and J. Wagoner, "Pseudo-isotopies of compact manifolds," Asterisque, 6 (1974)

[K] M. Karoubi, "Localisation des formes quadratiques I, II," Ann. Sc. Ec. Norm. Sup., Paris, 7 (1975), 359-404.

[M 1] A. Mishchenko, "Homotopy invariants of Manifolds," Math. of USSR-Izvestya, AMS Translation, 4 (1970), 506-519.

[M 2] __________, "Homotopy invariants of Manifolds III," Izvestya Akad. Nauk.

[Pa] W. Pardon, "The exact sequence of a localization in L-theory," Princeton University Thesis, 1974.

[R] A. Rainicki, "The algebraic theory of surgery," preprint.

[Sh] R. Sharpe, "Surgery on compact manifolds: the bounded even-dimensional case," Ann. Math.,

[Sp] E. Spanier, Algebraic Topology, McGraw-Hill, New York, 1966.

[SS] J. Shaneson, "Hermitian K-theory in topology," Springer Lectur Notes, No. 343.

[Sw] R.G. Swan, K-Theory of Finite Groups and Orders, Springer Lecture Notes, No. 149.

[W] C.T.C. Wall, Surgery on Compact Manifolds, Academic Press, New York, 1971.

[H] J.-C. Hausmann, "Homology sphere bordism and Quillen's plus-construction," these proceedings.

HIGHER WITT GROUPS : A SURVEY

by J.-L. Loday

In linear algebraic K-theory Quillen (and some others) defined higher algebraic K-functors K_n for all $n \geq 0$ such that they agree with the well-known Grothendieck functor K_0 and Bass-Whitehead functor K_1. From hermitian or quadratic modules an analogous Grothendieck-type functor has been known for a long time [Wi] : the Witt functor W_0. The purpose of this paper is a review of (essentially two) global* definitions of higher Witt functors W_n.

The first attempt follows closely the first definition of linear K_n given by Quillen, by means of the "plus" construction. This was done by Karoubi [Ka 1].

The second way, more geometric in spirit, is to consider chain complexes of modules with the symmetric (resp. quadratic) structure enjoyed by the chain complex of a manifold (resp. surgery kernel). This method was first suggested by Wall [Wl 2]. It was developed by Mischenko [Mi] in the symmetric case, and by Ranicki [Ra 1] in the quadratic case.

We do not want to discuss here the many different types of Witt groups in dimension 0 and refer to the article of Bak [Bk 1] for a general discussion on this subject (see also [Kn]). Nevertheless we try to follow the "uniform notation" adopted in [Bk 1].

In the first section we fix the notations and give the definitions of Witt groups, unitary and orthogonal groups.

The second section is devoted to the definitions of higher Witt groups

- by means of the "plus" construction,
- by means of algebraic complexes.

* "Global" means that these definitions give W_n for all n without assuming any periodicity a priori.

In the third section we give the properties of the higher Witt groups $W_n(A)$ when 2 is invertible in the ring A and when we neglect 2-torsion.

In the fourth section we look again at the properties of higher Witt groups cited before but without assuming 2 invertible and taking care of 2-torsion. This section is far from complete (especially in dimension 0 and 1) because of the complexity of the phenomena and because of lack of time in preparing this paper.

I should like to thank M. Karoubi and A. Ranicki for much help and useful comments.

1. NOTATIONS AND DEFINITIONS.

Let A be a hermitian ring i.e. a ring with involution $a \to \bar{a}$; thus $\bar{\bar{a}} = a$, $\overline{a+b} = \bar{a} + \bar{b}$, $\overline{ab} = \bar{b}\bar{a}$, for all $a, b \in A$.

Let M be a right A-module. The dual $\tilde{M}$ of M is the right A-module of additive functions $f : M \to A$ such that $f(ma) = \bar{a}\, f(m)$. A sesquilinear form on M is a A-homomorphism $\varphi : M \to \tilde{M}$. It is equivalent to give a bi-additive function $\Phi : M \times M \to A$ such that $\Phi(ma, nb) = \bar{a}\, \Phi(m,n)b$ (define Φ by $\Phi(m,n) = \varphi(n)(m)$). The form φ is called non degenerate (or non singular) if it is an isomorphism. Let $\mathrm{Sesq}(M)$ be the additive group of sesquilinear forms on the finitely generated projective (f.g.p.) module M . As $\tilde{\tilde{M}}$ is isomorphic to M (because M is f.g.p.) $\tilde{\varphi}$ is in $\mathrm{Sesq}(M)$. Let ε be in the center of A and $\varepsilon\bar{\varepsilon} = 1$. We put $S_\varepsilon : \mathrm{Sesq}(M) \to \mathrm{Sesq}(M)$ by $S_\varepsilon(\varphi) = \varphi + \bar{\varepsilon}\tilde{\varphi}$ (examples : $\varepsilon = +1$ or -1).

DEFINITION. An ε-hermitian module over A is a right f.g.p. A-module M together with a non-degenerate sesquilinear form φ such that $\tilde{\varphi} = \varepsilon\varphi$ (i.e. $\varphi \in \mathrm{Ker}\, S_{-\varepsilon}$).

DEFINITION. An ε-quadratic module over A is a right f.g.p. A-module M together with a class $[q] \in \mathrm{Coker}(S_{-\varepsilon})$ ($q \in \mathrm{Sesq}(M)$) such that the associated hermitian form $S_\varepsilon(q)$ is non degenerate.

A morphism $(M,\varphi) \to (M',\varphi')$ of ε-hermitian modules is a homomorphism $f : M \to M'$ such that $\varphi = \tilde{f}\varphi' f$.

A morphism $(M,[q]) \to (M',[q'])$ of ε-quadratic modules is a homomorphism $f : M \to M'$ such that $q - \tilde{f}q'f \in \mathrm{Im}(S_{-\varepsilon})$.

The category $\underline{\underline{H}}^\varepsilon(A)$ (resp. $\underline{\underline{Q}}^\varepsilon(A)$) of ε-hermitian modules (resp. ε-quadratic modules) and isomorphisms is additive with respect to the usual direct sum of modules and maps. The Grothendieck group of the additive category $\underline{\underline{H}}^\varepsilon(A)$ (resp. $\underline{\underline{Q}}^\varepsilon(A)$) is denoted $KH_0^\varepsilon(A)$ (resp. $KQ_0^\varepsilon(A)$).

Let $\overline{\underline{\underline{P}}(A)}$ be the category of f.g.p. A-modules and isomorphisms. The hyperbolic functor $H : P \to H(P) = (P \oplus \tilde{P}, \begin{pmatrix} 0 & 1 \\ \varepsilon & 0 \end{pmatrix})$ from $\overline{\underline{\underline{P}}(A)}$ to $\underline{\underline{H}}^\varepsilon(A)$ (resp. $H : P \to H(P) = (P \oplus \tilde{P}, [\begin{pmatrix} 0 & 1 \\ 0 & 0 \end{pmatrix}])$ from $\overline{\underline{\underline{P}}(A)}$ to $\underline{\underline{Q}}^\varepsilon(A)$) is additive and defines a map $K_0(A) \to KH_0^\varepsilon(A)$ (resp. $K_0(A) \to KQ_0^\varepsilon(A)$).

DEFINITION. The ε-symmetric Witt group of A is $W_0^\varepsilon(A) = \operatorname{Coker}(K_0(A) \to KH_0^\varepsilon(A))$. The ε-quadratic Witt group of A is $WQ_0^\varepsilon(A) = \operatorname{Coker}(K_0(A) \to KQ_0^\varepsilon(A))$.

Remark : The ε-hermitian form associated to the ε-quadratic form $[q]$ gives rise to a well-defined functor $\underline{\underline{Q}}^\varepsilon(A) \to \underline{\underline{H}}^\varepsilon(A)$ called the symmetrisation functor and hence a commutative square of abelian groups

$$\begin{array}{ccc} KQ_0^\varepsilon(A) & \longrightarrow & KH_0^\varepsilon(A) \\ \downarrow & & \downarrow \\ WQ_0^\varepsilon(A) & \longrightarrow & W_0^\varepsilon(A) \end{array}$$

If there exists λ in the center of A such that $\lambda + \bar{\lambda} = 1$, for instance if 2 is invertible in A , then $KQ_0^\varepsilon(A) = KH_0^\varepsilon(A)$, and $WQ_0^\varepsilon(A) = W_0^\varepsilon(A)$.

Orthogonal and unitary groups.(*) The automorphisms of the quadratic module M for a group $U(M)$ is called the unitary group of M . When M is hyperbolic, $M = H(A^n)$, then we denote this unitary group by $U_{2n}(A)$ (or $U_{2n}^\varepsilon(A)$ if we want to mention the choice of symmetry)

$$U_{2n}^\varepsilon(A) = \{\sigma \in GL_{2n}(A) \mid \bar{\sigma}\begin{pmatrix} 0 & 1_n \\ 0 & 0 \end{pmatrix}\sigma - \begin{pmatrix} 0 & 1_n \\ 0 & 0 \end{pmatrix} \in \operatorname{Im} S_{-\varepsilon}\} .$$

The usual stabilisation $A^n \to A^{n+1}$ leads to a monomorphism $U_{2n}(A) \to U_{2n+2}(A)$ and we put $U(A) = \varinjlim U_{2n}(A)$. Bak uses the notation $GQ_{2n}(A,\min)$ instead of $U_{2n}(A)$ [Bk 1].

Similarly the automorphisms of the hermitian module M form a group called the orthogonal group of M . When M is the hyperbolic module $H(A^n)$ we denote this orthogonal group by $O_{2n}(A)$ (or $O_{2n}^\varepsilon(A)$) and $O(A) = \varinjlim O_{2n}(A)$:

$$O_{2n}^\varepsilon(A) = \{\sigma \in GL_{2n}(A) \mid \bar{\sigma}\begin{pmatrix} 0 & 1_n \\ \varepsilon 1_n & 0 \end{pmatrix}\sigma = \begin{pmatrix} 0 & 1_n \\ \varepsilon 1_n & 0 \end{pmatrix}\} .$$

If one puts $\sigma = \begin{pmatrix} a & b \\ c & d \end{pmatrix}$ and $\sigma^* = \begin{pmatrix} \bar{d} & \varepsilon\bar{b} \\ \bar{c} & \bar{a} \end{pmatrix}$ then

$$O_{2n}^\varepsilon(A) = \{\sigma \in GL_{2n}(A) \mid \sigma^* . \sigma = 1_{2n}\} .$$

THEOREM [Ws, 1.4 and 1.4_0] . <u>The groups</u> $U(A)$ <u>and</u> $O(A)$ <u>are quasi-perfect i.e. their commutator subgroup</u> $[U(A),U(A)]$ <u>and</u> $[O(A),O(A)]$ <u>are perfect</u>.

Proofs of that result may also be found in [Bs 1, p. 156] [K-V] and [Bk 2, 5.5].

Note that there is a map $U^{\varepsilon}(A) \to O^{\varepsilon}(A)$ which is the identity when 2 is invertible in A . Moreover the hyperbolic functor induces a map

$$GL(A) \to U(A) \qquad \alpha \to \begin{pmatrix} \alpha & 0 \\ 0 & \bar{\alpha}^{-1} \end{pmatrix} .$$

(*) The group $U_{2n}(A)$ is called "unitary" here and in [Bs] [Sp] [Wl 2] "quadratic" in [Bk 1 and 2] and "orthogonal" in [Ws]. The word "hermitian" is frequently replaced by "symmetric".

2. DEFINITIONS OF HIGHER WITT GROUPS.

a) Homology of groups. In linear algebraic K-theory the commutator subgroup $E(A)$ of $GL(A)$ is perfect and has thus a universal central extension $St(A)$. Higher algebraic K-theory is defined by $K_1(A) = H_1(GL(A),\mathbb{Z})$, $K_2(A) = H_2(E(A),\mathbb{Z})$, $K_3(A) = H_3(St(A),\mathbb{Z})$ where $H_i(-,\mathbb{Z})$ denotes homology with trivial coefficients $\mathbb{Z}$. As the group $U^\varepsilon(A)$ is quasiperfect one can mimic the linear case and put :

$$KQ_1^\varepsilon(A) = H_1(U^\varepsilon(A),\mathbb{Z})$$

$$KQ_2^\varepsilon(A) = H_2(EU^\varepsilon(A),\mathbb{Z}) \quad \text{where} \quad EU^\varepsilon(A) = [U^\varepsilon(A),U^\varepsilon(A)]$$

$KQ_3^\varepsilon(A) = H_3(StU^\varepsilon(A),\mathbb{Z})$ where $StU^\varepsilon(A)$ is the universal central extension of $EU^\varepsilon(A)$ [Bk 1].

Finally one put $WQ_i^\varepsilon(A) = \mathrm{Coker}(K_i(A) \to KQ_i^\varepsilon(A))$ for $1 \leq i \leq 3$. This definition (with slight modifications) was first employed by Sharpe [Sp] who was able to compare the groups $KQ_2^\varepsilon(A)$ and $KQ_0^{-\varepsilon}(A)$ (see 4.2.a)).

b) With the "+" construction. The group $U(A)$ is quasi-perfect, so we can apply the "+" construction of Quillen [Ld, 1.1.1] to the classifying space $BU(A)$ relatively to the commutator subgroup and get the connected space $BU(A)^+$. The hyperbolic functor induces a continuous map $BGL(A)^+ \to BU(A)^+$.

DEFINITION. $KQ_n^\varepsilon(A) = \pi_n(BU^\varepsilon(A)^+)$, $WQ_n^\varepsilon(A) = \mathrm{Coker}(K_n(A) \to KQ_n^\varepsilon(A))$.

The same argument as in linear algebraic K-theory permits to prove that these definitions agree with those of a) for $1 \leq i \leq 3$.

The same construction can be done for the orthogonal group $O^\varepsilon(A)$ and gives rise to the (symmetric) higher Witt groups $W_n^\varepsilon(A)$, $n \geq 1$.
In [Ka 1] Karoubi always assume that 2 is invertible in A and denotes the group of orthogonal matrices of dimension 2n by ${}_\varepsilon O_{n,n}(A)$. Then he puts ${}_\varepsilon L_n(A) = \pi_n(BO(A)^+)$ and ${}_\varepsilon W_n(A) = \mathrm{Coker}(K_n(A) \to {}_\varepsilon L_n(A))$. Therefore when 2 is invertible in A we have

$$ {}_\varepsilon L_n(A) = KQ_n^\varepsilon(A) \quad \text{and} \quad {}_\varepsilon W_n(A) = WQ_n^\varepsilon(A) \ . $$

Let SA be the suspension of the ring A [K-V]. The proof of the following theorem is similar to the linear case (see [Ld, ch. I]).

THEOREM. <u>For</u> $n \geq 1$ <u>the group</u> $KQ_n^\varepsilon(SA)$ <u>is isomorphic to</u> $KQ_{n-1}^\varepsilon(A)$ <u>and</u> $WQ_n^\varepsilon(SA)$ <u>is isomorphic to</u> $WQ_{n-1}^\varepsilon(A)$.

Therefore it is natural to put $KQ_{-m}^\varepsilon(A) = KQ_0^\varepsilon(S^m A)$ and $WQ_{-m}^\varepsilon(A) = WQ_0^\varepsilon(S^m A)$ for $m > 0$.

c) <u>Algebraic Poincaré complexes.</u> To define higher Witt groups for all $n \geq 0$ Mischenko (resp. Ranicki) starts with n-dimensional chain complexes carrying a symmetric (resp. quadratic) structure.

Given a right A-module chain complex C define a $\mathbb{Z}[\mathbb{Z}_2]$-module chain complex

$$ C \otimes_A C = C \otimes_{\mathbb{Z}} C/\{xa \otimes y - x \otimes y\bar{a} \mid x, y \in C, a \in A\} \ , $$

with the generator $T \in \mathbb{Z}_2$ acting by

$$ T_\varepsilon : C \otimes_A C \to C \otimes_A C \ ; \ x \otimes y \mapsto (-)^{(\deg x)(\deg y)} y \otimes \varepsilon x \ . $$

Define the $\mathbb{Z}_2$- hypercohomology (resp. $\mathbb{Z}_2$- hyperhomology) groups

$$ Q^n(C,\varepsilon) = H_n(\mathrm{Hom}_{\mathbb{Z}[\mathbb{Z}_2]}(W, C \otimes_A C)) \qquad (\text{resp. } Q_n(C,\varepsilon) = H_n(W \otimes_{\mathbb{Z}[\mathbb{Z}_2]} (C \otimes_A C)) $$

with W a free $\mathbb{Z}[\mathbb{Z}_2]$-resolution of $\mathbb{Z}$, letting T act on $\mathbb{Z}$ by the identity. An element $\varphi \in Q^n(C,\varepsilon)$ (resp. $\psi \in Q_n(C,\varepsilon)$) is represented by a collection of chains $\varphi_s \in (C \otimes_A C)_{n+s}$ (resp. $\psi_s \in (C \otimes_A C)_{n-s}$) for $s \geq 0$ satisfying the relations

$$ d\varphi_s + (-)^{n+s-1}(\varphi_{s-1} + (-)^s T_\varepsilon \varphi_{s-1}) = 0 \in (C \otimes_A C)_{n+s-1} \qquad (\varphi_{-1} = 0) $$

$$ (\text{resp. } d\psi_s + (-)^{n-s-1}(\psi_{s+1} + (-)^{s+1} T_\varepsilon \psi_{s+1}) = 0 \in (C \otimes_A C)_{n-s-1}). $$

DEFINITION. ([Mi], resp. [Ra 1]). An n-dimensional ε-symmetric (resp. ε-quadratic) Poincaré complex over A (C,φ) (resp. (C,ψ)) is a f.g.p. A-module chain complex

$$C : C_n \to C_{n-1} \to C_{n-2} \to \ldots \to C_1 \to C_0$$

together with a class $\varphi \in Q^n(C,\varepsilon)$ (resp. $\psi \in Q_n(C,\varepsilon)$) such that evaluation of the slant product

$$\backslash : H_n(C \otimes_A C) \otimes_{\mathbb{Z}} H^r(C) \to H_{n-r}(C) \ ; \ (x \otimes y) \otimes f \mapsto x.\overline{f(y)}$$

on $\varphi_0 \in H_n(C \otimes_A C)$ (resp. $(1+T_\varepsilon)\psi_0 \in H_n(C \otimes_A C)$) induces A-module isomorphisms

$$H^r(C) \to H_{n-r}(C) \qquad (0 \leq r \leq n) \ .$$

In particular, a 0-dimensional ε-symmetric (resp. ε-quadratic) Poincaré complex is the same as an ε-hermitian (resp. ε-quadratic) module, as defined in § 1.

The higher Witt groups are obtained from such algebraic Poincaré complexes by passing to algebraic cobordism classes, using the appropriate abstraction of Poincaré-Lefschetz duality for cobordisms of manifolds. The algebraic cobordism group of n-dimensional ε-symmetric (resp. ε-quadratic) Poincaré complexes over A is denoted by $L^n(A,\varepsilon)$ (resp. $L_n(A,\varepsilon)$). Despite the notation both the functors

$$L^n(\ ,\varepsilon)\ ,\ L_n(\ ,\varepsilon) : (\text{hermitian rings}) \to (\text{abelian groups})$$

are covariant, and there is defined an ε-symmetrization map

$$(1+T_\varepsilon) : L_n(A,\varepsilon) \to L^n(A,\varepsilon)\ ;\ (C,\psi) \to (C,(1+T_\varepsilon)\psi)\ ,\ ((1+T_\varepsilon)\psi)_s = \begin{cases} (1+T_\varepsilon)\psi_0 & s=0 \\ 0 & s \geq 1 \end{cases} .$$

The groups $L^n(A,1)$ were denoted by $\Omega_n(A)$ in the work of Mischenko [Mi]. It is a theorem of Ranicki [Ra 1] that the groups $L_n(A,1)$ are isomorphic to the surgery obstruction groups $L_n(A)$ of Wall [Wl 2], which are 4-periodic by construction. Algebraic analogues of the n-ads of Wall [Wl 2] are used in [Ra 1] to construct infinite loop spaces $\mathbb{L}^0(A,\varepsilon)$, $\mathbb{L}_0(A,\varepsilon)$ (as simplicial sets) such that

$\pi_n(\mathbb{L}^0(A,\varepsilon)) = L^n(A,\varepsilon)$, $\pi_n(\mathbb{L}_0(A,\varepsilon)) = L_n(A,\varepsilon)$. Further, there are defined natural maps

$$BO^{\varepsilon}(A)^+ \to \mathbb{L}^0(A,\varepsilon)$$

$$BU^{\varepsilon}(A)^+ \to \mathbb{L}_0(A,\varepsilon)$$

such that composition with the hyperbolic maps

$$H : BGL(A)^+ \to BO^{\varepsilon}(A)^+$$

$$H : BGL(A)^+ \to BU^{\varepsilon}(A)^+$$

is null-homotopic, so that there are defined natural maps

$$W_n^{\varepsilon}(A) \to L^n(A,\varepsilon)$$

$$WQ_n^{\varepsilon}(A) \to L_n(A,\varepsilon) \ .$$

These maps are not isomorphisms in general : for example, both the maps

$$W_1^1(\mathbb{Z}) = \mathbb{Z}_2 \to L^1(\mathbb{Z},1) = \mathbb{Z}_2$$

$$WQ_1^1(\mathbb{Z}) = \mathbb{Z}_2 \to L_1(\mathbb{Z},1) = 0$$

are zero.

d) Comparison of notations.

	Grothendieck group of ε-quadratic modules	ε-Quadratic Witt group	(ε-Symmetric) Witt group
This paper	$KQ_0^{\varepsilon}(A)$	$WQ_0^{\varepsilon}(A)$	$W_0^{\varepsilon}(A)$
Bak [Bk 1]	$KQ_0^{\varepsilon}(A,\min)$	$WQ_0^{\varepsilon}(A,\min)$	
Bass [Bs 1]	$KU_0^{\varepsilon}(A,\min)$	$W_0^{\varepsilon}\ (A,\min)$	
Karoubi [Ka], [K-V]	${}_{\varepsilon}L_0(A)$	${}_{\varepsilon}W_0(A)$	
Ranicki [Ra 1]		$L_0(A,\varepsilon)$	$L^0(A,\varepsilon)$
Milnor-Husemoller [M-H]		$WQ(A)$ if $\varepsilon=1$	$W(A)$ if $\varepsilon=1$
Surgery groups [Wl 2]		$L_{2i}(\pi)$ if $\varepsilon=(-1)^i$ and $A=\mathbb{Z}[\pi]$	

Bak defines groups (for instance $WQ_i(A,\Lambda)$) depending on a "form parameter" Λ which is an additive subgroup of A such that

$$\{a - \varepsilon\bar{a} \mid a \in A\} \subset \Lambda \subset \{a \mid a \in A,\ a = -\varepsilon\bar{a}\}$$

and $a\Lambda\bar{a} \subset \Lambda$ for all $a \in A$. The minimum choice of Λ is denoted by min. For more complete comparisons see [Bk 1].

3. PROPERTIES OF $W_n(A)$ UP TO 2-TORSION.

Many of the theorems quoted in this section are consequences of more precise results (giving information on 2-torsion for instance) which are collected in section 4. The notation $\bar{W}_n$ is introduced in 3.2 and used till the end of this section.(*)

3.1. Hermitian and quadratic theories. The symmetrization map is an isomorphism up to 2-torsion :

$$WQ_n^{\varepsilon}(A) \otimes \mathbb{Z}[\tfrac{1}{2}] \simeq W_n^{\varepsilon}(A) \otimes \mathbb{Z}[\tfrac{1}{2}] .$$

3.2. Periodicity (and notations).

a) Karoubi's framework. The following result is called "weak periodicity theorem" and is proved in [Ka 1, th. 5.12] when 2 is invertible in A .

THEOREM. (See § 4.2.a) and (*)). The group ${}_{\varepsilon}W_n(A) \otimes \mathbb{Z}[\tfrac{1}{2}]$ is isomorphic to ${}_{-\varepsilon}W_{n+2}(A) \otimes \mathbb{Z}[\tfrac{1}{2}]$ and therefore isomorphic to ${}_{\varepsilon}W_{n+4}(A) \otimes \mathbb{Z}[\tfrac{1}{2}]$.

From the isomorphisms $WQ_n^{\varepsilon}(A) \otimes \mathbb{Z}[\tfrac{1}{2}] \simeq W_n^{\varepsilon}(A) \otimes \mathbb{Z}[\tfrac{1}{2}]$ we deduce that the groups $WQ_n^{\varepsilon}(A) \otimes \mathbb{Z}[\tfrac{1}{2}]$ are periodic of period 4 .

b) Ranicki's framework.

THEOREM. [Ra 1]. The group $L_n(A,\varepsilon)$ is isomorphic to $L_{n+2}(A,-\varepsilon)$ and therefore isomorphic to $L_{n+4}(A,\varepsilon)$.

When 2 is invertible in A the ε-symmetrization map $(1+T_{\varepsilon}) : L_n(A,\varepsilon) \to L^n(A,\varepsilon)$ is an isomorphism for all $n \geq 0$.

c) Comparison and notations. For $n = 0$ see [Bk 1].

PROPOSITION. $L_1(A,\varepsilon) \otimes \mathbb{Z}[\tfrac{1}{2}]$ is isomorphic to $W_1^{\varepsilon}(A) \otimes \mathbb{Z}[\tfrac{1}{2}]$. Therefore $L_n(A,\varepsilon) \otimes \mathbb{Z}[\tfrac{1}{2}]$ is isomorphic to $W_n^{\varepsilon}(A) \otimes \mathbb{Z}[\tfrac{1}{2}]$ for all n .

To simplify notations, till the end of this section we put

$$\bar{W}_0(A) = {}_1W_0(A) \otimes \mathbb{Z}[\tfrac{1}{2}] = L_0(A,\ 1) \otimes \mathbb{Z}[\tfrac{1}{2}]$$

$$\bar{W}_1(A) = {}_1W_1(A) \otimes \mathbb{Z}[\tfrac{1}{2}] = L_1(A,\ 1) \otimes \mathbb{Z}[\tfrac{1}{2}]$$

$$\bar{W}_2(A) = {}_{-1}W_0(A) \otimes \mathbb{Z}[\tfrac{1}{2}] = L_0(A,-1) \otimes \mathbb{Z}[\tfrac{1}{2}]$$

$$\bar{W}_3(A) = {}_{-1}W_1(A) \otimes \mathbb{Z}[\tfrac{1}{2}] = L_1(A,-1) \otimes \mathbb{Z}[\tfrac{1}{2}]$$

and for all $n \in \mathbb{Z}$, $\bar{W}_{4n+i}(A) = \bar{W}_i(A)$, $0 \leq i \leq 3$.

3.3. Change of rings. For every morphism of hermitian rings $f : A \to A'$ there are defined relative groups $\bar{W}_n(f)$ for $n(\bmod 4)$ which fit into a long exact sequence

$$\ldots \to \bar{W}_n(f) \to \bar{W}_n(A) \xrightarrow{f_*} \bar{W}_n(A') \to \bar{W}_{n-1}(f) \to \bar{W}_{n-1}(A) \to \ldots$$

3.4. Mayer-Vietoris sequence. For every cartesian square of hermitian rings

$$\begin{array}{ccc} A & \longrightarrow & A_1 \\ \downarrow & & \downarrow f_1 \\ A_2 & \xrightarrow{f_2} & A' \end{array}$$

where f_1 or f_2 is surjective one has the following exact sequence

$$\ldots \to \bar{W}_{n+1}(A') \to \bar{W}_n(A) \to \bar{W}_n(A_1) \oplus \bar{W}_n(A_2) \to \bar{W}_n(A') \to \bar{W}_{n-1}(A) \to \ldots$$

for all $n \in \mathbb{Z}$.

3.5. Multiplicative structure. The tensor product of modules and morphisms give an associative product

$$\bar{W}_n(A) \times \bar{W}_p(A') \to \bar{W}_{n+p}(A \otimes_{\mathbb{Z}} A')$$

for every hermitian rings A and A' .

3.6. Homotopy invariance. Let $A[t]$ the polynomial ring over A with $\bar{t} = t$. The natural inclusion $A \to A[t]$ induces a map

$$\bar{W}_n(A) \to \bar{W}_n(A[t] \ , \ n \in \mathbb{Z}$$

which is an isomorphism.

3.7. Laurent extensions. Let $A[t,t^{-1}]$ be the ring of polynomials in t and t^{-1} over A with the involution $\bar{t} = t^{-1}$. Then there is an isomorphism

$$\bar{W}_n(A[t,t^{-1}]) \simeq \bar{W}_n(A) \oplus \bar{W}_{n-1}(A) \ , \ n \in \mathbb{Z} .$$

The map $\bar{W}_n(A[t,t^{-1}]) \to \bar{W}_n(A)$ is induced by $t \to 1$.

The map $\bar{W}_n(A) \to \bar{W}_n(A[t,t^{-1}])$ is induced by the inclusion $A \to A[t,t^{-1}]$.

The map $\bar{W}_n(A[t,t^{-1}]) \to \bar{W}_{n-1}(A)$ is the composed homomorphism

$$\bar{W}_n(A[t,t^{-1}]) \to \bar{W}_n(SA) \simeq \bar{W}_{n-1}(A) .$$

The map $\bar{W}_{n-1}(A) \to \bar{W}_n(A[t,t^{-1}])$ is given by the product with a particular element of $\bar{W}_1(\mathbb{Z}[t,t^{-1}])$.

3.8. Localisation sequence. Let S be a multiplicative subset of non zero divisors s of A such that S lies in the center of A and is stable by the involution. Let A_S be the localised ring of A by S. We put $\hat{A} = \varprojlim_{s} A/sA$ and $\hat{A}_S = (\hat{A})_{i(S)}$ where i is the canonical map $A \to \hat{A}$.

THEOREM. The following sequence is exact

$$\ldots \to \bar{W}_n(A) \to \bar{W}_n(\hat{A}) \oplus \bar{W}_n(A_S) \to \bar{W}_n(\hat{A}_S) \to \bar{W}_{n-1}(A) \to \ldots \ , \ n \in \mathbb{Z} .$$

3.9. Topological K-theory. Let X be a compact space and $C_{\mathbb{R}}(X)$ be the ring of continuous functions from X to $\mathbb{R}$. Then we have an isomorphism

$$\bar{W}_n(C_{\mathbb{R}}(X)) \to KO^{-n}(X) \otimes \mathbb{Z}[\tfrac{1}{2}]$$

where KO denotes real topological K-theory.

Thus the periodicity theorem of 3.2. generalizes the "weak" Bott periodicity theorem.

3.10. <u>Surgery</u>. Let $A = \mathbb{Z}[\pi]$ be the group ring of π with involution $g \to g^{-1}$ for $g \in \pi$. Then we have isomorphisms

$$\bar{W}_n(\mathbb{Z}[\pi]) \simeq L_n(\pi) \otimes \mathbb{Z}[\tfrac{1}{2}]$$

where $L_n(\pi)$ denotes the surgery obstruction group defined by Wall [Wl 2].

(*) <u>Note about the proofs</u>. Karoubi claims that for every hermitian ring A one has isomorphisms ${}_\varepsilon\bar{W}_n(A) \approx {}_\varepsilon\bar{W}_n(A \otimes_{\mathbb{Z}} \mathbb{Z}[\tfrac{1}{2}])$. Therefore one can assume that 2 is invertible in A to prove all the results about ${}_\varepsilon\bar{W}_n(A)$ stated in § 3. References for them will be find in corresponding section of § 4.

4. PROPERTIES OF HIGHER WITT GROUPS.

Most often we state the assertions with the notations of their author. When the choice of symmetry is not important or well understood we omit ε in the formulas. As we are essentially interested in higher Witt theory many results only known in dimension 0 are not quoted here. No proofs, nor computations are given but we always try to indicate precise references.

4.1. Hermitian and quadratic theories. The symmetrization map $WQ_n(A) \to W_n(A)$ permits one to compare quadratic Witt groups and symmetric Witt groups. The generator E_8 of $WQ_0^1(\mathbb{Z})$ is the quadratic module $(\mathbb{Z}^8, [q])$ with

$$q = \begin{pmatrix} 1 & 1 & \cdot & \cdot & \cdot & \cdot & \cdot & \cdot \\ \cdot & 1 & 1 & \cdot & \cdot & \cdot & \cdot & \cdot \\ \cdot & \cdot & 1 & 1 & \cdot & \cdot & \cdot & \cdot \\ \cdot & \cdot & \cdot & 1 & 1 & \cdot & \cdot & \cdot \\ \cdot & \cdot & \cdot & \cdot & 1 & 1 & \cdot & \cdot \\ \cdot & \cdot & \cdot & \cdot & \cdot & 1 & 1 & 1 \\ \cdot & \cdot & \cdot & \cdot & \cdot & \cdot & 1 & \cdot \\ \cdot & \cdot & \cdot & \cdot & \cdot & \cdot & \cdot & 1 \end{pmatrix} .$$

Its image in $W_0^1(A)$ is eight times the generator $(\mathbb{Z},1)$. The product by E_8 defines $W_n(A) \to WQ_n(A)$ such that both the composites with the symmetrization map are multiplication by 8 .

Similarly, product with the generator $E_8 \in L_0(\mathbb{Z},1)$ defines a map $L^n(A,\varepsilon) \to L_n(A,\varepsilon)$ such both the composites with the ε-symmetrization map $(1+T_\varepsilon) : L_n(A,\varepsilon) \to L^n(A,\varepsilon)$ are multiplication by 8 ([Ra 1]).

4.2. Periodicity.

a) Sharpe's framework. [Sp]. Let $SU^\varepsilon(A)$ be the quasi-perfect subgroup of elements $\sigma \in U^\varepsilon(A)$ such that $\sigma \in E(A)$. Let $KU_1^\varepsilon(A)$ (resp. $KU_2^\varepsilon(A)$) be the homology group $H_1(SU^\varepsilon(A),\mathbb{Z})$ (resp. $H_2([SU^\varepsilon(A),SU^\varepsilon(A)],\mathbb{Z})$). Then if $KU_0^\varepsilon(A)$ denotes the Grothendieck group of stable, simple cogredience classes of simple matrices of the form $X+\varepsilon\bar{X}$, a detailed study of the universal central extension of the commutator subgroup $[SU^\varepsilon(A),SU^\varepsilon(A)]$ yields to the periodicity exact sequence :

$$KU_1(A) \to K_2(A)/\{c+\bar{c} \mid c \in K_2(A)\} \to KU_2^\varepsilon(A) \to KU_0^\varepsilon(A) \to \mathbb{Z}/(3-\varepsilon)\mathbb{Z} .$$

b) Karoubi's framework. In [Ka 1] Karoubi constructs two elements $v_2 \in {}_{-1}W_2(\mathbb{Z}[\frac{1}{2}])$ and $v_{-2} \in {}_{-1}W_2(\mathbb{Z}[\frac{1}{2}])$ the product of which is the generator of ${}_1W_0(\mathbb{Z}[\frac{1}{2}])$. The product by v_2 (resp. v_{-2}) induces a map $\beta : {}_\varepsilon W_n(A) \to {}_{-\varepsilon}W_{n+2}(A)$ (resp. $\beta' : {}_\varepsilon W_n(A) \to {}_{-\varepsilon}W_{n-2}(A)$) where A is a ring with 2 invertible (see 4.8).

THEOREM [Ka 1, 5.7]. For all hermitian rings A where 2 is invertible the composed maps $\beta \circ \beta'$ and $\beta' \circ \beta$ are both multiplication by 4. (*)

This result can be deduced from a refined periodicity theorem we describe now. We assume 2 invertible in A.

The hyperbolic functor (resp. the forgetful functor) induces a morphism $GL(A) \to {}_\varepsilon O(A)$ (resp. ${}_\varepsilon O(A) \to GL(A)$) and then a continuous map $BGL(A)^+ \to B_\varepsilon O(A)^+$ (resp. $B_\varepsilon O(A)^+ \to BGL(A)^+$) ; le ${}_\varepsilon\mathcal{U}(A)$ (resp. ${}_\varepsilon\mathcal{V}(A)$) be its homotopic-fiber and ${}_\varepsilon U_n(A) = \pi_n({}_\varepsilon\mathcal{U}(A))$ (resp. ${}_\varepsilon V_n(A) = \pi_n({}_\varepsilon\mathcal{V}(A))$) for $n \geq 1$. Ad hoc definitions for $n = 0$ can be given such that ${}_\varepsilon U_0(A) = {}_\varepsilon U_1(SA)$ (resp. ${}_\varepsilon V_0(A) = {}_\varepsilon V_1(SA)$) where SA denotes the suspension of the ring A.

"STRONG PERIODICITY" THEOREM [Ka 1, 5.15]. If 2 is invertible in A then the spaces $\Omega_\varepsilon\mathcal{U}(A)$ and ${}_{-\varepsilon}\mathcal{V}(A)$ are homotopy equivalent, thus ${}_\varepsilon U_{n+1}(A) \approx {}_{-\varepsilon}V_n(A)$.

This theorem is stated as a conjecture in [Ka 1] because it depends on a result [Ka 1, p. 312] which has been proved since [Ld, Cor. 2.3.6 and 3.1.7]. This strong theorem looks like the 8-periodicity theorems of Bott in real topological K-theory [Ka 1, p. 311].

c) Ranicki's framework. One has the following results.

THEOREM [Ra 1]. The product with the generator of $L^2(\mathbb{Z},-1)$ yields an isomorphism $L_n(A,\varepsilon) \to L_{n+2}(A,-\varepsilon)$, hence the groups $L_n(A,\varepsilon)$ are periodic of period 4.

THEOREM [Ra 1]. If A is noetherian of finite global dimension m then the product with the generator of $L^2(\mathbb{Z},-1)$ defines a map $L^n(A,\varepsilon) \to L^{n+2}(A,-\varepsilon)$ which is an isomorphism if $n \geq 2m$ and a monomorphism if $n = 2m-1$.

Note that no assumptions about the invertibility of 2 is made in Ranicki's results. Mischenko [Mi] claimed that the map $L^n(A,\varepsilon) \to L^{n+2}(A,-\varepsilon)$ is an isomorphism for all A, but his method of proof breaks down if 2 is not invertible in A ([Ra 1]).

4.3. Change of rings. For every higher Witt theory F_n and every morphism of hermitian rings $f : A \to A'$ there are defined relative groups $F_n(f)$, $n \in \mathbb{Z}$, which fit into a long exact sequence

$$\ldots \to F_n(f) \to F_n(A) \to F_n(A') \to F_{n-1}(f) \to F_{n-1}(A) \to \ldots$$

Taking $F_n = WQ_n$, $F_n(f)$ is the nth homotopy group of the homotopic-fiber of the map $BU(A)^+ \to BU(A')^+$ induced by f.

Taking $F_n = L_n(-,\varepsilon)$ then the groups $L_n(f,\varepsilon)$ are periodic of period 4.

Let F_n be $\overline{W}_n$ and suppose that 2 is invertible in A and A'. If f is surjective then $\overline{W}(f)$ is canonically isomorphic with $\overline{W}(\mathrm{Ker}\ f)$ [Ka 1, 5.9].

4.4. Mayer-Vietoris sequence. Let

(*)
$$\begin{array}{ccc} A & \xrightarrow{g_1} & A_1 \\ {\scriptstyle g_2}\downarrow & & \downarrow{\scriptstyle f_1} \\ A_2 & \xrightarrow{f_2} & A' \end{array}$$

be a cartesian square of hermitian rings and F_n be a higher Witt theory. A Mayer-Vietoris (abbreviated M-V) sequence for F_n is an exact sequence

$$\ldots \to F_n(A) \to F_n(A_1) \oplus F_n(A_2) \to F_n(A') \xrightarrow{\partial} F_{n-1}(A) \to F_{n-1}(A_1) \oplus F_{n-1}(A_2) \to \ldots$$

where the connected morphism ∂ is natural with respect to the square (*). The exactness of this sequence is equivalent to the "excision isomorphisms" of relative groups $F_n(g_2) \cong F_n(f_1)$ [Bs, p. 194].

a) A M-V sequence for $n = 0,1$ (resp. $n = 0,1,2$) is proved by Bass [Bs, p. 193] for general KU (= KQ)-theory when f_1 (resp. f_1 and f_2) is a surjection.

This sequence may also be found in [K-V] and in [Bk 1].

It is not known if such a sequence is valid for symmetric modules (KH-theory) when 2 is not invertible in A.

b) Various M-V sequences has been proved by Karoubi for the relative groups ${}_{\varepsilon}U_n$ and ${}_{\varepsilon}V_n$ discussed in 4.2.b) [Ka 1, ch. II].

c) Ranicki [Ra 1] claims that there is an M-V sequence in the groups $L_n(\ ,\varepsilon)$ associated to a cartesian square (*) with one of $A_1 \quad A'$, $A_2 \to A'$ onto, or if it is a localization-completion square.

d) Pardon [Pa 1] derived a M-V sequence for higher Witt groups from the M-V sequence of Bass and Sharpe's unitary periodicity.

In the particular case of surgery groups (i.e. $A = \mathbb{Z}[\pi]$) some M-V sequences may be found in the work of Cappell [Ca].

4.5. Multiplicative structure. The tensor product of hermitian modules is still an hermitian module and defines a pairing $KH_0^{\varepsilon}(A) \times KH_0^{\eta}(A') \to KH_0^{\varepsilon\eta}(A \otimes_{\mathbb{Z}} A')$. A basic remark is that the tensor product of an ε-quadratic module $(M,[q])$ over A and an η-hermitian module (N,φ) over A' defines a $\varepsilon\eta$-quadratic module $(M \otimes_{\mathbb{Z}} N, [q\otimes\varphi])$ over $A \otimes_{\mathbb{Z}} A'$. Indeed if q is in $\mathrm{Im}(S_{-\varepsilon})$ i.e. $q = \nu - \varepsilon\bar{\nu}$, then $q\otimes\varphi = \nu\otimes\varphi - \varepsilon\bar{\nu}\otimes\varphi$. As $\varphi \in \mathrm{Ker}\, S_{-\eta}$ i.e. $\varphi = \eta\bar{\varphi}$ one has $q\otimes\varphi = \nu\otimes\varphi - \varepsilon\eta\bar{\nu}\otimes\bar{\varphi} = S_{-\varepsilon\eta}(q\otimes\varphi)$ and $[q\otimes\varphi]$ is well defined.

Thus we get pairings $KQ_0^{\varepsilon}(A) \times KH_0^{\eta}(A') \to KQ_0^{\varepsilon\eta}(A\otimes A')$ and $KQ_0^{\varepsilon}(A) \times KQ_0^{\eta}(A') \to KQ_0^{\varepsilon\eta}(A\otimes A')$.

The tensor product of a hyperbolic module with any hermitian (or quadratic) module is still a hyperbolic module ; therefore we get pairings on Witt groups :

$$W_0^{\varepsilon}(A) \times W_0^{\eta}(A') \to W_0^{\varepsilon\eta}(A \otimes_{\mathbb{Z}} A')$$
$$WQ_0^{\varepsilon}(A) \times W_0^{\eta}(A') \to WQ_0^{\varepsilon\eta}(A \otimes_{\mathbb{Z}} A')$$
$$WQ_0^{\varepsilon}(A) \times WQ_0^{\eta}(A') \to WQ_0^{\varepsilon\eta}(A \otimes_{\mathbb{Z}} A') .$$

When A is commutative composing with maps induced by the product map $A \otimes_{\mathbb{Z}} A \to A$, $a\otimes b \to ab$ gives internal pairings. For instance $W_0^1(A)$ is a ring with unity.

Note that $WQ_0^1(\mathbb{Z})$ is a ring without unity : the product of the generator E_8 by himself is $8\ E_8$.

All these pairings extend to the higher Witt groups. In the "plus construction" framework we have the pairings of [Ka 1, p. 388] [Ld 1, ch. III]

$$W_n^{\varepsilon}(A) \otimes_{\mathbb{Z}} W_p^{\eta}(A') \to W_{n+p}^{\varepsilon\eta}(A \otimes_{\mathbb{Z}} A')$$
$$WQ_n^{\varepsilon}(A) \otimes_{\mathbb{Z}} W_p^{\eta}(A') \to WQ_{n+p}^{\varepsilon\eta}(A \otimes_{\mathbb{Z}} A')$$
$$WQ_n^{\varepsilon}(A) \otimes_{\mathbb{Z}} WQ_p^{\eta}(A') \to WQ_{n+p}^{\varepsilon\eta}(A \otimes_{\mathbb{Z}} A') .$$

There are corresponding pairings in the "chain complex" framework [Ra 1]

$$L^n(A,\varepsilon) \otimes_{\mathbb{Z}} L^p(A',\eta) \to L^{n+p}(A \otimes_{\mathbb{Z}} A', \varepsilon\eta)$$
$$L_n(A,\varepsilon) \otimes_{\mathbb{Z}} L^p(A',\eta) \to L_{n+p}(A \otimes_{\mathbb{Z}} A', \varepsilon\eta)$$
$$L_n(A,\varepsilon) \otimes_{\mathbb{Z}} L_p(A',\eta) \to L_{n+p}(A \otimes_{\mathbb{Z}} A', \varepsilon\eta) .$$

All these pairings are compatible with the symmetrization maps and with the natural maps $W_n \to L^n$, $WQ_n \to L_n$.

Remark : Many homomorphisms in K-theory are given via the product by special elements, for instance periodicity isomorphisms [Ka 1, th. 5.6] [Ra 1].

4.6. Homotopy invariance. Let A be a hermitian ring and $A[t]$ the polynomial ring with involution $\Sigma\, a_i t^i = \Sigma\, \bar{a}_i t^i$. A functor F from (hermitian) rings to abelian groups is homotopy invariant if the inclusion $A \to A[t]$ induces an isomorphism $F(A) \to F(A[t])$.

The forgetful functor from quadratic modules to f.g.p. modules induces a map (in Karoubi's notation) ${}_{\varepsilon}L_n(A) \to K_n(A)$ the kernel of which is denoted ${}_{\varepsilon}W'_n(A)$ ("co-Witt group").

THEOREM. If 2 is invertible in A the functors ${}_{\varepsilon}W_n(A)$ and ${}_{\varepsilon}W'_n(A)$ are homotopy invariant.

This theorem was proved in [Ka 1, 1.1] for ${}_{\varepsilon}W'_0$ $(= {}_{\varepsilon}L'_0)$ and in full generality by Cretin [Cr] up to a conjecture in linear algebraic K-theory which is proved by Quillen in [Qu 2].

This theorem fails when 2 is not invertible in A , namely for $\mathbb{Z}/2$ [Ka 1, p. 318].

When A is noetherian regular and 2 invertible in A the functors ${}_{\varepsilon}L_n$, ${}_{\varepsilon}U_n$ and ${}_{\varepsilon}V_n$ are also homotopy invariant [Ka 1, p. 312 and 5.21].

4.7. Laurent extensions. There are two ways of extending the involution on a ring A to the Laurent extension $A[t,t^{-1}]$. For $\bar{t} = t^{-1}$ we expect a higher Witt theory (F_n) to be such that

$$F_n(A[t,t^{-1}]) = F_n(A) \oplus F'_{n-1}(A)$$

for some other higher Witt theory (F'_n) differing from F_n in 2-torsion only. For $\bar{t} = t$ we expect a split exact sequence

$$0 \to F_n(A) \to F_n(A[t]) \oplus F_n(A[t^{-1}]) \to F_n(A[t,t^{-1}]) \to F'_n(A) \to 0 .$$

The first result of this type was the decomposition fo surgery obstruction groups

$$L^s_n(\pi \times \mathbb{Z}) = L^s_n(\pi) \oplus L^h_{n-1}(\pi)$$

obtained independently by Shaneson [Sh] and Wall [Wl 2] using geometric methods for finitely presented groups π . An algebraic proof of this was obtained by Novikov [No] modulo 2-torsion, and then by Ranicki [Ra 2] in complete generality, for any ring with involution. Another approach to such splitting theorems is due to Karoubi [Ka 1] [Ka 2]. The L-theory of $A[t,t^{-1}]$ for $\bar{t} = t$ has been studied in [Ka 3] and [Ra 3].

4.8. Localization. Let S be a multiplicative subset of non zero divisors s in the center of A such that $\bar{S} \subset S$; the localized ring is denoted by A_S .

a) Low dimension. If A is a Dedekind domain (trivial involution) with quotient field F then the sequence

$$0 \to W^1_0(A) \to W^1_0(F) \to \bigoplus_{\mathfrak{P}} W^1_0(A/\mathfrak{P})$$

is exact, where the direct sum extends over all maximal ideals $\mathfrak{P}$ of A [H-M, p. 93].

b) Karoubi's results. In [Ka 2] [Ka 3] and [Ka 4] the groups come all from quadratic modules that is to say ${}_{\varepsilon}L_n = KQ_n^{\varepsilon}$, ${}_{\varepsilon}W_n = WQ_n^{\varepsilon}$.

From the category of s-torsion A-modules one can construct a group ${}_{\varepsilon}U(A,S)$ [Ka 2, ch. II] such that the following sequence is exact

$$(*) \qquad {}_{\varepsilon}L_1(A) \to {}_{\varepsilon}L_1(A_S) \to {}_{\varepsilon}U(A,S) \to {}_{\varepsilon}L_0(A) \to {}_{\varepsilon}L_0(A_S) .$$

The comparison of this sequence with the analogue with $\hat{A}$ and $\hat{A}_S$ gives the exact sequence [Ka 2, th. 3.1]

$$(**) \qquad {}_{\varepsilon}L_1(\hat{A}) \oplus L_1(A_S) \to {}_{\varepsilon}L_1(\hat{A}_S) \to {}_{\varepsilon}L_0(A) \to {}_{\varepsilon}L_0(\hat{A}) \oplus {}_{\varepsilon}L_0(A_S) \to {}_{\varepsilon}L_0(A_S) .$$

Moreover Karoubi announces that the sequence (**) is still valid when one replace 1 by n and 0 by n-1 .

For particular rings A , namely Dedekind ring with 2 invertible, one can deduce the following exact sequence [Ka 3] :

$${}_{\varepsilon}L_{n+1}(A) \to {}_{\varepsilon}L_{n+1}(F) \to \bigoplus_{\mathfrak{P}} {}_{\varepsilon}U_n(A/\mathfrak{P}) \to {}_{\varepsilon}L_n(A) \to {}_{\varepsilon}L_n(F)$$

where the direct sum extends over all maximal ideals $\mathfrak{P}$ of A. The groups ${}_{\varepsilon}U_n(-)$ are the relative groups defined in 4.2.b) and F is the quotient field.

The situation for Witt groups is not so clear. However these results on ${}_{\varepsilon}L_n$ together with the localization sequence in linear algebraic K-theory enables one to deduce informations about the kernel and the cokernel of ${}_{\varepsilon}W_n(A) \to {}_{\varepsilon}W_n(A_S)$. For instance if $B = A/sA$ is a torsion ring (i.e. $B \otimes_{\mathbb{Z}} \mathbb{Q} = 0$) for all $s \in S$, then ${}_{\varepsilon}W_n(A) \otimes \mathbb{Z}[\frac{1}{2}]$ is isomorphic to ${}_{\varepsilon}W_n(A_S) \otimes \mathbb{Z}[\frac{1}{2}]$ for all $n \in \mathbb{Z}$ (cf. [Ka 4]).

c) Other results. There is a localization exact sequence of type (*) in the ε-symmetric (resp. ε-quadratic) higher Witt groups of Ranicki [Ra 1]

$$\ldots \to L^n(A,\varepsilon) \to L^n(A_S,\varepsilon) \to L^n(A,S,\varepsilon) \to L^{n-1}(A,\varepsilon) \to \ldots$$

$$(\text{resp. } \ldots \to L_n(A,\varepsilon) \to L_n(A_S,\varepsilon) \to L_n(A,S,\varepsilon) \to L_{n-1}(A,\varepsilon) \to \ldots).$$

The relative groups $L^n(A,S,\varepsilon)$ (resp. $L_n(A,S,\varepsilon)$) are algebraic cobordism groups of pairs (C,φ) (resp. (C,ψ)) consisting of an $(n+1)$-dimensional f.g.p. A-module chain complex C which becomes chain contractible over A_S together with a class $\varphi \in Q^{n+1}(C,-\varepsilon)$ (resp. $\psi \in Q_{n+1}(C,-\varepsilon)$). The ε-quadratic sequence is 4-periodic in n, and there is defined a 4-periodic sequence of type (**)

$$\dots \to L_n(A,\varepsilon) \to L_n(\hat{A},\varepsilon) \oplus L_n(A_S,\varepsilon) \to L_n(\hat{A}_S,\varepsilon) \to L_{n-1}(A,\varepsilon) \to \dots$$

Similar exact sequences have been obtained by Bak [Bk2,3] and Wall. Pardon [Pa 2] proves the exactness of (*) in full generality.

4.9. Topological K-theory. In linear algebraic K-theory the Grothendieck group $K_0(C_{\mathbb{C}}(X))$ is isomorphic to the topological K-group $KU^0(X)$. The ring $C_{\mathbb{C}}(X)$ of continuous functions of the compact space X in $\mathbb{C}$ is equipped with the complex conjugation. In Karoubi's notation we have [Ka 1]

$$_\varepsilon L_0(C_{\mathbb{C}}(X)) = KU^0(X) \oplus KU^0(X)\ ,$$

$$_\varepsilon W_0(C_{\mathbb{C}}(X)) = KU^0(X)\ ,$$

$$_\varepsilon U_0(C_{\mathbb{C}}(X)) = KU^{-1}(X)\ .$$

The ring $C_{\mathbb{R}}(X)$ is equipped with the trivial involution.

$$K_0(C_{\mathbb{R}}(X)) = KO^0(X)\ ,\ {}_1L_0(C_{\mathbb{R}}(X)) = KO^0(X) \oplus KO^0(X)\ ,$$

$$_1W_0(C_{\mathbb{R}}(X)) = KO^0(X)\ ,\ {}_1U_0(C_{\mathbb{R}}(X)) = KO^{-1}(X)\ ,\ {}_{-1}U_0(C_{\mathbb{R}}(X)) = KO^1(X)\ .$$

4.10. Surgery. Mischenko [Mi] defined a homotopy invariant "higher signature" $\sigma^*(M^n) \in L^n(\mathbb{Z}[\pi_1(M)],1)$ for any manifold M^n (or even a Poincaré complex M^n) by considering the symmetric structure on the chain complex of the universal cover $\widetilde{M}$. Ranicki [Ra 1] defines an analogous invariant $\sigma_*(f) \in L_n(\mathbb{Z}[\pi_1(X)]$ for a normal map $f: M^n \to X^n$ by considering the quadratic structure on the chain complex kernel, and identifies $\sigma_*(f)$ with the surgery obstruction in $L_n(\pi_1(X)) = L_n(\mathbb{Z}[\pi_1(X)],1)$ obtained by Wall [Wl 2] after surgery below the middle dimension. Surgery obstructions are related to higher signatures by

$$(1+T)\sigma_*(f) = \sigma^*(M) - \sigma^*(X) \in L^n(\mathbb{Z}[\pi_1(X)],1) .$$

The Mischenko higher signatures appear in the following product formula for surgery obstructions due to Ranicki [Ra 1]

$$\sigma_*(f \times g : M^n \times N^p \to X^n \times Y^p) = \sigma_*(f) \otimes \sigma_*(g) + \sigma^*(X) \otimes \sigma_*(g) + \sigma_*(f) \otimes \sigma^*(Y)$$

$$\text{in } L_{n+p}(\mathbb{Z}[\pi_1(X \times Y)],1) .$$

4.11. Miscellaneous remarks.

a) Change of theories. To define the higher Witt groups we started with f.g.p. A-modules ; many other choices are possible [Bk 1]. The comparison of the different theories gives rise to exact sequences like the Rothenberg exact sequence [Sh] [Ka 1, p. 374].

b) Let $f : A \to B$ a ring homomorphism such that B is an A-module of finite homological dimension. In linear algebraic K-theory there is defined a "transfer map" (= "Gysin map") $f^* : K_n(B) \to K_n(A)$ for all n [Qu 1, p. 111]. The definition of such a map seems to be more difficult in hermitian K-theory and has presently been solved in zero-dimension and for particular rings [Sc], [La] and [Ka 2, App. 4].

(*) When 2 is not invertible in A , then there exists an integer n and elements in ${}_{-1}WQ_2(\mathbb{Z})$ and ${}_{-1}WQ_{-2}(\mathbb{Z})$ such that the product is $2^n E_8 \in {}_1WQ_0(\mathbb{Z})$.

REFERENCES

[Bk 1] BAK A. Definitions and problems in surgery and related groups, Proceedings of the N.S.F. regional conference in topology, Knoxville, Tenn. (1974), Springer Lecture Notes (to appear).

[Bk 2] BAK A. K-theory of forms, Ann. Math. Studies, Princeton University Press (to appear).

[Bk 3] BAK A. Strong approximation for elementary groups (preprin

[Bs] BASS H. Unitary algebraic K-theory, Springer Lecture Notes 343 (1973) 57-265.

[Ca] CAPPELL S. Mayer-Vietoris sequences in Hermitian K-theory, Springer Lecture Notes 343 (1973) 478-512.

[Cr] CRETIN M. Invariance homotopique en K-théorie hermitienne, C.R.A.S. Paris 281 (1975) 1085-1087.

[Ka 1] KAROUBI M. Périodicité de la K-théorie hermitienne, Springer Lecture Notes 343 (1973) 301-411.

[Ka 2] KAROUBI M. Localisation des formes quadratiques I, Ann. Sc. Ec. Norm. Sup. Paris, 7 (1974) 359-404.

[Ka 3] KAROUBI M. Localisation des formes quadratiques II, Ann. Sc. Ec. Norm. Sup. Paris, 8 (1975) 99-155.

[Ka 4] KAROUBI M. Localisation des formes quadratiques III, (to appear).

[K-V] KAROUBI M. and VILLAMAYOR O. K-théorie algébrique et K-théorie topologique II Math. Scand. 28 (1971) 265-307.

[Kn] KNEBUSCH M. Grothendieck und Wittringe von nichtausgearteten symmetrischen Bilinearformen, Sitzungen b. Heidelberg, Springer Verlag, 1969/70.

[La] LANNES J. Formes quadratiques d'enlacement sur l'anneau des entiers d'un corps de nombres, Ann. Sc. Ec. Norm. Sup., t. 8 (1975) 535-579

[Ld] LODAY J.L. K-théorie algébrique et représentations de groupes, Ann. Sc. Ec. Norm. Sup., Fasc. 3, t. 9 (1976).

[M-H] MILNOR-HUSEMOLLER Symmetric bilinear forms, Ergebnisse d. Math., Springer Lecture Notes 73 (1973).

[Mi] MISCHENKO A.S. Homotopy invariants of non-simply-connected manifolds III, Higher signatures, Izv. Akad. Nauk. SSSR, 35 (1971) 1316-1355.

[No] NOVIKOV S.P. The algebraic construction and properties of hermitian analogues of K-theory..., Izv. Akad. Nauk. SSSR, 34 (1970) 253-288 and 475-500.

[Pa 1] PARDON W. Local surgery and applications to the theory of quadratic forms, to appear in Bull. A.M.S.

[Pa 2] PARDON W. (These proceedings).

[Qu 1] QUILLEN D. Higher algebraic K-theory I , Springer Lecture Notes, 341 (1973) 85-147.

[Qu 2] QUILLEN D. Higher algebraic K-theory II , these Proceedings.

[Ra 1] RANICKI A. The algebraic theory of surgery (to appear).

[Ra 2] RANICKI A. Algebraic L-theory IV . Polynomial extension rings, Comm. Math. Helv. 49 (1974) 137-167.

[Ra 3] RANICKI A. Algebraic L-theory II . Laurent extensions, Proc. Lond. Math. Soc. 3 (1973) 101-125.

[Sc] SCHARLAU W. Quadratic reciprocity laws, J. Number theory 4 (1972) 78-97.

THE EXACT SEQUENCE OF A LOCALIZATION FOR WITT GROUPS

by William Pardon*

§0 Introduction

It is often useful to compare the value of a functor defined on a "global" ring to its values on the "local" components of the ring. For example, the Hasse-Minkowski theorem [L,6.3.1] states that a quadratic form over a global field F is isotropic if and only if it is isotropic over its completions at the "places" of F. A straightforward consequence of this deep theorem is a local-global comparison of the above type, where the functor W of a ring A, $W(A)$, is a stable Grothendieck group on isometry classes of quadratic forms over A:

(0.1): There is an injection $W(F) \to \prod_{\mathfrak{p}} W(F_{\mathfrak{p}})$ where F is a global field and $F_{\mathfrak{p}}$ is its completion at the place (or prime, possibly infinite) $\mathfrak{p}$.

There are two observations to be made here. First, if $\mathfrak{p}$ is a finite prime, F is an algebraic number field and A is the ring of integers in F, then $W(F_{\mathfrak{p}}) \cong W(A/\mathfrak{p})$. $A/\mathfrak{p}$ is a finite field, so $W(A/\mathfrak{p})$ is easy to compute. Second, even though the components $W(F_{\mathfrak{p}}) \cong W(A/\mathfrak{p})$ of (0.1) give a classical, tractable list of invariants for elements of $W(F)$, (0.1) does not reveal which elements of $\prod_{\mathfrak{p}} W(F_{\mathfrak{p}})$ arise from global forms. That is, (0.1) does

*Partially supported by NSF Grant MPS 71-03442.

not compute $W(F)$, even in the simple case $F = \mathbb{Q}$. It turns out that if $F = \mathbb{Q}$, all finite collections of local invariants are realized, so that $W(\mathbb{Q}) \cong \coprod_p W(\mathbb{Z}/p\mathbb{Z}) + W(\mathbb{R})$. Since $W(\mathbb{R}) \cong \mathbb{Z} \cong W(\mathbb{Z})$ (via the signature), this fact may be expressed through a split exact sequence [MH, IV.2.1]

$$(0.2) \qquad W(\mathbb{Z}) \rightarrowtail W(\mathbb{Q}) \twoheadrightarrow \coprod_p W(\mathbb{Z}/p\mathbb{Z})$$

which amounts to the desired computation. It turns out that (0.2) generalizes to (see [MH, IV.3.4])

$$(0.3) \qquad W(A) \rightarrowtail W(F) \longrightarrow \coprod_{\mathfrak{p}\ \text{finite}} W(A/\mathfrak{p}) \twoheadrightarrow \mathcal{C}/\mathcal{C}^2$$

where $\mathcal{C}$ is the ideal class group of A. Taking into account Quillen's localization sequence for algebraic K_i-groups (for low dimensions [S, 8.4]) and the fact that there exists a comparable sequence of Witt groups W_i, it is reasonable to expect that (0.3) extends to a "long exact" sequence. Theorem (2.1) of this paper exhibits the localization sequence for Witt groups.

The above remarks are hindsight since (2.1) was first proved (in the special case where A is the integral group ring of a finite group) using surgery theory; the algebraic constructions in this paper are thus geometrically "realizable". Further, in the applications of (2.1) (or (0.3)) one calculates $W_i^\lambda(A)$ (or $W(A)$) from knowledge of the rest of the exact sequence. Max Karoubi [K] has independently produced a part of this exact sequence (2.1), but some of his arguments required conditions on the ring A in (2.1) (e.g., $1/2 \in A$ or A Dedekind), which made the applications I had in mind inaccessible. Theorem (2.1) contains strong (but removeable) restrictions on the localized ring B, but is general enough to make calculations of Witt groups and surgery obstruction groups. These could not appear here due to space limitations and will be in another paper (see [P2]). Andrew Ranicki has also announced a version of (2.1) in [R'].

I have followed the definitions and notational conventions of Bass' foundational paper [B] as closely as possible. In an effort to make the paper self-contained, §1 integrates into a summary of background material from [B] the basic definitions and facts concerning quadratic forms in the category of torsion modules with short free resolution. A statement of the main theorem (2.1) and a summary of the rest of the paper is the content of §2.

Proposition (1.17) and the idea of an "integral lattice" ((1.10),(3.1)) are the basic tools in the proof of (2.1). I am aware that (2.1) can be proved much more generally, but it seemed that in making the most general formulation the paper would be too long and technical and that the prominence of these two ideas would be obscured.

Notational Conventions

$+$	means	direct sum
$\rightarrowtail$	"	1 - 1 homomorphism
$\twoheadrightarrow$	"	onto homomorphism
[*]	"	bibliographical reference to *
(*)	"	reference to (*) in this paper.

If M_i and N_j are modules, and $f_{ij}: M_i \to N_j$ are module homomorphisms, $1 \leq i \leq n$, $1 \leq j \leq m$, then the matrix (f_{ji}) denotes the obvious homomorphism $M_1 + M_2 + \ldots + M_n \to N_1 + N_2 + \ldots + N_m$.

§1 The category $\mathfrak{D}_F^1$; basic definitions and results for quadratic forms

(1.1) The localization. Let A be a ring-with-involution containing 1, where the involution is denoted "$\overline{}$": $\overline{a+b} = \bar{a} + \bar{b}$, $\overline{ab} = \bar{b}\bar{a}$, $\bar{1} = 1$, for all $a,b \in A$. All A-modules will be right A-modules unless otherwise specified. Let $\Sigma \subset A$ be a central multiplicative subset containing 1 such that Σ contains no zero-divisors ($as = 0 \Rightarrow a = 0$, $a \in A$, $s \in \Sigma$), $\bar{s} = s$ for all $s \in \Sigma$, and the ring of quotients $B = A_\Sigma$ is an Artin, semi-simple, A-injective ring containing 1/2. Thus the localization map ℓ: $A \to B$ is an injection of rings-with-involution and $- \otimes_A B$ is exact. All A-modules considered in this paper, with the exception of B and B/A, will be finitely generated.

Let V be a two-sided A-module (in practice, $V = A, B$, or B/A) and M an A-module. If M_V^* denotes $\mathrm{Hom}_A(M,V)$ then M_V^* is in a natural way a left A-module, but is always taken to be a right A-module by setting $(fa)(m) = \bar{a}f(m)$ for $a \in A$, $f \in M_V^*$, and $m \in M$. If h: $M \to N$ is an A-module homomorphism, there is a natural homomorphism h_V^*: $N_V^* \to M_V^*$. If $V = A$, then set $M_V^* = \bar{M}$ and $h_V^* = \bar{h}$; we make the same conventions for B-modules M and set $M_B^* = \bar{M}$, $h_B^* = \bar{h}$. A homomorphism μ: $F \to G$ between free A-modules (B-modules) with chosen bases will be identified (when convenient) with its matrix, also denoted μ; conversely, any matrix μ will be identified with a homomorphism of free modules. If we write $F = A^n$ ($F = B^n$) we mean F has a chosen basis. The induced homomorphism $\bar{\mu}$: $\bar{G} \to \bar{F}$ has matrix $\bar{\mu}$ [B,I.2.7], the conjugate transpose of the matrix μ. If $V = B/A$, set $M_V^* = M^\wedge$ and $h_V^* = h^\wedge$, where M is an A-module and h: $M \to N$ an A-module homomorphism.

(1.2) <u>Σ-torsion</u>. An element m in the A-module M is said to be Σ-torsion or a <u>torsion element</u> if there is $s \in \Sigma$ such that $ms = 0$. Since Σ is central, the subset <u>tM</u> $\subseteq M$ consisting of torsion elements is an A-submodule, the <u>torsion submodule</u>. M is called a <u>torsion module</u> or <u>torsion</u> if $tM = M$. A standard argument [c, 0.6.1] shows that the sequence

$$(1.3) \qquad tM \rightarrowtail M \longrightarrow M \otimes_A B$$

is exact. Hence M is a torsion module if and only if $M \otimes_A B = 0$.

Let $\mathfrak{D}^1_{\underline{F}}$ be the category of torsion modules having short free resolution; hence $S \in \mathfrak{D}^1_F$ if $S \otimes B = 0$ and there is an exact sequence $A^n \rightarrowtail A^n \twoheadrightarrow S$. Here is a fundamental proposition.

(1.4) <u>Proposition</u>: If M is a torsion module with short free resolution $F \overset{\mu}{\rightarrowtail} G \overset{j}{\twoheadrightarrow} M$, there is a natural isomorphism

$$k\colon \operatorname{Ext}^1_A(M,A) \cong M^{\wedge}$$

and hence an exact sequence $\bar{G} \overset{\bar{\mu}}{\rightarrowtail} \bar{F} \twoheadrightarrow M^{\wedge}$.

<u>Proof</u>: From the short free resolution, we obtain the exact sequence

$$\bar{G} \overset{\bar{\mu}}{\rightarrowtail} \bar{F} \overset{j'}{\twoheadrightarrow} \operatorname{Ext}^1_A(M,A)$$

since G is A-free and $tM = M$ implies $\bar{M} = 0$. Identifying $\operatorname{Ext}^1_A(M,A)$ with $\operatorname{cok}(\bar{\mu})$, define k as follows. Let $f \in \bar{F}$. The bottom horizontal map can be defined to make the following diagram commute because B is A-injective

$$\begin{array}{ccc} F & \xrightarrow{\ f\ } & A \\ \big\downarrow{\scriptstyle \mu} & & \big\downarrow \\ G & \xrightarrow{\ f(\mu^{-1})\ } & B \end{array} .$$

The map $f(\mu^{-1})$ is unique because $\mu \otimes B$ is an isomorphism. If $m \in M$ and $x \in \mathrm{Ext}^1_A(M,A) \equiv \mathrm{cok}(\bar{\mu})$ define k by the formula

$$(1.5) \qquad k(x)(m) = r\cdot f(\mu^{-1})(j^{-1}(m))$$

where $r\colon B \to B/A$ is the quotient map, $j'(f) = x$, and $j^{-1}(m) \in G$ is any element for which $j(j^{-1}(m)) = m$. Thus,

$$(1.6) \qquad \tilde{j}(f)(m) = r\cdot f(\mu^{-1})(j^{-1}(m)).$$

Details are left to the reader.

The following is well-known:

(1.7) Proposition: Let $(R) = (A^n \overset{\alpha}{\rightarrowtail} A^n \overset{j}{\twoheadrightarrow} S)$ and $(R') = (A^m \overset{\alpha'}{\rightarrowtail} A^m \overset{j'}{\twoheadrightarrow} S)$ be short free resolutions of $S \in \mathfrak{D}^1_F$. Then there are non-negative integers k, ℓ with $n + k = m + \ell$ and automorphisms $\mu, \mu'\colon A^{m+\ell} \to A^{m+\ell}$ such that $\mu(\alpha + I_k) = (\alpha' + I_\ell)\mu'$ and $(j,0_k) = (j',0_\ell)\mu$, where $(j,0_k)\colon A^{n+k} \to S$ is j on the first n factors and zero on the rest; $(j',0_\ell)$ is defined analogously.

(1.8) Examples: (a) Let π be a finite group, $A = \mathbb{Z}[\pi]$, $B = \mathbb{Q}[\pi]$, $\overline{\Sigma n_g g} = \Sigma n_g g^{-1}$, $n_g \in \mathbb{Z}$ or $\mathbb{Q}$, $g \in \pi$. (b) Let B be a number field with $\mathbb{Z}_2 \subseteq \mathrm{Gal}(B/\mathbb{Q})$ furnishing the involution (or B could have trivial involution), and A = the ring of integers. More generally, B could be a semi-simple algebra-with-involution, $1/2 \in B$, and A any order taken to itself by the involution. (c) Let $A = \mathbb{Z}[t_1,t_1^{-1},\ldots,t_n,t_n^{-1}]$, "$\overline{}$" induced by $\bar{t}_i = t_i^{-1}$, $B = \mathbb{Q}(t_1,\ldots,t_n)$. (d) Making (c) more exotic, let Φ be a Bieberbach group. This means there is an extension $\mathbb{Z}^n \rightarrowtail \Phi \twoheadrightarrow \pi$, where $\mathbb{Z}^n$ is free abelian, normal and maximal abelian in Φ, and π is finite. Φ is "classified" by $\theta \in H^2(\pi;\mathbb{Z}^n)$ where $\mathbb{Z}^n$ has a π-action arising from the extension. From θ construct, by extension of coefficients, $\theta' \in H^2(\pi; \mathbb{Q}(t_1,\ldots,t_n)^{\cdot})$ determining a division ring B such that $\mathbb{Z}[\phi]$ imbeds in B as a localization.

(1.9) <u>The setting for quadratic forms</u>: We will recall from [B] the context and definitions of unitary K-theory. Many of our definitions will be stated only for A-modules, but when they make sense, they are understood to apply to B-modules as well. Let (A,λ,Λ) be a unitary ring with involution ([B,I.4.1]) where we require special values for λ and Λ. Thus, A is a ring-with-involution (1.1), $\lambda = \pm 1$, and $\Lambda = \underline{S_\lambda(A)} = \{a + \lambda\bar{a} \mid a \in A\}$. If $\alpha \in M_n(A)$, the set of $(n\times n)$-matrices over A, $\bar{\alpha}$ denotes its conjugate transpose, thus making $M_n(A)$ a ring-with-involution. Set $\underline{S_\lambda(A^n)} = S_\lambda(M_n(A)) = \{\alpha \in M_n(A) \mid \alpha = \lambda\bar{\alpha},\ \alpha_{ii} \in S_\lambda(A)\}$. If $1/2 \in A$, $S_\lambda(A) = \{a \in A \mid a = \lambda\bar{a}\}$.

(1.10) <u>Forms</u>: If M,N are A-modules and $V = A$ or $V = B/A$, a function $g\colon N \times M \to V$ is a <u>sesquilinear form</u> if it is biadditive and satisfies $g(na,mb) = \bar{a}g(n,m)b$, for all $a,b \in A$, $n \in N$, $m \in M$. If $L \subseteq N$ we define its <u>orthogonal complement</u>, $\underline{L^\perp}$ to be $\{m \in M \mid g(L,m) = 0\}$. The <u>natural form</u> $\langle\ ,\ \rangle_M\colon M_V^* \times M \to V$, defined by $\langle f,m\rangle_M = f(m)$, where $f \in M_V^*$, $m \in M$, is a sesquilinear form. If $h\colon M \to N$ is a homomorphism, $\langle h_V^* f,m\rangle_M = \langle f,hm\rangle_N$. ([B,I.2.1].) A sesquilinear form $g\colon N \times M \to V$ is <u>nonsingular</u> if the <u>adjoints</u>

$$(1.11)\qquad \begin{aligned} {}_g d\colon N &\longrightarrow M_V^*\ , & \langle {}_g d(n),m\rangle &= g(n,m)\\ d_g\colon M &\longrightarrow N_V^*\ , & \langle d_g(m),n\rangle &= \overline{g(n,m)} \end{aligned}$$

are isomorphisms. If $N = M$ is A-free with basis $\{e_1,\ldots,e_n\}$, the map $d_g\colon A^n \to (A^n)^*_V$ has matrix $(g(e_i,e_j))$ ([B,I.2.7]). We sometimes use the matrix $(g(e_i,e_j))$ to denote d_g; on the other hand ${}_gd$ is more convenient for some purposes since no "$\overline{}$" appears in its definition. From the natural form we obtain the adjoint map $d_M = d_{\langle\ ,\ \rangle_M}\colon M \to (M^*_V)^*_V$. M is called <u>V-reflexive</u> if d_M is an isomorphism. If M is A-projective (B-projective) then it is A-reflexive (B-reflexive); if $M \in \mathfrak{D}^1_F$, then by (1.4) it is B/A-reflexive. If M and N are V-reflexive then [B,I.2.4] we need only verify one of the conditions (1.11) for nonsingularity.

An A-submodule $L \subset B^n$ is a <u>lattice</u> if the inclusion induces $L \otimes B \cong B^n$. If $g\colon B^n \times B^n \to B$ is sesquilinear and $L \subset B^n$ is a lattice, then L inherits a sesquilinear form $g_L\colon L \times L \to B$. The <u>dual lattice</u> $L' = \{x \in B^n \mid g(x,L) \subseteq A\}$. If g is nonsingular, then the induced sesquilinear form $\ell\colon L' \times L \to A$ is nonsingular and there is a natural identification of L' with $\bar{L}$. If L is A-free with basis $\{e_1,\ldots,e_n\}$ then there is a "dual basis" $\{e^*_1,\ldots,e^*_n\}$ such that $\ell(e^*_i,e_j) = \delta_{ij}$. If L has a basis, it will be assumed L' has the dual basis.

(1.12) <u>λ-Forms</u>: If a sesquilinear form $g\colon M \times M \to V$ satisfies $g(m,m') = \lambda\overline{g(m',m)}$, for all $m,m' \in M$, it is called <u>λ-hermitian</u>. It is <u>even λ-hermitian</u> if, in addition, $g(m,m) \in S_\lambda(A)$, for all $m \in M$. If M is free and $V = A$, g is even if and only if $d_g \in S_\lambda(A^n)$. Now let $r\colon B \to B/A$ and $r'_\lambda\colon B \to B/S_\lambda(A)$ be the projections.

(1.12a) If $V = A$, M is A-free and $q\colon M \to A/S_{-\lambda}(A)$ is a function, the triple (M,g,q) is a <u>λ-form</u> if $g\colon M \times M \to A$ is λ-hermitian and for all $m,m' \in M$, $a \in A$,

(i) $q(ma) = \bar{a}q(m)a$

(ii) $q(m+m') - q(m) - q(m') = r'_{-\lambda}g(m,m')$

(iii) $q(m) + \lambda\overline{q(m')} = r'_{-\lambda}g(m,m)$.

$K \subseteq M$ is <u>totally isotropic</u> if $g \mid K \times K \equiv 0 \equiv q \mid K$. <u>$F_0^\lambda(A)$</u> denotes the set of isometry classes of nonsingular λ-forms (M,g,q) where M is free of <u>even rank</u>; $F_0^\lambda(A)$ is an abelian semigroup under <u>orthogonal sum</u>, denoted "$\perp$". We define λ-forms (B^n,g,q) in a similar way.

<u>Remark</u>: Given λ-hermitian g: $M \times M \to A$, M projective, the condition "$g(m,m) \in S_\lambda(A)$ for all $m \in M$" guarantees the existence of a function q with the properties above [B,I.3.4]. In our definition we require a specific choice of q. The reader should compare (1.12(a)) with the setting of [B,I,4.4]. If $1/2 \in A$, g determines q.

(1.12b): If $V = B/A$, $M \in \mathfrak{D}_F^1$, φ: $M \times M \to B/A$ is λ-hermitian, and ψ: $M \to B/S_\lambda(A)$ is a function, the triple (M,φ,ψ) is called a <u>λ-form</u> if for all $m,m' \in M$, $a \in A$

(i) $\psi(ma) = \bar{a}\psi(m)a$

(ii) $\psi(m+m') - \psi(m) - \psi(m') = \varphi(m,m') + \varphi(m',m)$

(the right-hand side is well-defined in $B/S_\lambda(A)$)

(iii) $r\psi(m) = \varphi(m,m)$

$K \subseteq M$ is called <u>totally isotropic</u> if $\varphi \mid K \times K \equiv 0 \equiv \psi \mid K$. <u>$F_0^\lambda(B/A)$</u> denotes the set of isometry classes of nonsingular λ-forms (M,φ,ψ), $M \in \mathfrak{D}_F^1$; $F_0^\lambda(B/A)$ is an abelian semigroup under the operation of <u>orthogonal sum</u>, denoted "$\perp$".

<u>Remark</u>: (<u>i</u>) We have defined "λ-form" in two quite different ways; the context makes clear which of (1.12a) or (1.12b) we mean and

in any case we always use Latin characters f,g,h,q,p, etc. in (M,g,q) for (1.12a), and Greek characters in [1.12b). (ii) We are unable to show that an intrinsic condition like "$\varphi(m,m) \in S_\lambda(B/A)$" allows us to conclude the existence of ψ as an analogous condition does in the category of projectives (remark after (1.12a)). This is what keeps us from a neat formalism for λ-forms over $\mathfrak{D}_F^1$ analogous to that in [B,I.4.4] for projectives; (1.17) and (1.18) might be thought of as a substitute.

(1.13) $W_0^\lambda(A)$, $W_0^\lambda(B)$, and $W_0^\lambda(B/A)$: (a) Let P be A-free. The λ-form $(P + \bar{P}, g_h, q_h)$ is hyperbolic, and is denoted $\mathcal{H}(P)$ if P and $\bar{P}$ are totally isotopic and $g_h \mid \bar{P} \times P$ is the natural form. The quotient of the Grothendieck group on $F_0^\lambda(A)$ by the subgroup generated by $\mathcal{H}(A^n)$, $n \in \mathbb{Z}^+$, is denoted $W_0^\lambda(A)$, the Witt group. We define $W_0^\lambda(B)$ similarly.

(b) Let $S \in \mathfrak{D}_F^1$. The λ-form $\mathcal{H}(S) = (S + S^\wedge, \varphi_h, \psi_h)$ is hyperbolic if S and $S^\wedge$ are totally isotropic and $\varphi_h \mid S^\wedge \times S$ is the natural form. If $T \in \mathfrak{D}_F^1$, (T,φ,ψ) is a λ-form, and $K \subset T$, $K \in \mathfrak{D}_F^1$, is a totally isotropic submodule, (T,φ,ψ) is a kernel and K a subkernel if the induced sesquilinear form $(T/K) \times K \to B/A$ is nonsingular. Clearly $\mathcal{H}(S)$ is a kernel with subkernels S and $S^\wedge$. $W_0^\lambda(B/A)$ is the Grothendieck group on $F_0^\lambda(B/A)$ modulo the subgroup generated by kernels.

(1.14) Before we define the W_1^λ-functors, we make constructions (1.17) and (1.18) fundamental to this paper.

(1.15) Lemma [C2,1.4]: Let (S,φ,ψ) be a λ-form. For each $t \in S$ there is $b \in B$ such that $r'b = \psi(t)$ and $b = \lambda\,\bar{b}$, where

$r'\colon B \to B/S_\lambda(A)$.

<u>Proof</u>: By (1.12b(i)) $\psi(-t) = \psi(t)$ and by (1.12b(ii)) $\psi(t+(-t)) - \psi(t) - \psi(-t) = -(\varphi(t,t) + \lambda\overline{\varphi(t,t)})$ so $2\psi(t) = \varphi(t,t) + \lambda\overline{\varphi(t,t)}$. Hence if $b,b' \in B$ are such that $r'b = \psi(t)$ and $rb' = \varphi(t,t)$, $r\colon B \to B/A$, then $2b = b' + \lambda\bar{b}' + c$, $c \in S_\lambda(A)$. Hence $b = 1/2(b' + \lambda\bar{b}') + 1/2c$ so $b = \lambda\bar{b}$.

(1.16) <u>Corollary</u>: $\psi(t) = \lambda\overline{\psi(t)}$ for each $t \in S$.

With (S,φ,ψ) as above and $A^n \overset{\mu}{\rightarrowtail} A^n \overset{j}{\twoheadrightarrow} S$ a resolution of S, let $\{e_1,\ldots,e_n\}$ be a basis for A^n and choose (by (1.15)) $\tau_{ik} \in B$, $1 \le i,k \le n$, such that $\varphi(je_i,je_k) = r\tau_{ik}$, $\psi(je_i) = r'\tau_{ii}$ and $\tau_{ik} = \lambda\overline{\tau_{ki}}$. We use $\tau = (\tau_{ik}) \in M_n(B)$ to define in the obvious way a λ-hermitian form, also denoted τ,

$$\tau\colon A^n \times A^n \longrightarrow B.$$

τ is called a <u>covering</u> of (S,φ,ψ) with respect to the resolution $A^n \overset{\mu}{\rightarrowtail} A^n \overset{j}{\twoheadrightarrow} S$. This gives part (a) of

(1.17) <u>Proposition</u>: Given a λ-form (S,φ,ψ), $S \in \mathfrak{D}_F^1$, and any resolution $A^n \overset{\mu}{\rightarrowtail} A^n \overset{j}{\twoheadrightarrow} S$, we may find a λ-hermitian form $\tau\colon A^n \times A^n \to B$ such that if $r\colon B \to B/A$ and $r'\colon B \to B/S_\lambda(A)$ are the projections and $m,m' \in A^n$,

(a) $$\varphi(j(m),j(m')) = r\tau(m,m')$$

and $$\psi(j(m)) = r'\tau(m,m).$$

(b) If τ also denotes the matrix of $\tau\colon A^n \times A^n \to B$, then $\bar{\mu}\tau \in M_n(A)$, $\bar{\mu}\tau\mu \in S_\lambda(A)$ and the diagram

$$\begin{array}{ccc} A^n & \xrightarrow{\bar{\mu}\tau} & \bar{A}^n \\ \Big\downarrow j & & \Big\downarrow \tilde{j} \\ T & \xrightarrow[d_\varphi]{} & T^\wedge \end{array}$$

commutes ((1.6) for $\tilde{j}$).

<u>Proof</u> (of (b)): Consider the sesquilinear form $g\colon A^n \times A^n \to B$ where $g(m,n) = \tau(\mu(m),n)$, $m,n \in A^n$. Since $r\cdot g(m,n) = r\cdot\tau(\mu(m),n) = \varphi(j\mu(m),n)$ and $j\mu = 0$, g takes values in A. We claim $d_g = \bar{\mu}\cdot d_\tau$; this gives $\bar{\mu}\tau \in M_n(A)$ since the matrix of $d_\tau : A^n \to (A^n)^*_B$ is (τ_{ij}) by (1.10). Let $m,n \in A^n$: $(\bar{\mu}d_\tau(m))(n) = \langle\bar{\mu}d_\tau(m),n\rangle_{B^n} = \langle d_\tau(m),\mu(n)\rangle_{B^n}$ $= \overline{\tau(\mu(n),m)} = d_g(m)(n)$, which verifies the claim. Similarly, if we consider the λ-hermitian form $f\colon A^n \times A^n \to A$, where $f(m,n) = \tau(\mu(m),\mu(n))$ we find its matrix is $\bar{\mu}\tau\mu$; it is in $S_\lambda(A^n)$ because $r'f(m,m) = r'\tau(\mu(m),\mu(m)) = \psi(j\mu(m)) = 0$. Finally, let $m \in A^n$, $t \in S$, and $j^{-1}(t) \in A^n$ any element of A^n such that $j(j^{-1}(t)) = t$. Then $\{\tilde{j}(\bar{\mu}d_\tau(m))\}(t) = r\langle\bar{\mu}d_\tau(m),\mu^{-1}(j^{-1}(t))\rangle_{B^n} = r\langle d_\tau(m),j^{-1}(t)\rangle_{B^n}$ $= \overline{r\tau(j^{-1}(t),m)} = \overline{\varphi(t,jm)} = d_\varphi(jm)(t)$. This completes the proof.

The following "converse" is left to the reader.

(1.18) <u>Proposition</u>: Let $S \in \mathfrak{D}^1_F$ have short free resolution $A^n \overset{\mu}{\rightarrowtail} A^n \twoheadrightarrow S$ and let $\tau\colon A^n \times A^n \to B$ be a λ-hermitian form such that $\bar{\mu}\tau \in M_n(A)$ and $\bar{\mu}\tau\mu \in S_\lambda(A^n)$. Then the equations of (1.17a) define a λ-form (S,φ,ψ).

(1.19) <u>The unitary group $U^\lambda_{2n}(A)$</u>: The set of isometries of the hyperbolic λ-form $\mathcal{H}(A^n) = (A^n + \bar{A}^n, g_h, q_h)$ (1.13) form a group, $U^\lambda_{2n}(A)$, whose elements σ may be written ([B,II.4.1.2])

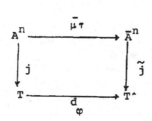

$$\sigma = \begin{pmatrix} \alpha & \beta \\ \gamma & \delta \end{pmatrix} \in GL_{2n}(A)$$

where

(1.20):

(i) $\alpha,\beta,\gamma,\delta \in M_n(A)$

(ii) $\bar{\delta}\alpha + \lambda\bar{\beta}\gamma = I_n$, the (n×n) identity matrix

(iii) $\bar{\beta}\delta,\ \bar{\gamma}\alpha \in S_{-\lambda}(A^n)$ ((1.9) for $S_{-\lambda}(A^n)$).

If $\sigma \in U_{2n}^{\lambda}(A)$ as above and $\sigma' = \begin{pmatrix} \alpha' & \beta' \\ \gamma' & \delta' \end{pmatrix}$ set

$$\sigma \perp \sigma' = \begin{bmatrix} \alpha & 0 & \beta & 0 \\ 0 & \alpha' & 0 & \beta' \\ \gamma & 0 & \delta & 0 \\ 0 & \gamma' & 0 & \delta' \end{bmatrix} \in U_{2(m+n)}^{\lambda}(A).$$

If $I_2 = \begin{pmatrix} 1 & 0 \\ 0 & 1 \end{pmatrix} \in U_{2n}^{\lambda}$, we obtain a sequence of <u>stabilization</u> homomorphisms $U_{2n}^{\lambda}(A) \to U_{2(n+1)}^{\lambda}(A)$, $\sigma \mapsto \sigma \perp I_2$. Letting $n \to \infty$ we obtain $U^{\lambda}(A)$, whose abelianization is denoted <u>$KU_1^{\lambda}(A)$</u>.

(1.21): Here are some special elements of $U_{2n}^{\lambda}(A)$.

(a) $w_1^{\lambda} = \begin{pmatrix} 0 & 1 \\ \lambda & 0 \end{pmatrix} \in U_2^{\lambda}(A)$; $w_n^{\lambda} = w_{n-1}^{\lambda} \perp w_1^{\lambda}$.

(b) The homomorphism $H: GL_n(A) \to U_{2n}^{\lambda}(A)$ sends $\alpha \in GL_n(A)$ to $H(\alpha) = \begin{pmatrix} \alpha & 0 \\ 0 & \bar{\alpha}^{-1} \end{pmatrix}$. This stabilizes to $H: GL(A) \to U^{\lambda}(A)$, whose abelianization is a homomorphism $H_1: K_1(A) \to KU_1^{\lambda}(A)$. <u>$W_1^{\lambda}(A)$</u> is the cokernel of H_1.

(c) Let $\rho,\tau \in S_{-\lambda}(A^n)$ and set $X_+(\rho) = \begin{pmatrix} I_n & 0 \\ \rho & I_n \end{pmatrix}$ $X_-(\tau) = \begin{pmatrix} I_n & 0 \\ \tau & I_n \end{pmatrix} \in U_{2n}^{\lambda}(A)$. If $\langle Z \rangle$ denotes the smallest subgroup containing Z, then ([B,II.5.2,II.4.1.3])

(1.22) $KU_1^{\lambda}(A) = U^{\lambda}(A)/\langle\{X_+(\rho),X_-(\tau) \mid \rho,\tau \in S_{-\lambda}(A^n), n \in \mathbb{Z}^+\}\rangle$

(1.23) $W_1^\lambda(A)=U^\lambda(A)/\langle\{X_+(\rho),X_-(\tau),H(\alpha)\mid \rho,\tau\in S_{-\lambda}(A^n),\alpha\in GL_n(A),n\in\mathbb{Z}^+\}\rangle.$

(1.24) The matrices α,γ (1.20) define an injective map $(\alpha,\gamma)\colon A^n \to A^n + \bar{A}^n$, whose image has a complementary summand equal to $\mathrm{im}((\beta,\delta)\colon \bar{A}^n \to A^n + \bar{A}^n)$. The relation $\bar{\gamma}\alpha \in S_{-\lambda}(A^n)$ in (1.20) is equivalent to the condition that g_h and q_h annihilate $\mathrm{im}(\alpha,\gamma)$. Indeed, let a sesquilinear form $[\ ,\ ]\colon A^n \times A^n \to A$ be defined by

$$(1.25) \qquad [x,y] = g_h(\gamma(x),\alpha(y)), \qquad x,y \in A^n.$$

Then $[\ ,\ ]^d = \bar{\gamma}\alpha$, so $\bar{\gamma}\alpha \in S_{-\lambda}(A^n)$ if and only if $[\ ,\ ]$ is an even $(-\lambda)$-hermitian form. But $g_h(\mathrm{im}(\alpha,\gamma)) \equiv 0$ if and only if $g_h(\gamma(x),\alpha(y)) + g_h(\alpha(x),\gamma(y)) = 0$ for all x,y; and $q_h(\mathrm{im}(\alpha,\gamma)) \equiv 0$ if and only if $q_h(\alpha(x) + \gamma(x)) = g_h(\alpha(x),\gamma(x)) \in S_{-\lambda}(A)$. The condition $\bar{\beta}\delta \in S_{-\lambda}(A^n)$ may be interpreted analogously.

In a similar way the condition $\bar{\delta}\alpha + \lambda\bar{\beta}\gamma = I_n$ (1.20(ii)) means that $\mathrm{im}(\alpha,\gamma) + \mathrm{im}(\beta,\delta) = A^n + \bar{A}^n$ and that the form induced by g_h on $\bar{A}^n \times A^n$--identified with $\mathrm{im}((\beta,\delta)\colon \bar{A}^n \to A^n + \bar{A}^n) \times \mathrm{im}((\alpha,\gamma)\colon A^n \to A^n + \bar{A}^n)$-- is the natural form (1.10).

(1.26): Using the remarks in (1.25), suppose conversely that a split injection $(\alpha,\gamma)\colon A^n \to A^n + \bar{A}^n$ is given, with totally isotropic image. Then [B,I.3.10] shows we may find $(\beta,\delta)\colon \bar{A}^n \to A^n + \bar{A}^n$ such that $\mathrm{im}(\beta,\delta)$ is totally isotropic, $\mathrm{im}(\alpha,\gamma) + \mathrm{im}(\beta,\delta) = A^n + \bar{A}^n$, and $\mathrm{im}(\beta,\delta) \times \mathrm{im}(\alpha,\gamma)$ has induced on it the natural form. By (1.25), this means

$$\sigma = \begin{pmatrix} \alpha & \beta \\ \gamma & \delta \end{pmatrix} \in U_{2n}^\lambda(A).$$

In fact it is easy to show that (α,γ) determines the class of σ in

$W_1^\lambda(A)$. In this setting stabilization of σ, addition of w_1^λ, the right action of $X_+(\rho)$, $X_-(\tau)$ and $H(\epsilon)$ (1.21) on σ translate, respectively, to

(1.27):

(i) $(\alpha,\gamma) \perp I_2 = (\alpha + 1_A, \gamma + 0_A)$.

(ii) $(\alpha,\gamma) \perp w_1^\lambda = (\alpha + 0_A, \gamma + \lambda 1_A)$.

(iii) $(\alpha,\gamma)X_+(\rho) = (\alpha + \gamma\rho,\gamma)$.

(iv) $(\alpha,\gamma)X_-(\tau) = (\alpha,\gamma + \alpha\tau)$.

(v) $(\alpha,\gamma)H(\epsilon) = (\alpha\epsilon,\gamma\bar{\epsilon}^{-1})$.

The operations of (1.27) (i), (iii), (iv), (v) generate an equivalence relation on split injections (α,γ) with totally isotropic image, having a group of equivalence classes equal to $W_1^\lambda(A)$. (See [R].)

(1.28) <u>$W_1^\lambda(B/A)$:</u> The purpose of the remarks in (1.24) and (1.26) was to motivate the following definitions. Suppose $H,K \in \mathfrak{D}_F^1$, $\mathcal{H}(H) = (H + H^\wedge,\varphi_h,\psi_h)$ is the hyperbolic λ-form and Δ: $K \to H + H^\wedge$ is an injection with totally isotropic image. Then if $\Delta = (\xi,\zeta)$, ξ: $K \to H$, ζ: $K \to H^\wedge$, define a $(-\lambda)$-hermitian form $[\ ,\ ]$: $K \times K \to B/A$ for $k,k' \in K$ by

(1.29) $$[k,k'] = \varphi_h(\zeta(k),\xi(k'))$$ (compare (1.25)).

(1.30) <u>Definition:</u> A <u>λ-formation</u> is a 4-tuple $(K,H,\Delta,\varkappa)$ where K and $H \in \mathfrak{D}_F^1$, Δ: $K \to H + H^\wedge$ is an injection whose image is a subkernel of $\mathcal{H}(H)$ and $(K, [\ ,\ ], \varkappa)$ is a $(-\lambda)$-form, where $[\ ,\ ]$ is given by (1.29). Let $F_1^\lambda(B/A)$ denote the set of isomorphism classes of λ-formations (isomorphisms induced by isomorphisms of K and H

preserving Δ and $\varkappa$). $F_1^\lambda(B/A)$ is an abelian semigroup in the obvious way, with zero element the λ-formation where $K = H = 0$, (the "zero formation").

(1.31) Remarks: (a) $[\ ,\ ]^d = \xi^\wedge\zeta$. Indeed, if $k,k' \in K$, then $\{\xi^\wedge\zeta(k)\}(k') = \langle\xi^\wedge\zeta(k),k'\rangle_K = \langle\zeta(k),\xi(k')\rangle_K = \varphi_h(\zeta(k),\xi(k'))$ $= \{[\ ,\ ]^d(k)\}(k')$. (b) The $(-\lambda)$-form $(K,[\ ,\ ],\varkappa)$ corresponds by (a), (1.24), and (1.25) to the condition $\bar{\alpha}_\gamma \in S_{-\lambda}(A^n)$ (1.20). We have required additionally in (1.30) a choice $\varkappa$ of splitting, corresponding to a choice of splitting for $\bar{\alpha}_\gamma$ (or a choice of q-form for $[\ ,\ ]$ of (1.25)--c.f. Remark (1.12a).) Thus $\varkappa: K \to B/S_{-\lambda}(A)$ is extra structure and is comparable to Sharpe's idea of the "split unitary group" ([Sh, §3]).

(1.32) Operations on $F_1^\lambda(B/A)$: Let $\theta = (K,H,\Delta,\varkappa) \in F_1^\lambda(B/A)$ be given and let E denote the short exact sequence of elements of $\mathfrak{D}_F^1$, $(E) = (J \rightarrowtail H_1 \twoheadrightarrow H)$. We define a λ-formation $\sigma_E\theta = (K_1,H_1,\Delta_1,\varkappa_1)$ as follows. Let K_1 be the pullback in the diagram

(1.33)

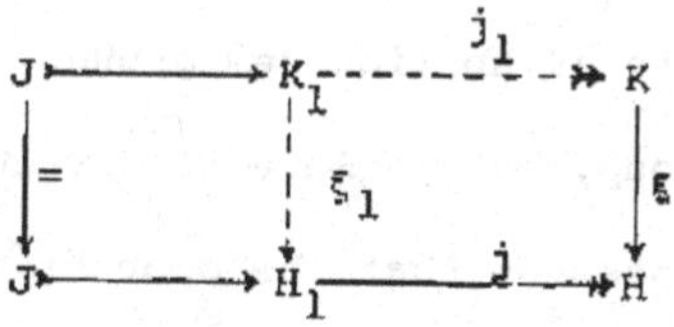

Define $\zeta_1 = j^\wedge\zeta j_1$, and $\varkappa_1 = \varkappa j_1$. Then one verifies that if $\Delta_1 = (\xi_1,\zeta_1)$, then $\sigma_E\theta = (K_1,H_1,\Delta_1,\varkappa_1)$ is a λ-formation, well-defined up to isomorphism by θ and the isomorphism class of (E). In a similar way, given $(E) = (L \to H_1^\wedge \overset{\ell}{\to} H^\wedge)$ let K_1 be the pullback in

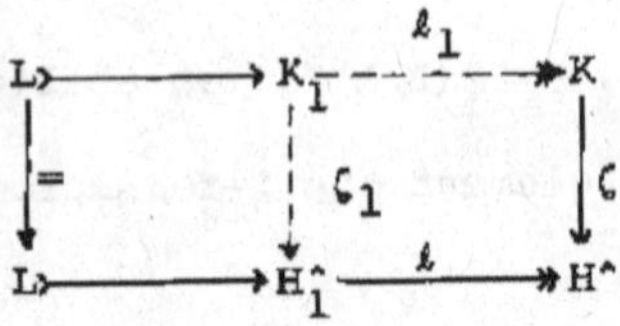

and set $\xi_1 = \ell^{\wedge} \cdot \xi \cdot \ell_1$, $\varkappa_1 = \varkappa \ell_1$ and $\Delta_1 = (\xi_1, \zeta_1)$. Then $_E\sigma\theta = (K_1, H_1, \Delta_1, \varkappa_1)$ is a λ-formation.

(b) If (H, φ, ψ) is a $(-\lambda)$-form, let $\chi_-(H,\varphi,\psi)\theta = (K, H, \Delta', \varkappa')$ where $\Delta' = (\xi, \zeta + d_\varphi \cdot \xi)$ and $\varkappa' = \varkappa + \psi \cdot \xi$. If $(H^{\wedge}, \varphi, \psi)$ is a $(-\lambda)$-form, let $\chi_+(H^{\wedge},\varphi,\psi)\theta = (K, H, \Delta', \varkappa')$ where $\Delta' = (\xi + d_\varphi \cdot \zeta, \zeta)$ and $\varkappa' = \varkappa + \psi \cdot \zeta$.

(1.34) Definition: $W_1^\lambda(B/A)$ is the semigroup $F_1^\lambda(B/A)$ modulo the equivalence relation, $\approx$, generated by the following four operations (notation as in (1.32 (a)(b)):

(i) $\theta \to \sigma_E\theta$, $(E) = (J \to H_1 \to H)$.

(ii) $\theta \to {_E\sigma\theta}$, $(E) = (L \to H_1^{\wedge} \to H)$.

(iii) $\theta \to \chi_-(H,\varphi,\psi)\theta$, (H,φ,ψ) a $(-\lambda)$-form.

(iv) $\theta \to \chi_+(H,\varphi,\psi)\theta$, $(H^{\wedge},\varphi,\psi)$ a $(-\lambda)$-form.

(1.35) Remark: $W_1^\lambda(B/A)$ is an abelian semigroup. The main theorem (2.1) implies it is a group, but we have no direct proof of this. The notation has been chosen so that the operations of (1.34) correspond to the following operations on $\sigma \in U_{2n}^\lambda(A)$ (see (1.27))

(i) $\sigma \to I_{2m} \perp \sigma$.

(ii) $\sigma \to w_{2m}^\lambda \perp \sigma$.

(iii) $\sigma \to X_-(\rho)\sigma$, $\rho \in S_{-\lambda}(A^n)$.

(iv) $\sigma \to X_+(\tau)\sigma$, $\tau \in S_{-\lambda}(A^n)$.

It is customary to call $W_1^\lambda(A)/\langle w_n, n \in Z^+\rangle$ the Wall group, $L_\lambda^h(A)$.

Hence our $W_1^\lambda(B/A)$ is essentially a Wall group, not a Witt group.

§2 The main theorem

(2.1) Theorem: Let A be a ring-with-involution, B a ring of quotients as in (1.1). Then there is a long exact sequence of abelian groups, $\lambda = \pm 1$,

$$\cdots \longrightarrow W_1^\lambda(A) \xrightarrow{\mathcal{K}_1^\lambda} W_1^\lambda(B) \xrightarrow{\mathcal{L}_1^\lambda} W_1^\lambda(B/A) \xrightarrow{\partial_1^\lambda} W_0^\lambda(A) \xrightarrow{\mathcal{K}_0^\lambda} W_0^\lambda(B)$$

$$\xrightarrow{\mathcal{L}_0^\lambda} W_0^\lambda(B/A) \xrightarrow{\partial_0^\lambda} W_1^{-\lambda}(A) \xrightarrow{\mathcal{K}_1^{-\lambda}} W_1^{-\lambda}(B) \longrightarrow \cdots$$

The proof of this theorem occupies the rest of this paper. We first define $\mathcal{L}_0^\lambda$ and ∂_0^λ (§3,4) and prove exactness of the last five terms (§5). Then we define $\mathcal{L}_1^\lambda$ and ∂_1^λ (§6,7) and prove exactness of the first five terms (§8). The maps $\mathcal{K}_i^\lambda$, $i = 0,1$, are induced by tensoring with B ("change of rings," [B, I.6.3]). The theorem implies $W_1^\lambda(B/A)$ is a group (cf. (1.35)).

§3 $\mathcal{L}_0^\lambda\colon W_0^\lambda(B) \to W_0^\lambda(B/A)$

(3.1): This homomorphism is classical for example (1.8b). It is easily shown that, when A is the ring of integers in a number field B, $\mathcal{L}_0^\lambda$ is essentially the direct sum over all primes $\mathfrak{p} \subseteq A$ of the "second residue class map", denoted ∂_2 in [L], ψ_2 in [MH](c.f (0.3))

Let a λ-form (B^n,g,q) be given (since $1/2 \in B$, g determines q, but we keep it in the notation for completeness). Find an "integral" lattice $L \subset B^n$; i.e., find $L \cong A^n$ such that $g(L \times L) \subseteq A$, $q(L) \subseteq A/S_{-\lambda}(A)$ (clear denominators in a matrix representation for g). If L' denotes the dual lattice (1.10), then $L \subseteq L'$ and if S denotes L'/L, $S \in \mathfrak{D}_F^1$. We construct a non-singular λ-form (S,φ,ψ) as follows. Since $g(L' \times L) = g(L \times L') \subseteq A$ and $q(L) \subseteq A/S_{-\lambda}(A)$, $g_{L'} = g|L'$ defines a λ-form (S,φ,ψ) (1.12b)

$$(3.2)\qquad \begin{aligned} &\varphi\colon (L'/L) \times (L'/L) \longrightarrow B/A, && \varphi(j\ell,j\ell') = rg(\ell,\ell') \\ &\psi\colon L'/L \longrightarrow B/S_\lambda(A), && \psi(j\ell) = r'g(\ell,\ell) \end{aligned}$$

where $S = L'/L$, $j\colon L' \to L'/L$, $r\colon B \to B/A$, $r'\colon B \to B/S_\lambda(A)$, and $\ell,\ell' \in L'$. That ${}_\varphi d$ is injective follows from the definition of dual lattice. To see it is surjective, if $f \in S^\wedge = (L'/L)^\wedge$ there is $\tilde f \in (L')^*_B$ such that $r\tilde f(\ell) = fj(\ell)$, $\ell \in L'$. Since g is nonsingular, there is $\tilde y \in B^n$ such that ${}_g d(\tilde y) = \tilde f$. Since $\tilde f(L) \subseteq A$ and $\tilde f(L) = g(\tilde y,L)$, we have $\tilde y \in L'$. By definition ${}_\varphi d(j\tilde y) = f$. Thus ${}_\varphi d$ is an isomorphism, so (S,φ,ψ) is nonsingular. Let $L_0^\lambda(B,g,q;L)$ denote the isometry class of $(S,\varphi,\psi) \in F_0^\lambda(B/A)$ and $\mathcal{L}_0^\lambda(B,g,q;L)$ its class in $W_0^\lambda(B/A)$.

(3.3) __Proposition:__ $\mathcal{L}_0^\lambda(B^n,g,q;L)$ depends only on the class of (B,g,q) in $W_0^\lambda(B)$.

__Proof:__ To show independence of $\mathcal{L}_0^\lambda(B^n,g,q;L)$ from L, it suffices to show $\mathcal{L}_0^\lambda(B^n,g,q;L) = \mathcal{L}_0^\lambda(B^n,g,q;I)$ where $I \subseteq L$, since any two integral lattices L and M contain a common sublattice I. From $I \subseteq L$ we obtain

$$I \subseteq L \subseteq L' \subseteq I'.$$

If $L_0^\lambda(B^n,g,q;I) = (T,\varphi_I,\psi_I)$ and $K = L/I$ then $K \in \mathfrak{D}_F^1$, $K \subseteq T$ and K is totally isotropic. By definition of dual lattice, $K^\perp = L'/I$ and under the identification $K^\perp/K = (L'/I)/(L/I) \cong L'/L$ we have

$\varphi_I | (K^{\perp}/K) \times (K^{\perp}/K) = \varphi$ and $\psi_I | (K^{\perp}/K) = \psi$. Hence to show $\mathcal{L}_0^{\lambda}(B^n, g, q; L) = \mathcal{L}_0^{\lambda}(B^n, g, q; I)$ it suffices to prove the following lemma.

(3.4) Lemma: Let a nonsingular λ-form (T, φ, ψ) be given, $T \in \mathfrak{D}_F^1$. Suppose there is $K \in \mathfrak{D}_F^1$, $K \subseteq T$, and K is totally isotropic. Then if $S = K^{\perp}/K$, the naturally induced λ-form (S, φ', ψ') equals (T, φ, ψ) in $W_0^{\lambda}(B/A)$, provided (S, φ', ψ') is nonsingular.

Proof: $(S, \varphi', \psi') \perp (S, (-\varphi'), (-\psi')) = 0$ in $W_0^{\lambda}(B/A)$ (the diagonal submodule is a subkernel) so it suffices to show $(T, \varphi, \psi) \perp (S, (-\varphi'), (-\psi')) = 0$ in $W_0^{\lambda}(B/A)$. Hence it suffices to take $K = K_1$ in the following sublemma.

(3.5): Suppose given $(S, \varphi, \psi) \in F_0^{\lambda}(B/A)$ and $K_1 \subset S$ totally isotropic such that the induced sesquilinear form $(S/K_1^{\perp}) \times K_1 \to B/A$ is nonsingular and the induced λ-form $(K_1^{\perp}/K_1, \varphi_1, \psi_1)$ is nonsingular and is a kernel. Then (S, φ, ψ) is a kernel (see (1.13b)).

Proof: Let K_2 be a subkernel for $(K_1^{\perp}/K_1, \varphi_1, \psi_1)$. Let J' be the pullback in

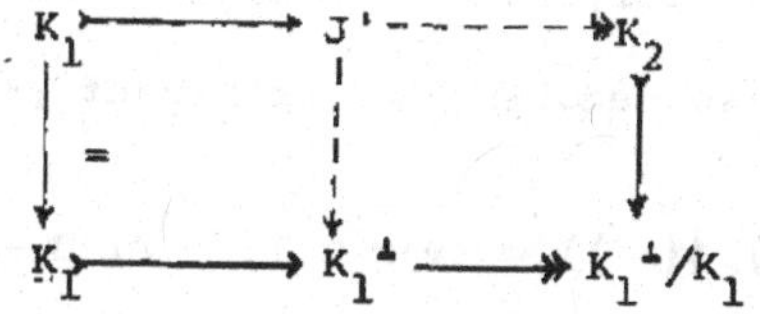

and let J be its image in $K_1^{\perp} \subseteq S$; it is easy to see J is totally isotropic. Further, the diagram shows that $K_1^{\perp}/J \cong (K_1^{\perp}/K_1)/K_2 \cong K_2^{\wedge}$. The sequence of injections $J \to K_1^{\perp} \to S$ gives rise to the short exact sequence of cokernels

$$K_2^{\wedge} \cong K_1^{\perp}/J \rightarrowtail S/J \twoheadrightarrow S/K_1^{\perp} \cong K_1^{\wedge},$$

which by construction is $\mathrm{Hom}_A(-,B/A)$ of the exact sequence $K_1 \to J \to K_2$. This shows $S/J \cong J^\wedge$ so the induced form $(S/J) \times J \to B/A$ is nonsingular. Hence J is a subkernel.

Continuing the proof of (3.3), suppose (B^n,g,q) is isometric to (B^n,g',q'). Then there is an automorphism $\alpha\colon B^n \to B^n$ such that $d_g = \bar{\alpha} d_{g'} \alpha$. If $L \subseteq B^n$ is integral with respect to g, then $I = \alpha(L)$ is integral with respect to g' and $\bar{\alpha}(I') = L'$. It follows easily that $L_0^\lambda(B^n,g,q;L) = L_0^\lambda(B^n,g',q';I)$. Using this we may take $\mathcal{H}(B^n)$ to be represented by $\mathcal{H}(A^n) \otimes B$ and clearly $\mathcal{L}_0^\lambda(\mathcal{H}(A^n) \otimes B) = 0$ (we may find a lattice $L = A^n + \bar{A}^n$ with $L = L'$). This completes the proof of (3.3) and shows we have a well-defined homomorphism $\mathcal{L}_0^\lambda\colon W_0^\lambda(B) \to W_0^\lambda(B/A)$.

§4 $\partial_0^\lambda\colon W_0^\lambda(B/A) \to W_1^{-\lambda}(A)$

(4.1): Let a nonsingular λ-form (S,φ,ψ) be given. Choose a short free resolution for S, $(R) = (A^n \overset{\alpha}{\rightarrowtail} A^n \overset{j}{\twoheadrightarrow} S)$ and let $(R^\wedge) = (\bar{A}^n \overset{\bar{\alpha}}{\rightarrowtail} \bar{A}^n \overset{\tilde{j}}{\twoheadrightarrow} S^\wedge)$ be the dual resolution (1.4). Choose a λ-hermitian form $\tau\colon A^n \times A^n \to B$ satisfying the conditions of (1.17). Hence for $m,m' \in A^n$, and τ also denoting the matrix of $\tau\colon A^n \times A^n \to B$,

(4.2)
$$\begin{aligned}
&\varphi(j(m),j(m')) = r\tau(m,m'), \qquad r\colon B \longrightarrow B/A\\
&\psi(j(m)) = r'\tau(m,m) \qquad , \qquad r'\colon B \longrightarrow B/S_\lambda(A)\\
&\bar{\alpha}\tau \in M_n(A) \qquad , \qquad \bar{\alpha}\tau\alpha \in S_\lambda(A^n)\\
&\tilde{j}\cdot\bar{\alpha}\tau = d_\varphi \cdot j\colon A^n \longrightarrow S^\wedge.
\end{aligned}$$

Setting $\gamma = \tau\alpha$, we have $\bar{\gamma}\alpha = \lambda\bar{\alpha}\tau\alpha \in S_\lambda(A^n)$; further, we claim $(\alpha,\gamma)\colon A^n \to A^n + \bar{A}^n$ is a split injection with totally isotropic image. Once we verify this the discussion of (1.26) shows how to construct

$$(4.3) \qquad \sigma = \begin{pmatrix} \alpha & \beta \\ \gamma & \delta \end{pmatrix} \in U_{2n}^{-\lambda}(A),$$

which, as an element of $W_1^{-\lambda}(A)$, depends only on the pair (α,γ). Since α is injective, (α,γ) is injective. We will show that $t(\operatorname{cok}(\alpha,\gamma))$, the torsion submodule of $\operatorname{cok}(\alpha,\gamma)$, is zero and conclude that $\operatorname{cok}(\alpha,\gamma)$ is free. Hence (α,γ) is a split injection. Suppose $(e,f) \in A^n + \bar{A}^n$ and $(\alpha(m),\gamma(m)) = (e,f)s = (es,fs)$, for some $m \in M$, $s \in \Sigma$. Since $\gamma = \tau\alpha$, we get $\alpha(m) = es$ and $\tau\alpha(m) = fs$; hence $\tau(e) = f$. If $e \notin \operatorname{im}(\alpha)$, then $j(e) \neq 0$ so $d_\varphi j(e) = \tilde{j}\bar{\alpha}\tau(e) \neq 0$ since d_φ is bijective. But $\tau(e) = f \in \bar{A}^n$ and $\tilde{j}\alpha$ is the zero map. Thus $\alpha(m') = e$, for some $m' \in A^n$ and $\gamma(m')s = \tau\alpha(m')s = fs$, so $\gamma(m') = f$. Thus $(e,f) \in \operatorname{im}(\alpha,\gamma)$. Denote $J = \operatorname{im}(\alpha,\gamma)$, let $h\colon (A^n + \bar{A}^n) \times (A^n + \bar{A}^n) \to A$ denote the hyperbolic λ-hermitian form and let $g\colon J \times ((A^n + \bar{A}^n)/J) \to A$ denote the induced sesquilinear form. We have the exact sequence $J \overset{i}{\rightarrowtail} A^n + \bar{A}^n \overset{j}{\twoheadrightarrow} (A^n + \bar{A}^n)/J$, and the commutative diagram

$$\begin{array}{ccc} J & \xrightarrow{\quad (\alpha,\gamma) \,=\, i \quad} & A^n + \bar{A}^n \\ \Big\downarrow {}_g d & & \cong \Big\downarrow h^d \\ (A^n + \bar{A}^n)/J & \xrightarrow{\quad \bar{j} \quad} & A^n + \bar{A}^n \end{array} .$$

Since h is nonsingular, ${}_g d$ is injective; clearly ${}_g d \otimes B$ is an isomorphism so $\operatorname{cok}({}_g d)$ is torsion (1.2). But $\operatorname{cok}(i)$ is torsion-free, so ${}_g d$ is an isomorphism. Thus $\operatorname{cok}(\alpha,\gamma) = (A^n + \bar{A}^n)/J \cong \bar{J} \cong \bar{A}^n$ is free, so J is a summand of $A^n + \bar{A}^n$ as desired.

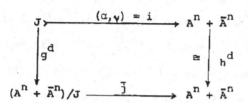

We denote $I_0^\lambda(S,\varphi,\psi;R,\tau) = \sigma \in U_{2n}^{-\lambda}(A)$ (4.3) and let $\vartheta_0^\lambda(S,\varphi,\psi;R,\tau)$ be the class of σ in $W_1^{-\lambda}(A)$.

(4.4) Proposition: $\vartheta_0^\lambda(S,\varphi,\psi;R,\tau)$ is independent of the choices of R and τ.

Proof: Suppose first that $\tau'\colon A^n \times A^n \to B$ also satisfies (4.2) with respect to the resolution (R), so that $\rho = \tau - \tau' \in S_\lambda(A^n)$. Then if $I_0^\lambda(S,\varphi,\psi;R,\tau') = \begin{pmatrix}\alpha' & \beta' \\ \gamma' & \delta'\end{pmatrix}$, we have $\alpha = \alpha'$ and $(\alpha,\gamma) = (\alpha,\tau\alpha)$ $= (\alpha,\tau'\alpha + \rho\alpha) = (\alpha,\tau'\alpha' + \rho\alpha') = (\alpha',\gamma' + \rho\alpha')$. Hence $I_0^\lambda(S,\varphi,\psi;R,\tau) = X_+(\rho)I_0^\lambda(S,\varphi,\psi;R,\tau')$ so $\vartheta_0^\lambda(S,\varphi,\psi;R,\tau) = \vartheta_0^\lambda(S,\varphi,\psi;R,\tau')$.

Stabilizing $(R) = (A^n \overset{\alpha}{\rightarrowtail} A^n \twoheadrightarrow S)$ to $(R + I_k) = (A^{n+k} \xrightarrow{\alpha+I_k} A^{n+k} \twoheadrightarrow S)$ we get $I_0^\lambda(S,\varphi,\psi;R + I_k, \tau \perp O_k) = I_0^\lambda(S,\varphi,\psi;R,\tau) \perp I_{2k}$ (stabilization (1.19)) where $O_k\colon A^k \times A^k \to B$ is the zero form. If $(R') = (A^m \overset{\alpha'}{\rightarrowtail} A^m \twoheadrightarrow S)$ is another resolution for S, then by (1.7) there are integers k, ℓ with $n + k = m + \ell$ and automorphisms $\mu,\mu'\colon A^{n+k} \to A^{n+k}$ such that $(\alpha' + I_\ell)\mu' = \mu(\alpha + I_k)$. If $\tau'\colon A^{m+\ell} \times A^{m+\ell} \to B$ is defined by $\tau'(m,m') = (\tau \perp O_k)(\mu^{-1}(m),\mu^{-1}(m'))$ we have conditions (4.2) with $\tau',\alpha' + I_\ell$ replacing τ,α and $\tau' = \bar{\mu}^{-1}(\tau + O_k)\mu^{-1}$. Thus, if $O_k\colon A^k \to A^k$ is the zero map, $(\alpha + I_k, \gamma + O_k) = (\alpha + I_k,(\tau + O_k)(\alpha + I_k))$ $= (\mu^{-1}(\alpha' + I_\ell)\mu', (\bar{\mu}\tau'\mu)(\mu^{-1}(\alpha' + I_\ell)\mu')) = (\mu^{-1}(\alpha' + I_\ell)\mu', \bar{\mu}\gamma'\mu')$ where $\gamma' = \tau'(\alpha' + I_\ell)$. Hence if $\tau''\colon A^m \times A^m \to B$ is any covering of (S,φ,ψ) with respect to (R') (i.e., satisfying (4.2)),

$I_0^\lambda(S,\varphi,\psi;R,\tau) \perp I_{2k} = I_0^\lambda(S,\varphi,\psi;R + I_k,\tau \perp O_k)$

$= H(\mu^{-1})I_0^\lambda(S,\varphi,\psi;R' + I_\ell,\tau')H(\mu')$

$= H(\mu^{-1})X_+(\tau' - (\tau'' \perp O_\ell))I_0^\lambda(S,\varphi,\psi;R' + I_\ell,\tau'' + O_\ell)H(\mu')$

$= H(\mu^{-1})X_+(\tau' - (\tau'' \perp O_\ell))(I_0^\lambda(S,\varphi,\psi;R',\tau'') \perp I_{2\ell})H(\mu')$. This completes the proof.

(4.5) <u>Corollary</u>: Given $I_0^\lambda(S,\varphi,\psi;R,\tau) \in U_{2n}^{-\lambda}(A)$, $\rho \in S_\lambda(A^n)$ and $\mu' \in GL_n(A)$, we may find (R,τ') and $(R",\tau")$ satisfying (4.2) such that $I_0^\lambda(S,\varphi,\psi;R",\tau")H(\mu') = I_0^\lambda(S,\varphi,\psi;R,\tau) = X_-(\rho)I_0^\lambda(S,\varphi,\psi;R,\tau')$.

<u>Proof</u>: This follows from the proof of (4.4).

(4.6): To complete the construction of $\mathfrak{s}_0^\lambda$, it remains only to show that if $(S,\varphi,\psi) = 0$ in $W_0^\lambda(B/A)$, then $\mathfrak{s}_0^\lambda(S,\varphi,\psi) = 0$. Thus we assume $(S,\varphi,\psi) \perp (U,\varphi',\psi') = (V,\varphi",\psi")$ where (U,φ',ψ') and $(V,\varphi",\psi")$ are kernels. The following lemma suffices.

(4.7) <u>Lemma</u>: Let (S,φ,ψ) be a kernel. Then $\mathfrak{s}_0^\lambda(S,\varphi,\psi) = 0$.

<u>Proof</u>: By assumption there is $K \in \mathfrak{D}_F^1$ such that K is totally isotropic and there is a short exact sequence $K \rightarrowtail S \twoheadrightarrow S/K \cong K^\wedge$. If $A^n \overset{\mu}{\rightarrowtail} A^n \overset{j}{\twoheadrightarrow} K$ is a short free resolution for K, then (1.4) we have $\bar{A}^n \overset{\bar{\mu}}{\rightarrowtail} \bar{A}^n \overset{\tilde{j}}{\twoheadrightarrow} K^\wedge$ and consequently a short free resolution (R) for S:

$$(R) = (A^n + \bar{A}^n \xrightarrow{\begin{pmatrix} \mu & \theta \\ 0 & \bar{\mu} \end{pmatrix}} A^n + \bar{A}^n \twoheadrightarrow S + S^\wedge)$$

where $\theta: \bar{A}^n \to A^n$ is a homomorphism. We claim there is $\tau: (A^n + \bar{A}^n) \times (A^n + \bar{A}^n) \to B$ satisfying (4.2) with matrix

$$\tau = \begin{bmatrix} 0 & \lambda\bar{\mu}^{-1} \\ \mu^{-1} & \nu \end{bmatrix} .$$

Since $\varphi|K \times K = 0$ and, under the identification $S/K \cong K^\wedge$, $\varphi|(S/K) \times K$ is the natural form, $\langle\ ,\ \rangle_K$, it suffices to show that $r\langle x,\mu^{-1}y\rangle_{B^n} = \langle \tilde{j}(x), j(y)\rangle_K$ where $x,y \in A^n$ and $r: B \to B/A$. But this follows directly from the definition (1.6) of $\tilde{j}$. So τ has the form claimed. Thus

$$I_0^{\lambda}(S,\varphi,\psi;R,\tau) = \left(\begin{array}{cc|c} \mu & \theta & \beta \\ 0 & \bar{\mu} & \\ \hline 0 & \lambda I_n & \delta \\ I_n & \eta & \end{array}\right) \in U_{4n}^{-\lambda}(A),$$

where $\eta = \mu^{-1}\theta + \nu\bar{\mu}$. Consider the element $w_1^{\lambda} = \begin{pmatrix} 0 & 1 \\ \lambda & 0 \end{pmatrix} \in GL_2(A)$ (see 1.20--w_1^{λ} is not in $U_2^{-\lambda}(A)$) and the element $T_4^{\lambda} \in U_4^{-\lambda}(A)$.

(4.9) $$T_4^{\lambda} = X_-(-w_1^{\lambda})X_+(\bar{w}_1^{-\lambda})H(w_1^{\lambda})X_-(w_1^{\lambda})$$

introduced by Sharpe in [Sh]. Then T_4^{λ} represents zero in $W_1^{-\lambda}(A)$ and

(4.10) $$T_4^{\lambda} = \begin{pmatrix} 0 & 0 & 1 & 0 \\ 0 & 0 & 0 & 1 \\ \lambda & 0 & 0 & 0 \\ 0 & \lambda & 0 & 0 \end{pmatrix} \quad ; \quad \text{set } T_{4n}^{\lambda} = T_4^{\lambda} \perp \cdots \perp T_4^{\lambda}$$

(n terms)

Thus,

$$T_{4n}^{\lambda}L_0^{\lambda}(S,\varphi,\psi;R,\tau) = \left(\begin{array}{cc|c} 0 & \lambda I_n & \delta \\ I_n & \eta & \\ \hline \lambda\mu & \lambda\theta & \lambda\beta \\ 0 & \lambda\bar{\mu} & \end{array}\right)$$

The upper left-hand $(2n \times 2n)$ block is invertible so by [B,II.2.5(b)], (4.11) has the form $X_-(\gamma_1)H(\alpha)X_+(\gamma_2)$, hence represents zero in $W_1^{-\lambda}(A)$.

§5 Exactness of the last five terms of (2.1)

(5.1) Proposition: The following sequence of groups and homomorphisms is exact, where $\mathcal{K}_i^{\pm\lambda}$ is induced by $\otimes$ B:

$$W_0^{\lambda}(A) \xrightarrow{\mathcal{K}_0^{\lambda}} W_0^{\lambda}(B) \xrightarrow{\mathcal{L}_0^{\lambda}} W_0^{\lambda}(B/A) \xrightarrow{\partial_0^{\lambda}} W_1^{-\lambda}(A) \xrightarrow{\mathcal{K}_1^{-\lambda}} W_1^{-\lambda}(B).$$

<u>Proof:</u> (a) Exactness at $W_0^\lambda(B)$. It is clear that $\mathcal{L}_0^\lambda \mathcal{K}_0^\lambda = 0$. Let (B^n,g,q) be given with $L_0^\lambda(B^n,g,q;L) \perp (U,\varphi,\psi) = (V,\varphi,\psi)$ in $F_0^\lambda(B/A)$ (notation as in (3.1)) where (U,φ,ψ) and (V,φ',ψ') are kernel λ-forms. We need the following lemma.

(5.2) <u>Lemma:</u> Let (U,φ,ψ) be a kernel λ-form, $U \in \mathfrak{D}_F^1$. Then there is a hyperbolic λ-form (B^n,g,q) and an integral lattice $L \subset B^n$ such that $L_0^\lambda(B^n,g,q;L) = (U,\varphi,\psi)$.

<u>Proof:</u> In (4.7) we showed there was a resolution R of U and a covering τ of (U,φ,ψ) (4.2) so that

$$I_0^\lambda(U,\varphi,\psi;R,\tau) = \left[\begin{array}{cc|c} \mu & \theta & \beta \\ 0 & \bar{\mu} & \\ \hline 0 & \lambda I_n & \delta \\ I_n & \eta & \end{array}\right]$$

Hence $I_0^\lambda(U,\varphi,\psi;R,\tau)H\begin{pmatrix}-\eta & \lambda I_n \\ I_n & 0\end{pmatrix} = \left[\begin{array}{cc|c} -\mu\eta+\theta & \lambda\mu & \beta \\ \bar{\mu} & 0 & \\ \hline I_n & 0 & \delta' \\ 0 & I_n & \end{array}\right], -\mu\eta + \theta = -\mu\nu\bar{\mu}$

$\in S_\lambda(A^n) = I_0^\lambda(U,\varphi,\psi;R',\tau')$, for some choice of R', τ' by (4.5). By definition this implies there is a λ-form $\theta = (L,g,q)$, $L = A^n \times \bar{A}^n$ where g has matrix $\begin{pmatrix}-\mu\nu\bar{\mu} & \lambda\mu \\ \bar{\mu} & 0\end{pmatrix}$ such that $\theta \otimes B = (B^n + \bar{B}^n, g_B q_B)$ has integral lattice equal to $L \subset B^n + \bar{B}^n$. Thus $L_0^\lambda(B^n + \bar{B}^n, g_B, q_B; L) = (S,\varphi,\psi)$. The matrix for θ above implies $(B^n + \bar{B}^n, g_B, q_B)$ is hyperbolic, so the proof of (5.2) is complete.

Returning to exactness, (5.2) allows us to assume $L_0^\lambda(B^n,g,q;L) = (V,\varphi,\psi) =$ a kernel λ-form. We thus have a resolution $L \rightarrowtail L' \overset{j}{\twoheadrightarrow} V$ and $\tau = g|\, L'$ covering (V,φ,ψ) as in (4.2). By (5.2) again, we have a hyperbolic λ-form (B^n,h,p), an integral lattice $I \subset B^n$ with corresponding resolution $I \rightarrowtail I' \overset{k}{\twoheadrightarrow} V$, and a covering $\theta\colon I' \times I' \to B$ $(\theta = h|I')$ of (V,φ,ψ) satisfying (4.2). Let P be the pullback $\{(x,y) \in L' \times I' \mid j(x) = k(y)\}$,

$$\begin{array}{ccc} P & \dashrightarrow & I' \\ \Big\downarrow & & \Big\downarrow k \\ L' & \xrightarrow{\;j\;} & V \end{array} .$$

Clearly P is stably free. Define f in a λ-form (P,f,s) by setting $f((x,y),(x',y')) = \tau(x,x') - \theta(y,y')$ where $(x,y),(x',y') \in P$. This f is clearly λ-hermitian and, since τ and θ each cover (S,φ,ψ), f takes values in A and $f((x,y),(x,y)) \in S_\lambda(A)$, for all $(x,y) \in P$. To define s, take P to be free with basis $\{e_1,\ldots,e_n\}$, $f(e_i,e_i) = a_i + \lambda\bar{a}_i \in S_\lambda(A)$, $a_i \in A$. Define $s(e_i) \equiv a_i \bmod S_{-\lambda}(A)$, and extend to $s\colon P \to A/S_{-\lambda}(A)$ using the conditions (1.12a).

By definition of P, there are split exact sequences

$$L \overset{i_1}{\rightarrowtail} P \twoheadrightarrow I', \quad i_1(x) = (x,0)$$

and

$$I \overset{i_2}{\rightarrowtail} P \twoheadrightarrow L', \quad i_2(y) = (0,y).$$

Clearly $f(L \times I) \equiv 0 \equiv f(I \times L)$ from which we obtain induced forms

$$L \times (P/I) \longrightarrow A \qquad \text{and} \qquad I \times (P/L) \longrightarrow A \quad .$$

Identifying P/I with L' and P/L with I' these become the nonsingular forms pairing a lattice with its dual lattice. It is easy to verify that this implies f is nonsingular. But since

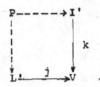

$(L + I) \otimes B \xrightarrow{(i_1+i_2)\otimes B} P \otimes B$ is an isomorphism the construction of f shows $(P \otimes B,\ f \otimes B,\ s \otimes B)$ is isometric to $(B^n,g,q) \perp (B^n,h,p)$. Since (B^n,h,p) is hyperbolic $\mathcal{K}_0^\lambda((P,f,s)) = (B^n,g,q)$ in $W_0^\lambda(B)$.

(b) Given (B^n,g,q) representing an element of $W_0^\lambda(B)$, let $L \subset B^n$ be an integral lattice and let $L_0^\lambda(B^n,g,q;L) = (S,\varphi,\psi)$. Then $(R) = (L \overset{\alpha}{\rightarrowtail} L' \twoheadrightarrow S)$ gives a resolution and $\tau\colon L' \times L' \to B$ covers the λ-form (S,φ,ψ) as in (4.2), $\tau = g|L' \times L'$. Recalling that if L has basis $\{e_i\}$, we give L' the basis $\{e_i{}^*\}$ satisfying $g(e_i{}^*,e_j) = \delta_{ij}$, it is easily verified that $\tau\alpha = I_n$ so that

$$I_0^\lambda(S,\varphi,\psi;R,\tau) = \begin{pmatrix} \alpha & \beta \\ I_n & \delta \end{pmatrix} = \sigma.$$

Since n is even (1.13a) we may multiply σ on the left by T_{2n}^λ as in (4.11) to see that σ represents zero in $W_1^{-\lambda}(A)$. Thus $\partial_0^\lambda \mathcal{L}_0^\lambda(B,g,q) = 0$.

Now suppose given $(S,\varphi,\psi) \in F_0^\lambda(B/A)$, a resolution $(R) = (A^{2n} \overset{\alpha}{\rightarrowtail} A^{2n} \twoheadrightarrow S)$, and a covering $\tau\colon A^{2n} \times A^{2n} \to B$ of (S,φ,ψ) (4.2) such that $I_0^\lambda(S,\varphi,\psi;R,\tau) = \sigma \in U_{4n}^{-\lambda}(A)$ and σ represents zero in $W_1^{-\lambda}(A)$. By [Sh, 5.5,5.6] there exist $\rho,\tau_1,\tau_2 \in S_\lambda(A^{2n})$, $\alpha \in GL_{2n}(A)$ such that

$$\sigma = X_-(\tau_1)T_{4n}^\lambda X_-(\tau_2)H(\alpha)X_+(\bar\rho),$$

where T_{4n}^λ is defined in (4.10). (Sharpe works in the "unitary Steinberg group", but his matrix calculations show that the above form is valid.) Hence $X_-(-\tau_1)\sigma X_+(-\bar\rho)H(\alpha^{-1}) = T_{4n}X_-(\tau_2) = \begin{pmatrix} \tau_2 & * \\ -\lambda I_{2n} & * \end{pmatrix}$.

Right multiplication by $X_+(-\bar\rho)$ does not change the first column of $(2n \times 2n)$ blocks in σ while left multiplication by $X_-(-\tau_1)$ and right multiplication by $H(\alpha^{-1})$ are realized by changes in R and

τ (4.5). Hence there are R', τ' so that $I_0^\lambda(S,\varphi,\psi;R',\tau') = \begin{pmatrix} \tau_2 & \beta \\ -\lambda I_{2n} & \delta \end{pmatrix}$. As we saw in the proof of (5.2), this means (S,φ,ψ) is in the image of L_0^λ.

(5.3) Remark: In proving that $\partial_0^\lambda \mathcal{L}_0^\lambda = 0$, we needed n even in the λ-form (B^n,g,q). This is the only place we will use this condition. If $\lambda = 1$, it is unnecessary: if n is odd we may add $\mathcal{H}(B)$ to (B^n,g,q), observe that $[1] \perp [-1] \perp (B^n,g,q) \cong \mathcal{H}(B) \perp (B^n,g,q)$ (where [b] denotes the unary form on B with matrix (b)), that [1] is sent to zero by $\partial_0^\lambda \mathcal{L}_0^\lambda$ and that $[-1] \perp [B^n,g,q]$ is a λ-form on B^{n+1}, n + 1 even. If $\lambda = -1$ and B has simple component acted upon trivially by the involution (e.g., $B = \mathbb{Q}\pi$, π finite), then each (B^n,g,q) has n even. The author does not know whether the assumption that n be even is necessary in general.

(c) Exactness at $W_1^{-\lambda}(A)$. If $(S,\varphi,\psi) \in F_0^\lambda(B/A)$, choose $(R) = (A^n \overset{\alpha}{\rightarrowtail} A^n \twoheadrightarrow S)$ and τ as usual so that $I_0^\lambda(S,\varphi,\psi;R,\tau) = \begin{pmatrix} \alpha & \beta \\ \gamma & \delta \end{pmatrix} = \sigma$.

Since $\alpha \otimes B$ is invertible, we may apply [B.II.2.5b] to conclude $\mathcal{K}_1^{-\lambda}(\sigma) = 0$.

Next suppose $\sigma \in U_{2n}^{-\lambda}(A)$ is such that $\mathcal{K}_1^{-\lambda}(\sigma) = 0$ in $W_1^{-\lambda}(B)$. By stabilizing (if necessary) we may assume n is even. If $\sigma = \begin{pmatrix} \alpha & \beta \\ \gamma & \delta \end{pmatrix}$, we showed in [P3, 3.8] that $[\sigma] = 0$ in $W_1^{-\lambda}(B)$ implies there is $\tau \in S_\lambda(B^n)$ such that $\gamma + \tau\alpha \in GL_n(B)$; moreover, τ is required to be the adjoint of the orthogonal sum of a λ-form nonsingular on $\alpha(\ker \gamma)$ with the zero form on some complement of $\alpha(\ker \gamma)$ in A^n. Replacing τ by $a\tau\bar{a}$, where $a\tau\bar{a} \in S_\lambda(A^n)$, $a \in \Sigma$, the λ-form $a\tau\bar{a}$ still has the above mentioned properties. Hence we may find

$\tau \in S_\lambda(A^n)$ such that $\gamma + \tau\alpha \in GL_n(B)$. If $\sigma' = T^\lambda_{2n}X_-(\tau)\sigma$ and $\sigma' = \begin{pmatrix}\alpha' & \beta' \\ \gamma' & \delta'\end{pmatrix}$ then α' is a unit in $M_n(B)$. Taking $\tau = \gamma'\alpha'^{-1}$ and $\mu = \alpha'$ in (1.18) we obtain a λ-form (S,φ,ψ) for which (by definition-- see (4.2))

$$I_0^\lambda(S,\varphi,\psi;R,\tau) = \sigma',$$

where $(R) = (A^n \overset{\alpha'}{\rightarrowtail} A^n \twoheadrightarrow S)$. This completes the proof of (5.1).

§6 $\underline{\varkappa_1^\lambda\colon W_1^\lambda(B) \to W_1^\lambda(B/A)}$

(6.1): Let $\sigma = \begin{pmatrix}\alpha & \beta \\ \gamma & \delta\end{pmatrix} \in U_{2n}^\lambda(B)$ be given; we want to construct an element $(K,H,\Delta,\varkappa) \in F_1^\lambda(B/A)$. We may find $\nu,\eta \in M_n(A)$, invertible as elements of $M_n(B)$, such that $\alpha\nu$, $\gamma\nu$, $\beta\eta$, $\delta\eta \in M_n(A)$. Let

(6.2) $$\sigma(\nu,\eta) = \begin{pmatrix}\alpha\nu & \beta\eta \\ \gamma\nu & \delta\eta\end{pmatrix}, \qquad \sigma'(\nu,\eta) = \begin{pmatrix}\overline{\delta\eta} & \lambda\overline{\beta\eta} \\ \overline{\lambda\gamma\nu} & \overline{\alpha\nu}\end{pmatrix}$$

and consider the sequence of injections,

(6.3) $$A^n + \bar{A}^n \xrightarrow{\sigma(\nu,\eta)} A^n + \bar{A}^n \xrightarrow{\sigma'(\nu,\eta)} A^n + \bar{A}^n.$$

Denote the terms of (6.3) L,I,L', respectively, keeping in mind that each has a basis (it turns out below that L' is the dual lattice to L). We can thus consider the sequence of lattices in $B^n + \bar{B}^n$,

(6.4) $$L \subseteq I \subseteq L' \subset B^n + \bar{B}^n,$$

where $I \otimes B \subseteq B^n + \bar{B}^n$ is the <u>identity</u>. Hence it makes sense to endow $I = A^n + \bar{A}^n$ with the hyperbolic structure (1.13a) so that $I \otimes B = \mathcal{H}(A^n) \otimes B \to \mathcal{H}(B^n)$ is the identity. L and L' inherit λ-hermitian forms h_L and $h_{L'}$ using the injections $\sigma(\nu,\eta)$ and $\sigma'(\nu,\eta)$; $h_{L'}$ may take values in B. Calculation using the relations (1.20) defining σ as an element of $U^{\lambda}_{2n}(B)$ shows that h_L has matrix $\begin{pmatrix} 0 & \lambda\bar{\nu}\eta \\ \bar{\eta}\nu & 0 \end{pmatrix}$ and $h_{L'}$ has matrix $\begin{pmatrix} 0 & \lambda(\bar{\nu}\eta)^{-1} \\ (\bar{\eta}\nu)^{-1} & 0 \end{pmatrix}$, where inverses are taken in $GL_n(B)$. It follows that L and L' are dual lattices in $\mathcal{H}(B^n)$. If $D\colon A^n + \bar{A}^n \xrightarrow{\cong} A^n + \bar{A}^n$ denotes the adjoint of the hyperbolic λ-form on $A^n + \bar{A}^n$, another calculation shows

$$\overline{\sigma(\nu,\eta)}D = D\sigma'(\nu,\eta) \tag{6.5}$$ (see [B,II.1.2])

and

$$\sigma'(\nu,\eta)\sigma(\nu,\eta) = \begin{pmatrix} \bar{\eta}\nu & 0 \\ 0 & \bar{\nu}\eta \end{pmatrix}. \tag{6.6}$$

(6.7) <u>Construction of K, H and Δ in (K,H,Δ,κ):</u> Taking $\tau = \begin{pmatrix} 0 & \lambda(\bar{\nu}\eta)^{-1} \\ (\bar{\eta}\nu)^{-1} & 0 \end{pmatrix}$ (= the matrix of $h_{L'}\colon L' \times L' \to B$) and $\mu = \sigma'(\nu,\eta)\sigma(\nu,\eta)$ in (1.18) we obtain the hyperbolic form $\mathcal{H}(H) = (H + H^{\wedge}, \varphi_h, \psi_h)$, $H = \operatorname{cok}(\bar{\eta}\nu)$. If K denotes I/L, $\sigma'(\nu,\eta)$ induces an injection $\Delta\colon K \to H + H^{\wedge} = L'/L$. Explicitly, if $\Delta = (\xi,\zeta)$, $\xi\colon K \to H$, $\zeta\colon K \to H^{\wedge}$, then the diagrams

$$\begin{array}{ccc} A^n + \bar{A}^n & & A^n \\ \downarrow{\scriptstyle \sigma'\sigma} & & \downarrow{\scriptstyle \bar{\eta}\nu} \\ I = A^n + \bar{A}^n & \xrightarrow{(\overline{\delta\eta},\,\lambda\overline{\beta\eta})} & A^n \\ \downarrow{\scriptstyle j_K} & & \downarrow{\scriptstyle j_H} \\ K & \xrightarrow{\quad \xi \quad} & H \end{array} \quad \text{and} \quad \begin{array}{ccc} A^n + \bar{A}^n & & \bar{A}^n \\ \downarrow{\scriptstyle \sigma'\sigma} & & \downarrow{\scriptstyle \bar{\nu}\eta} \\ I = A^n + \bar{A}^n & \xrightarrow{(\lambda\overline{\gamma\nu},\,\overline{\alpha\nu})} & \bar{A}^n \\ \downarrow{\scriptstyle j_K} & & \downarrow{\scriptstyle \tilde{j}_H} \\ K & \xrightarrow{\quad \zeta \quad} & H^{\wedge} \end{array} \tag{6.8}$$

commute.

(6.9) <u>Lemma:</u> $\operatorname{Im}(\Delta) \subset H + H^{\wedge}$ is a subkernel (1.13b).

<u>Proof:</u> To see $\operatorname{im}(\Delta)$ is totally isotropic, observe that the

λ-hermitian form $h_{L'}$ convering $\mathcal{H}(H)$ (in the sense of (1.17)) is A-valued and even (in fact hyperbolic) on I; hence, by construction, the forms φ_h and ψ_h restricted to the image of $I/L = K$ are identically zero in B/A and $B/S_\lambda(A)$. It remains to verify that the sesquilinear form $g\colon ((H + H^\wedge)/K) \times K \to B/A$ induced by φ_h is nonsingular, where $K \equiv \mathrm{im}(\Delta)$. There is a commutative diagram of exact sequences

$$\begin{array}{ccccc}
\mathrm{cok}(\sigma(\nu,\eta)) & \rightarrowtail & \mathrm{cok}(\sigma(\nu,\eta)\sigma'(\nu,\eta)) & \longrightarrow & \mathrm{cok}(\sigma'(\nu,\eta)) \\
\downarrow = & & \downarrow = & & c \downarrow \cong \\
K & \overset{\Delta}{\rightarrowtail} & H + H^\wedge & \xrightarrow{\Delta^\wedge\cdot(\varphi_h)d} & K^\wedge \\
\downarrow = & & \downarrow = & & \downarrow g^d \\
K & \overset{\Delta}{\rightarrowtail} & H + H^\wedge & \twoheadrightarrow & (H + H^\wedge)/K
\end{array}$$

where c is an isomorphism from (6.5) and (1.4). $\Delta^\wedge\cdot(\varphi_h)d$ is a surjection by commutativity, so g^d is an isomorphism.

(6.10) <u>Construction of $\varkappa\colon K \to B/S_{-\lambda}(A)$</u>: To construct and study the $(-\lambda)$-form $(K, [\ ,\ ],\varkappa)$ (1.30) we need an explicit expression for $[\ ,\ ]\colon K \times K \to B/A$, where $[k,\ell] = \varphi_h(\zeta(k),\xi(\ell))$. Let $k,\ell \in K$ and $x,y \in A^n + \bar{A}^n$, with $j_K(x) = k$, $j_K(y) = \ell$ (6.8); let $x = x_1 + x_2$, $y = y_1 + y_2$, $x_1,y_1 \in A^n$, $x_2,y_2 \in \bar{A}^n$. Then by construction and (6.8), if $r\colon B \to B/A$ is the projection,

(6.11) $$[k,\ell] = \varphi_h(\zeta(k),\xi(\ell)) = rh_{L'}(\lambda\overline{\gamma\nu}(x_1)+\overline{\alpha\nu}(x_2),\overline{\delta\eta}(y_1)+\lambda\overline{\beta\eta}(y_2)).$$

Expanding the term on the right, we find that if $\tau\colon (A^n + \bar{A}^n) \times (A^n + \bar{A}^n) \to B$ is the sesquilinear form with matrix

(6.12) $$\tau = \begin{pmatrix} \lambda\gamma\bar{\delta} & \gamma\bar{\beta} \\ \alpha\bar{\delta} & \lambda\alpha\bar{\beta} \end{pmatrix}$$

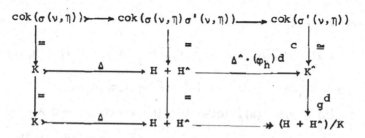

then $[k,\ell] = r\tau(x,y)$, where r,k,ℓ,x,y are as above. (τ is not λ-hermitian, so does not fit into the context of (1.17). But $\tau + \lambda\tau = \begin{pmatrix} 0 & I_n \\ \lambda I_n & 0 \end{pmatrix} \in M_{2n}(A)$ which is sufficient for τ to induce the $(-\lambda)$-hermitian form $[\ ,\ ]: K \times K \to B/A$ using the first equation of (1.17a).)

Let $k \in K$, $x = x_1 + x_2 \in A^n + \bar{A}^n$, $j_K(x) = k$. Define

$$(6.13)\quad \varkappa(k) = r'\tau(x,x) - q_h(x)$$
$$= r'h_{L'}(\lambda\overline{\gamma\nu}(x_1) + \overline{\alpha\nu}(x_2), \overline{\delta\eta}(x_1) + \lambda\overline{\beta\eta}(x_2)) - q_h(x)$$

where $r': B \to B/S_{-\lambda}(A)$, $\mathcal{H}(A^n) = (A^n + \bar{A}^n = I, g_h, q_h)$, $q_h(x) \in A/S_{-\lambda}(A) \subset B/S_{-\lambda}(A)$. Clearly $\varkappa(k) \equiv r\tau(x,x) \bmod A$ and $r\tau(x,x) = [k,k]$ by (6.11).so 1.12 b(iii) is satisfied; it is routine to verify the rest of (1.12b) so $[K, [\ ,\ ],\varkappa)$ is a $(-\lambda)$-form. Now define

$$(6.14)\qquad L_1^\lambda(\sigma;\nu,\eta) = (K,H,\Delta,\varkappa) \in F_1^\lambda(B/A)$$

as constructed in (6.7), (6.10), and (6.13), and let $\mathcal{L}_1^\lambda(\sigma;\nu,\eta)$ denote its class in $W_1^\lambda(B/A)$.

(6.15) <u>Proposition</u>: $\mathcal{L}_1^\lambda(\sigma;\nu,\eta)$ depends only on the class of σ in $W_1^\lambda(B)$.

<u>Proof</u>: We need to show independence of $\mathcal{L}_1^\lambda(\sigma;\nu,\eta)$ from the choices made in its construction. Denote $\sigma = \begin{pmatrix} \alpha & \beta \\ \gamma & \delta \end{pmatrix}$.

(i) <u>Choice of ν,η</u>. Given $\nu \in M_n(A)$, invertible as an element of $M_n(B)$, there is $\nu' \in M_n(A)$ and $s \in \Sigma$ such that $\nu\nu' = \nu'\nu = sI_n$. Hence it suffices to show that $\mathcal{L}_1^\lambda(\sigma;\nu\nu',\eta) = \mathcal{L}_1^\lambda(\sigma;\nu,\eta) = \mathcal{L}_1^\lambda(\sigma;\nu,\eta\eta')$. But it is easy to verify that if $(E) = (\mathrm{cok}(\nu') \rightarrowtail \mathrm{cok}(\bar{\eta}\nu\nu') \twoheadrightarrow \mathrm{cok}(\bar{\eta}\nu) = H)$, then (1.32) $L_1^\lambda(\sigma;\nu\nu',\eta) = \sigma_E L_1^\lambda(\sigma;\nu,\eta)$.

Similarly, $L_1^\lambda(\sigma;\nu,\eta\eta') = {}_E\sigma L_1^\lambda(\sigma;\nu,\eta)$ if $(E) = (\mathrm{cok}(\eta') \to \mathrm{cok}(\bar{\nu}\eta\eta') \to \mathrm{cok}(\bar{\nu}\eta) = H^*)$. Thus we have $\mathcal{L}_1^\lambda\colon U_{2n}^\lambda(B) \to W_1^\lambda(B/A)$.

(ii) <u>Stabilizing σ</u>. Clearly $L_1^\lambda(\sigma;\nu,\eta) = L_1^\lambda(\sigma \perp I_2;\nu + I_1,\eta + I_1)$. Hence $\mathcal{L}_1^\lambda\colon U^\lambda(B) \to W_1^\lambda(B/A)$.

(iii) It remains to show that $\mathcal{L}_1^\sigma(\sigma;\nu,\eta) = \mathcal{L}_1^\lambda(\sigma E;\nu',\eta')$ for each $E \in X = \langle\{H(\epsilon),X_+(\rho),X_-(\tau) \mid \epsilon \in GL_n(B),\rho,\tau \in S_{-\lambda}(B^n),n \in \mathbb{Z}^+\}\rangle$ (see 1.23).

First let $E = H(\epsilon)$. Choose $\nu,\eta \in M_n(A)$, invertible in $M_n(B)$ such that $\epsilon\nu$, $\alpha\epsilon\nu$, $\gamma\epsilon\nu$, $\bar{\epsilon}^{-1}\eta$, $\beta\bar{\epsilon}^{-1}\eta$ and $\delta\bar{\epsilon}^{-1}\eta$ are in $M_n(A)$. Then $\sigma(\epsilon\nu,\bar{\epsilon}^{-1}\eta) = \{\sigma H(\epsilon)\}(\nu,\eta)$ so $L_1^\lambda(\sigma;\epsilon\nu,\bar{\epsilon}^{-1}\eta) = L_1^\lambda(\sigma H(\epsilon);\nu,\eta)$.

Next if $E = X_-(\rho)$, $\rho \in S_{-\lambda}(B^n)$, choose $a' \in \Sigma$ so that if $a = a'\bar{a}'$, then αa^2, γa^2, βa, δa, $\rho a \in M_n(A)$. Then

$$\{\sigma X_-(\rho)\}(a^2 I_n,aI_n) = \begin{bmatrix} \alpha a^2 + (\beta a)(\rho a) & \beta a \\ \gamma a^2 + (\delta a)(\rho a) & \delta a \end{bmatrix}.$$

We claim that $L_1^\lambda(\sigma X_-(\rho);\, a^2 I_n,aI_n) = \chi_-(\varphi,\psi)L_1^\lambda(\sigma;a^2 I_n,aI_n)$, where (H,φ,ψ) is induced as in (1.18) where we take $\mu = a^3 I_n$, $\tau = \lambda\bar{\rho}a^{-2}$. Since $\{\sigma X_-(\rho)\}(a^2 I_n,aI_n) = \{\sigma H(a^2)X_-(a^4\rho)\}(I_n,aI_n)$ this is a special case of the following lemma, where $\nu = a^2 I_n$, $\eta = aI_n$, $P = a\rho$.

(6.16) <u>Lemma:</u> Let $\sigma \in U_{2n}^\lambda(B)$, $\nu,\eta \in M_n(A)$, invertible in $M_n(B)$ as usual, and let $P\colon A^n \to \bar{A}^n$ be such that $\bar{P}\bar{\eta}\nu \in S_{-\lambda}(A^n)$ and the λ-form (H,φ,ψ) is induced as in (1.18) for $\mu = \bar{\eta}\nu$ and $\tau = \tau_H = \lambda(\bar{\nu}\eta)^{-1}\bar{P}$. Then $L_1^\lambda(\sigma H(\nu)X_-(\bar{\nu}\eta P),I_n,\bar{\nu}\eta) = \chi_-(\varphi,\psi)L_1^\lambda(\sigma;\nu,\eta)$.

<u>Proof:</u> We have

$$(6.17)\qquad \{\sigma H(\nu)X_-(\bar{\nu}\eta P)\}(I_n,\bar{\nu}\eta) = \begin{bmatrix}\alpha\nu+(\beta\eta)P & \beta\eta\\ \gamma\nu+(\delta\eta)P & \delta\eta\end{bmatrix} = \sigma(\nu,\eta)\cdot\begin{bmatrix} I_n & 0\\ P & I_n\end{bmatrix}.$$

Denote $\sigma H(\nu)X_-(\bar{\nu}\eta P) = \theta$. Then a calculation shows

$$(6.18)\qquad \theta'(I_n,\bar{\nu}\eta)\theta(I_n,\bar{\nu}\eta) = \begin{pmatrix}\bar{\eta}\nu & 0\\ -0 & \bar{\nu}\eta\end{pmatrix} = \sigma'(\nu,\eta)\sigma(\nu,\eta)$$

where

$$\theta'(I_n,\bar{\nu}\eta) = \begin{bmatrix}\overline{\delta\eta} & \lambda\overline{\beta\eta}\\ \lambda(\overline{\gamma\nu}+\bar{P}\,\overline{\delta\eta}) & \overline{\alpha\nu}+\bar{P}\,\overline{\beta\eta}\end{bmatrix}.$$

Denote by L_σ and L_σ' the lattices L' and L' obtained for $\sigma(\nu,\eta)$ in (6.4) and by L_θ, L_θ' the lattices constructed for $\theta(I_n,\bar{\nu}\eta)$. Let τ_σ and τ_θ be the corresponding sesquilinear forms (6.12). From (6.17) and (6.18) we obtain (using the same I for σ and θ) the following commutative diagram where the top two horizontal maps are isometries

(6.19)

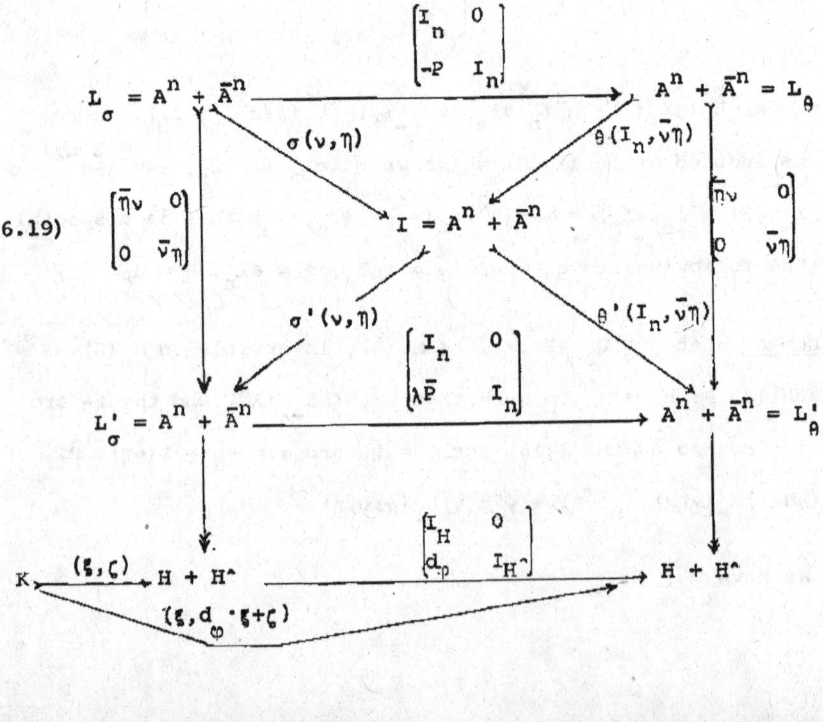

If $L_1^\lambda(\theta;I_n,\bar{\nu}\eta) = (K',H,\Delta',\kappa')$ then (6.19) shows that $\mathrm{im}(L_\sigma) = \mathrm{im}(L_\theta) \subseteq I$ so that $K = K'$ and, using (6.8) and the bottom horizontal composite in (6.18), that $\Delta' = (\xi, d_\varphi \cdot \xi + \zeta)$.

Now let us compute κ'. Let $k \in K$, $x = x_1 + x_2 \in A^n + \bar{A}^n$, $j_K(x_1 + x_2) = k$. By definition (6.13)

$$\kappa'(k) = r'h_{L'_\theta}(\lambda(\overline{\gamma\nu+\delta\eta P})(x_1) + (\overline{\alpha\nu+\beta\eta P})(x_2), \overline{\delta\eta}(x_1) + \lambda\overline{\beta\eta}(x_2)) - q_h(x)$$

$$= r'\tau_\theta(x,x) - q_h(x).$$

Since

$$h_{L'_\sigma} = h_{L'_\theta}, \quad r'\tau_\theta(x,x) = r'h_{L'_\sigma}(\lambda\overline{\gamma\nu}(x_1) + \overline{\alpha\nu}(x_2), \overline{\delta\eta}(x_1) + \lambda\overline{\beta\eta}(x_2))$$

$$+ r'h_{L'_\sigma}(\lambda\bar{P}(\overline{\delta\eta}(x_1) + \lambda\overline{\beta\eta}(x_2)), \overline{\delta\eta}(x_1) + \lambda\overline{\beta\eta}(x_2)).$$

Denoting $\overline{\delta\eta}(x_1) + \lambda\overline{\beta\eta}(x_2) = z$, the last expression is $r'\tau_\sigma(x,x) + r'\langle\lambda\bar{P}z, (\bar{\eta}\nu)^{-1}z\rangle_{B^n} = r'\tau_\sigma(x,x) + r'\langle\lambda(\bar{\nu}\eta)^{-1}\bar{P}z, z\rangle_{B^n}$ $= r'\tau_\sigma(x,x) + r'\tau_H(z,z)$, since $\lambda(\bar{\nu}\eta)^{-1}\bar{P} = \tau_H$ by assumption. By (6.8) and the definition of τ_H and ψ, the last expression is $r'\tau_\sigma(x,x) + \psi(\xi(k))$. This completes the proof of (6.16) since, from above, $\kappa'(k) = r'\tau_\theta(x,x) - q_h(x) = r'\tau_\sigma(x,x) - q_h(x) + \psi(\xi(k))$ $= \kappa(k) + \psi(\xi(k))$.

Returning finally to the proof of (6.15), an argument similar to that just given shows $L_1^\lambda(\sigma X_+(\rho); aI_n, a^2 I_n) = \chi_+(\varphi,\psi) L_1^\lambda(\sigma; aI_n, a^2 I_n)$ for suitable $a \in \Sigma$ and $(-\lambda)$-form (H,φ,ψ). This completes the proof of (6.15) and shows we have a well-defined map $\mathscr{L}_1^\lambda: W_1^\lambda(B) \to W_1^\lambda(B/A)$ which is easily seen to be a homomorphism.

For §8 we will need the following proposition, which "reverses" (6.15).

(6.20) <u>Proposition</u>: Let $L_1^\lambda(\sigma;\nu,\eta) = \theta \approx \theta'$ in $F_1^\lambda(B/A)$, for some $\sigma \in U_{2n}^\lambda(B)$. Then there is a nonnegative integer r, $D \in \langle\{X_+(\rho), X_-(\tau), H(\epsilon) \mid \rho,\tau \in S_{-\lambda}(B^n),\ \epsilon \in GL_n(B),\ n \in \mathbb{Z}^+\}\rangle$, and $\nu',\eta' \in M_{n+r}(A)$ such that

$$L_1^\lambda((\sigma \perp I_{2r})D;\nu',\eta') = \theta'.$$

<u>Proof</u>: Suppose $\theta' = \sigma_E\theta$, where E is the extension $(E) = (J \overset{i}{\rightarrowtail} H \twoheadrightarrow H)$. By construction we have a resolution $A^n \overset{\bar{\eta}\nu}{\rightarrowtail} A^n \twoheadrightarrow H$. We need to find $\mu: A^n \rightarrowtail A^n$, $\operatorname{cok}(\mu) \cong J$, such that the sequence of cokernels,

$$\operatorname{cok}(\mu) \rightarrowtail \operatorname{cok}(\bar{\eta}\nu\mu) \twoheadrightarrow \operatorname{cok}(\bar{\eta}\nu)$$

is isomorphic to the extension E. Then it is clear that $\theta' = L_1^\lambda(\sigma;\nu\mu,\eta)$; actually, we work stably. Precisely, let $A^m \overset{\mu_1}{\rightarrowtail} A^m \overset{j}{\twoheadrightarrow} H'$ be any resolution of H' and let F be the pullback of j and i in

(6.21)
$$\begin{array}{ccccc} A^m & \overset{\mu_3}{\rightarrowtail} & F & \dashrightarrow & J \\ \Big\| & & \Big\downarrow{\scriptstyle \mu_2} & & \Big\downarrow{\scriptstyle i} \\ A^m & \overset{\mu_1}{\rightarrowtail} & A^m & \overset{j}{\twoheadrightarrow} & H' \end{array}$$

Then F is stably free, so we may assume it is free, $F \cong A^m$. Since $\operatorname{cok}(\mu_2) \cong H$, by (1.7) there are integers k, ℓ with $m + \ell = n + k$ and automorphisms $k,h: A^{m+\ell} \to A^{n+k}$ such that $k(\mu_2 + I_\ell)h^{-1} = (\bar{\eta}\nu) + I_k$. Then we have the commutative diagram

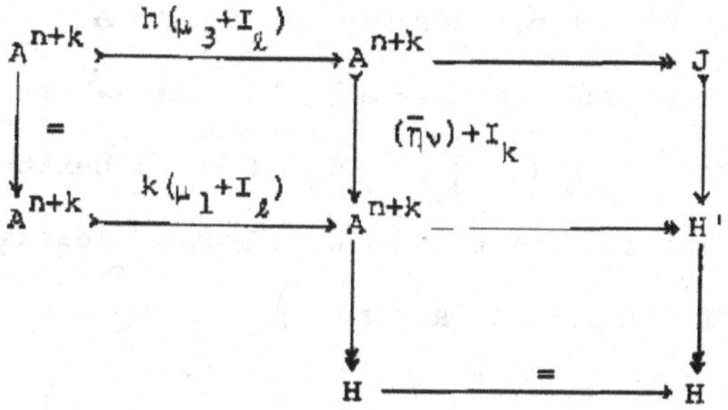

where the right vertical extension is isomorphic to E. Denoting $\mu = h(\mu_3 + I_\ell)$, we find $\theta' = \sigma_E\theta = L_1^\lambda((\sigma \perp I_{2\ell});(\nu + I_k)\mu, \eta + I_k)$. A similar argument applies in the cases $\sigma_E\theta' = \theta$, $\theta' = {}_E\sigma\theta$ and ${}_E\sigma\theta' = \theta$.

To complete the proof of (6.20) suppose $\theta' = \chi_-(\varphi,\psi)\theta$ and $L_1^\lambda(\sigma;\eta,\nu) = \theta$. Construct $P: A^n \to \bar{A}^n$ so that if in (1.17) we take $\mu = \bar{\eta}\nu$ and $\tau = \lambda(\bar{\nu}\eta)^{-1}\bar{P}$, the $(-\lambda)$-form (H,φ,ψ) is produced. Let $Q = \bar{\nu}\eta P$. Then $Q \in S_{-\lambda}(A^n)$ and by (6.16) $L_1^\lambda(\sigma H(\nu)X_-(Q);I_n,\bar{\nu}\eta)$ $= \chi_-(\varphi,\psi)L_1^\lambda(\sigma;\nu,\eta) = \theta'$. A similar argument applies in case $\theta' = \chi_+(\varphi,\psi)\theta$. Since $\chi_{\pm}(-\varphi,-\psi)\chi_{\pm}(\varphi,\psi)\theta = \theta$, the proof of (6.20) is complete.

§7 $\partial_1^\lambda: W_1^\lambda(B/A) \to W_0^\lambda(A)$

(7.1): Let $\theta = (K,H,\Delta,\kappa) \in F_1^\lambda(B/A)$ be given, let $(R) = (A^n \overset{\alpha}{\rightarrowtail} A^n \twoheadrightarrow H)$ be a short free resolution of H and consider the short exact sequence $A^n + \bar{A}^n \xrightarrow{\alpha+\bar{\alpha}} A^n + \bar{A}^n \overset{j}{\to} H + H^\wedge$. Setting $P = j^{-1}(K)$ we have the commutative diagram of injections

(7.2)

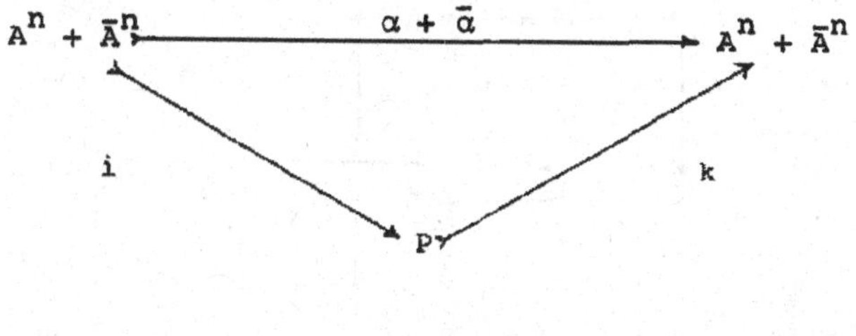

Let L' = the range of $\alpha + \bar{\alpha}$, identify the domain $A^n + \bar{A}^n$ of $\alpha + \bar{\alpha}$ with its image $L \subset L'$ and give $L(\equiv A^n + \bar{A}^n)$ the λ-hermitian form $h_L: L \times L \to A$ with matrix $\begin{pmatrix} 0 & \lambda\bar{\alpha} \\ \alpha & 0 \end{pmatrix}$. Then $L \otimes B$ inherits a (hyperbolic) λ-hermitian form and (as the notation already indicates) L' is the dual lattice to L in $L \otimes B$. Replacing (7.2) we have the sequence of lattices

(7.2)' $\qquad L \subseteq P \subseteq L' \qquad$ (compare (6.4))

in which h_L induces a λ-hermitian form $g: P \times P \to B$ and $h_{L'}: L' \times L' \to B$, the latter having matrix $\begin{pmatrix} 0 & \lambda\bar{\alpha}^{-1} \\ \alpha^{-1} & 0 \end{pmatrix}$.

(7.3) <u>Proposition:</u> g takes values in A and is nonsingular.

<u>Proof:</u> The hyperbolic λ-form $(H + H^{\wedge}, \varphi_h, \psi_h)$ (in the definition of $\theta \in F_1^{\lambda}(B/A)$) is constructed by taking $\tau = h_{L'}$ and $\mu = \alpha + \bar{\alpha}$ in (1.18). Since the image of P under j in $H + H^{\wedge}$ is K, K is totally isotropic by assumption, and $g = h_{L'} \mid P \times P$, we have $g(P \times P) \subseteq A$. To show g is nonsingular it suffices to show that if P' is the dual lattice to P, then $P = P'$. Since g is A-valued on P it suffices to show $P' \subseteq P$. If not, let $x \,\bar{\in}\, P$ and $g(P,x) \subseteq A$. Then $j(x) \,\bar{\in}\, K$, but $\varphi_h(K, j(x)) = 0$. This contradicts the nonsingularity of $((H + H^{\wedge})/K) \times K \to B/A$ induced by φ_h (1.13b). Hence $P = P'$.

(7.4) <u>Corollary:</u> We have the commutative diagram

$$\begin{array}{ccc} P & \xrightarrow{\quad k \quad} & A^n + \bar{A}^n \\ \Big\downarrow{\scriptstyle g^d} & & \Big\downarrow{\scriptstyle D} \\ P & \xrightarrow{\quad \tilde{i} \quad} & \overline{A^n + \bar{A}^n} \end{array}$$

where k and i are in (7.2), and $D = {}_{(g_h)}d$, $\mathcal{H}(A^n) = (A^n + \bar{A}^n, g_h, q_h)$.

Proof: Left to the reader.

We have constructed P and g in a proposed λ-form (P,g,q). To construct q, let $i_1: P \to A^n$ and $i_2: P \to \bar{A}^n$ be inclusion $P \subset L' = A^n + \bar{A}^n$ followed by the coordinate projections. Let $x \in P$, $r: B \to B/A$, $r': B \to B/S_{-\lambda}(A)$ and set

$$q(x) = r'h_{L'}(i_1x, i_2x) - \varkappa(j(x)) \qquad \text{(compare (6.13))}. \tag{7.5}$$

q takes values in $A/S_{-\lambda}(A)$ since $rh_{L'}(i_1x,i_2x) = [jx,jx]$ (see (6.11) and the first sentence of the proof of (7.3)). We compute $r'g(x,x) = r'g(i_1x + i_2x, i_1x + i_2x) = r'h_{L'}(i_1x,i_2x) + r'\lambda\overline{h_{L'}(i_1x,i_2x)} = (r'h_{L'}(i_1x,i_2x) - \varkappa(jx)) + \lambda\overline{(r'h_L(i_1x,i_2x) - \varkappa(j(x))}$ ((1.16)) $= q(x) + \lambda q(x)$. The other properties (1.12a) are verified similarly so (P,g,q) is a λ-form. Set $I_0^\lambda(\theta;R) = (P,g,q) \in F_0^\lambda(A)$ and let $\mathfrak{s}_1^\lambda(\theta;R)$ denote its class in $W_0^\lambda(A)$.

(7.6) Proposition: For $\theta \in F_1^\lambda(B/A)$, $\mathfrak{s}_1^\lambda(\theta;R) \in W_0^\lambda(A)$ depends only on the equivalence class of θ in $W_1^\lambda(B/A)$.

Proof: (a) Choice of R, $(R) = (A^n \overset{\alpha}{\rightarrowtail} A^n \twoheadrightarrow H)$. If $(R') = (A^m \overset{\beta}{\rightarrowtail} A^m \twoheadrightarrow H)$ is another resolution of H then by (1.7) there are integers k and ℓ with $n + k = m + \ell$ and automorphisms $k_1: A^{n+k} \to A^{n+k}$, $k_2: A^{m+\ell} \to A^{m+\ell}$ with

$$k_1(\alpha + I_k)k_2 = \beta + I_\ell. \tag{7.7}$$

Then if $(R + A^k)$ denotes $(A^{n+k} \xrightarrow{\alpha+I_k} A^{n+k} \twoheadrightarrow H)$, $I_1^\lambda(\theta;R + A^k) = I_1^\lambda(\theta;R) \perp \mathcal{H}(A^k)$ while (7.7) implies $I_1^\lambda(\theta;R + A^k) = I_1^\lambda(\theta;R' + A^\ell)$

in $F_0^\lambda(A)$.

(b) <u>Choice of θ</u> within its equivalence class in $W_1^\lambda(B/A)$. Let $\theta = (K,H,\Delta,\kappa)$.

(i) Let (H,φ,ψ) be a $(-\lambda)$-form and let $\chi_-(\varphi,\psi)\theta = \theta'$ $= (K,H,\Delta',\kappa')$. We claim $I_1^\lambda(\theta';R) = I_1^\lambda(\theta;R)$ in $F_0^\lambda(A)$. Using (1.17) find $\tau\colon A^n \times A^n \to B$ covering (H,φ,ψ) with respect to $(R) =$ $(A^n \overset{\alpha}{\rightarrowtail} A^n \twoheadrightarrow H)$. Then if $\lambda\bar{\rho} = \bar{\alpha}_\tau$ and $(P',g',q') = I_1^\lambda(\theta';R)$ we have the diagram (compare (6.19)) in which the top two horizontal maps are isometries ([B,II.1.2])

(7.8)

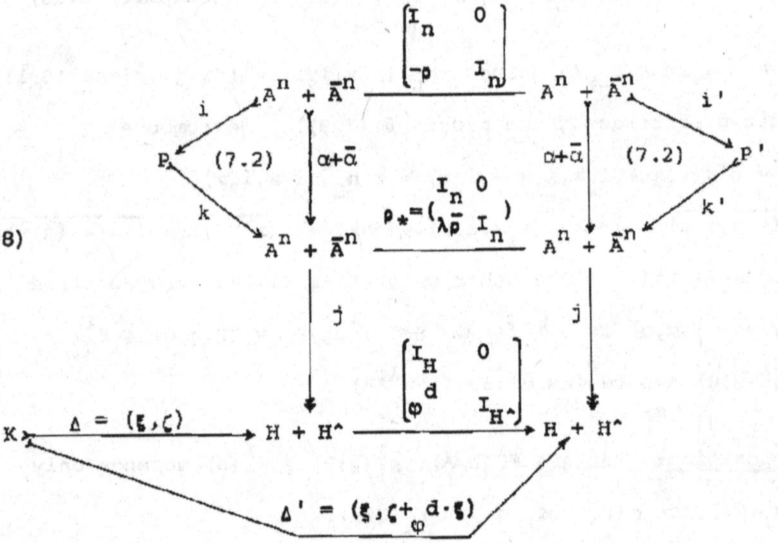

In (7.8) $P = j^{-1}(\Delta K)$, $P' = j^{-1}(\Delta' K)$. By commutativity $\rho_*(P) = (P')$; but ρ_* is an isometry, so $I_1^\lambda(\theta';R) = I^\lambda(\theta;R)$. A similar argument shows that $I_1^\lambda(\chi_+(\varphi,\psi)\theta;R) = I_1^\lambda(\theta;R)$, for any $(-\lambda)$-form $(H^\wedge,\varphi,\psi)$.

(ii) <u>$\mathfrak{s}_1^\lambda(\sigma_E\theta;R) = \mathfrak{s}_1^\lambda(\theta;R)$</u>. Let $\theta' = \sigma_E\theta = (K,H,\Delta,\kappa)$ where $(E) = (J \rightarrowtail H' \twoheadrightarrow H)$. Since $\mathfrak{s}_1^\lambda(\theta';R)$ is independent of the resolution R (part (a)), we assume by the argument of (6.20) that if $(R) = (A^n \overset{\alpha}{\rightarrowtail} A^n \twoheadrightarrow H)$, there is $\beta\colon A^n \rightarrowtail A^n$, $\mathrm{cok}(\beta) \cong J$, such that the extension $\mathrm{cok}(\beta) \rightarrowtail \mathrm{cok}(\alpha\beta) \twoheadrightarrow \mathrm{cok}(\alpha)$ is isomorphic to E. Let $(R') = (A^n \overset{\alpha\beta}{\rightarrowtail} A^n \twoheadrightarrow H')$; then $I_1^\lambda(\theta;R) = I_1^\lambda(\sigma_E\theta; R')$. Indeed, let $L \subseteq P \subseteq L'$ be the lattices of (7.2)' associated to θ and R. Let $M = \beta(A^n) + \bar{A}^n \subseteq A^n + \bar{A}^n = L$, and let M' be the dual lattice. Then $M \subseteq P \subseteq M'$ is the chain of lattices associated to $\sigma_E\theta$ and R', $M \subseteq L \subseteq P \subseteq L' \subseteq M$, and so $I_1^\lambda(\sigma_E\theta;R') = I_1^\lambda(\theta;R)$. Invariance of $\mathscr{s}_1^\lambda$ under other stabilizations is proved similarly. This completes the proof of (7.6).

§8 Exactness of the first five terms of (2.1)

(8.1): The following sequence of homomorphisms is exact:

$$W_1^\lambda(A) \xrightarrow{\mathcal{K}_1^\lambda} W_1^\lambda(B) \xrightarrow{\mathscr{L}_1^\lambda} W_1^\lambda(B/A) \xrightarrow{\mathscr{s}_1^\lambda} W_0^\lambda(A) \xrightarrow{\mathcal{K}_0^\lambda} W_0^\lambda(B).$$

Consequently, $W_1^\lambda(B/A)$ is a group.

Proof: (a) Exactness at $W_1^\lambda(B)$ $\mathscr{L}_1^\lambda\mathcal{K}_1^\lambda = 0$ since if $\sigma \in U_{2n}^\lambda(A)$ we take $\nu = \eta = I_n$ in the construction of $\mathscr{L}_1^\lambda(\sigma)$, thus obtaining the zero formation. If $\sigma \in U_{2n}^\lambda(B)$ and $\mathscr{L}_1^\lambda(\sigma;\nu,\eta) = 0$ for some (hence any) choice of $\nu,\eta \in M_n(A)$, $L_1^\lambda(\sigma;\nu,\eta)$ may be converted to the zero formation by the operations of (1.34). By (6.20) each operation on $L_1^\lambda(\sigma;\nu,\eta)$ is realized by changes in ν, η and σ within its class in $W_1^\lambda(B/A)$. Hence we may find ν', η' and σ' with $[\sigma'] = [\sigma]$ in $W_1^\lambda(B)$ such that $L_1^\lambda(\sigma';\nu',\eta')$ = the zero formation. This easily implies $\nu',\eta' \in GL(A)$ so $\sigma' \in U_{2m}^\lambda(A)$ for some m.

(b) Exactness at $W_1^\lambda(B/A)$ The lattice I in (6.4) becomes P in (7.2). I has the hyperbolic structure by construction so $\mathscr{s}_1^\lambda\mathscr{L}_1^\lambda = 0$. Suppose on the other hand that $I_1^\lambda(\theta;R) = \mathcal{H}(A^n) =$

$(A^n + \bar{A}^n, g_h, q_h)$, where $\theta = (K,H,\Delta,\kappa) \in F_1^\lambda(B/A)$ and $(R) = (A^n \overset{\mu}{\rightarrowtail} A^n \twoheadrightarrow H)$

In the notation of (7.2) there are form-preserving injections $(L,h_L) \overset{i}{\rightarrowtail} (A^n + \bar{A}^n, g_h, q_h) \overset{k}{\rightarrowtail} (L',h_{L'})$ so that under the identification $L \equiv A^n + \bar{A}^n$, i is given by a matrix

$$\begin{bmatrix} \alpha' & \beta \\ \gamma' & \delta \end{bmatrix}, \quad \alpha', \beta, \gamma', \delta \in M_n(A).$$

By (7.4), in which ${}_g d$ now equals D, k has matrix $\begin{pmatrix} \bar{\delta} & \lambda\bar{\beta} \\ \lambda\bar{\gamma}' & \bar{\alpha}' \end{pmatrix}$ and so

(7.2) $\begin{pmatrix} \bar{\delta} & \lambda\bar{\beta} \\ \lambda\bar{\gamma}' & \bar{\alpha}' \end{pmatrix} \begin{pmatrix} \alpha' & \beta \\ \gamma' & \delta \end{pmatrix} = \begin{pmatrix} \mu & 0 \\ 0 & \bar{\mu} \end{pmatrix}$, $(R) = (A^n \overset{\mu}{\rightarrowtail} A^n \twoheadrightarrow H)$. Hence

$\bar{\delta}(\alpha'\mu^{-1}) + \lambda\bar{\beta}(\gamma'\mu^{-1}) = I_n$; and $(\alpha'\mu^{-1})(\gamma'\mu^{-1})$, $\bar{\beta}\delta \in S_{-\lambda}(B^n)$. Setting $\alpha = \alpha'\mu^{-1}$, $\gamma = \gamma'\mu^{-1}$ (1.2) shows that

$$\sigma = \begin{bmatrix} \alpha & \beta \\ \gamma & \delta \end{bmatrix} \in U_{2n}^\lambda(B),$$

By definition $L_1^\lambda(\sigma;\mu,I_n) = \theta$.

(c) <u>Exactness at $W_0^\lambda(A)$</u> In the construction of ∂_1^λ (7.1), the lattice L supports a form h_L which becomes hyperbolic after tensoring up to B. Since for $\theta = (K,H,\Delta,\kappa) \in F_1^\lambda(B/A)$ and R a resolution of H, $(L,h_L) \otimes B \cong L_1^\lambda(\theta;R) \otimes B$ is an isometry, $\mathcal{K}_1^\lambda\partial_1^\lambda(\theta) = 0$. Now let (P,g,q) be a λ-form, P a free A-module such that $(P,g,q) \otimes B$ is isometric to $\mathcal{H}(B^n)$. This implies there is an inclusion of $L \cong A^n + \bar{A}^n$ in P, preserving forms, where L supports a form h_L with matrix $\begin{pmatrix} 0 & \lambda\bar{\alpha} \\ \alpha & 0 \end{pmatrix}$, for some $\alpha \in M_n(A)$. Referring to the construction I_1^λ, we find $\theta = (K,H,\Delta,\kappa) \in F_1^\lambda(B/A)$ such that $I_1^\lambda(\theta;R) = (P,g,q)$ where $(R) = (A^n \overset{\alpha}{\rightarrowtail} A^n \twoheadrightarrow H)$. Details are left to the reader.

Bibliography

[B] H. Bass, "Unitary Algebraic K-theory," Springer Lecture Notes, No. 343 (1973), 57-265.

[C1] P.M. Cohn, Free Rings and Their Relations, Academic Press (1974).

[C2] F. Connolly, "Linking Numbers and Surgery," Topology 12 (1973), 389-412.

[K] M. Karoubi, "Localisation des formes quadratriques I, II, Ann. Sc. Ec. Norm. Sup., Paris, 7 (1975), 359-404; 8 (1975), 99-155.

[L] T.-Y. Lam, The Algebraic Theory of Quadratic Forms, W.A. Benjamin (1973).

[MH] J. Milnor and D. Husemoller, Symmetric Bilinear Forms, Ergebnisse d. Math., Springer Verlag, B.73 (1973).

[P1] W. Pardon, "The exact sequence of a localization in L-theory," Princeton Ph.D. Thesis, 1974.

[P2] ________, "Local surgery and the theory of quadratic forms," Bull. A.M.S., 82 (1976), 131-133.

[P3] ________, "An invariant determining the Witt class of a unitary transformation over a semi-simple ring," to appear in J. of Algebra.

[R] A. Ranicki, "Algebraic L-theory, I," Proc. Lond. Math. Soc. 27 (1973), 101-125.

[R'] ________, "The algebraic theory of surgery," preprint.

[Sh] R. Sharpe, "On the structure of the unitary Steinberg group," Ann. Math. 96 (1972), 444-479.

[S] R. Swan and E.G. Evans, K-theory of Finite Groups and Orders, Springer Lecture Notes, No. 149 (1970).

Columbia University

Orthogonal Representations on Positive Definite Lattices.

An application of Kneser's strong approximation theorem towards the computation of defect groups in orthogonal representation theory.

Andreas W.M. Dress, Bielefeld

AMS(MOS) subject classification (1970), Primary 20 c 10, Secondary 10 c 05, 10 J 05.

In [c] I have studied systematically the problem as to what extent an integral linear or orthogonal representation of a given finite group π is determined by its restriction to the various proper subgroups γ of π. It turned out that - essentially - for any category $\mathcal{C}$, which - like the category F of finite sets, the category $\mathcal{P}(R)$ of finitely generated, projective modules over some commutative ring R or the category $\mathcal{B}(R)$ of inner product spaces over R (see [M] for its definition) - is endowed with two coherently commutative and associative compositions, say "$\perp$" and "o", such that $\perp$, the "sum", is coherently distributive with respect to o , the "product" [1], there exists a unique minimal class $\mathcal{D}(\mathcal{C})$ of finite groups - the defect class of C - which is closed with respect to subgroups and epimorphic images and which, in a way, controls the representation theory for any finite group π over $\mathcal{C}$. $\mathcal{D}(\mathcal{C})$ can be defined as the class of all finite groups γ, such that there exist two γ-objects $(X, u : \gamma \to \mathrm{Aut}_{\mathcal{C}}(x))$ and $(Y, v : \gamma \to \mathrm{Aut}_{\mathcal{C}}(Y))$ in $\mathcal{C}$, which restrict to isomorphic γ'-objects for any proper subgroup γ' of γ, but are not even "stably isomorphic" as γ-objects, i.e. one cannot even find some $k \in \mathbb{N}$ and some γ-object $(Z, w : \gamma \to \mathrm{Aut}_{\mathcal{C}}(Z)$ with $\underbrace{(X,u)\perp \dots \perp(X,u)}_{k \text{ times}} \perp (Z,w) \cong \underbrace{(Y,v)\perp \dots \perp(Y,v)}_{k \text{ times}} \perp (Z,w)$.

With this definition [2] it is almost obvious that for an arbitrary finite group π two π-objects in $\mathcal{C}$ are stably isomorphic if and only if there restrictions to all subgroups $\gamma \leq \pi$ with $\gamma \in \mathcal{D}(\mathcal{C})$ are stably isomorphic, whereas it is less obvious and indeed quite remarkable that $\mathcal{D}(\mathcal{C})$ is closed with respect to subgroups, - a fact, which was used repeatedly in the actual computation of $\mathcal{D}(\mathcal{P}(R))$ and $\mathcal{D}(\mathcal{B}(R))$ in the last sections of [c]. Here the result was $\mathcal{D}(\mathcal{P}(R)) = \mathcal{D}(\mathcal{B}(R)) = \{\gamma \mid \gamma$ cyclic or there exists a prime p with $p\cdot R \neq R$ and a normal p-subgroup $\gamma_0 \trianglelefteq \gamma$ with γ / γ_0 cyclic $\} =: \mathcal{D}(R)$. This was proved by combining the actual computation of

---

[1] Thus $\mathcal{C}$ or more precisely $(\mathcal{C},\perp,o)$ is a "distributive" category in the sense of [c].

[2] Another definition can be given within the frame works of J.A. Green's abstract representation theory (cf. |G|. That's also where the name "defect-class" comes from.

$\mathcal{D}(P(R))$ and $\mathcal{D}(B(R))$ for a local ring R with a weak local global principle for integral representations which implied $\mathcal{D}(P(R)) = \bigcup_{\mathfrak{m}} \mathcal{D}(P(R_{\mathfrak{m}}))$ and $\mathcal{D}(B(R)) = \bigcup_{\mathfrak{m}} \mathcal{D}(B(R_{\mathfrak{m}}))$, where $\mathfrak{m}$ runs through all maximal ideals of R (cf. [c], §9,10; see also [LGP]).

The fact that $P(R)$ and $B(R)$ led to the same, relatively small class of groups was quite surprising in view of the fact that not only $\mathcal{D}(F)$ consists of all finite groups, but also $\mathcal{D}(B^+(\mathbb{Z}))$, where $B^+(\mathbb{Z})$ denotes the category of positive definite inner product spaces over $\mathbb{Z}$. So I tried to compute $\mathcal{D}(B^+(\mathbb{Z}(\frac{1}{p})))$, hoping that perhaps this would lead to a new class somewhere in between $\mathcal{D}(\mathbb{Z}(\frac{1}{p}))$ and the class $\mathcal{D}(F)$ of all finite groups, but the result was, that $\mathcal{D}(B^+(\mathbb{Z}(\frac{1}{p})))$ again equals $\mathcal{D}(\mathbb{Z}(\frac{1}{p}))$ for any prime p.

To clarify this striking difference between the representation theoretic behaviour of $B^+(\mathbb{Z})$ on the one hand and $B(\mathbb{Z})$ or $B^+(\mathbb{Z}(\frac{1}{p}))$ on the other hand I confronted myself with the following problem: let $\mathcal{O} \subsetneq K$ be a ring of algebraic numbers with the algebraic number field K as field of fractions. Let δ be a set of orderings of K and let $B^{\delta}(\mathcal{O})$ denote the category of inner product spaces over $\mathcal{O}$ which are positive definite with respect to any ordering in δ. What can be said about its defect class? The answer is given in the following theorem, whose proof is the purpose of this note.

<u>Theorem:</u> One has always $\mathcal{D}(B^{\delta}(\mathcal{O})) = \mathcal{D}(\mathcal{O})$ unless $\mathcal{O}$ is an order in K (i.e. $\mathcal{O}$ is finite over $\mathbb{Z}$), K is totally real and δ consists of all ((K : $\mathbb{Q}$)) orderings of K, in which case $\mathcal{D}(B^{\delta}(\mathcal{O})) = \mathcal{D}(F)$.

The proof of this theorem consists essentially in a simple application of Kneser's strong approximation theorem (cf. [K 1/2]), which is used to replace the weak local global principle, referred to above, since the latter one does not easily seem to extend to the situation one has to consider in $B^{\delta}(\mathcal{O})$.

The proof is given in several steps. (1) From [c] we'll use the following facts: For any finite group π there exist finite left π-sets S_0 and T_0 with $S_0 \not\cong T_0$, but $S_0|_{\gamma} \cong T_0|_{\gamma}$ for all proper subgroups $\gamma \subsetneq \pi$. For any π-set S let $(\mathcal{O}[S], f_S)$ denote the π-object in $B^{\delta}(\mathcal{O})$, which as an inner product space over $\mathcal{O}$ is just the free $\mathcal{O}$-module $\mathcal{O}[S] = \{\sum_{s\in S} a_s s \mid a_s \in \mathcal{O}\}$, generated by S with $f_S(\sum_s a_s s, \sum_s b_s s) = \sum_{s\in S} a_s b_s$, on which π acts by the $\mathcal{O}$-linear extension of its action on S. Then for any local ring $\mathcal{O}$ with $\pi \notin \mathcal{D}(\mathcal{O})$ there exists a π-set X with $(\mathcal{O}[S_0 \dot\cup X], f_{S_0 \dot\cup X}) \cong (\mathcal{O}[T_0 \dot\cup X], f_{T_0 \dot\cup X})$. Moreover for an arbitrary ring $\mathcal{O}$ one has $\pi \notin \mathcal{D}(B^{\delta}(\mathcal{O}))$ if and only if $(\mathcal{O}[S_0], f_{S_0})$ and $(\mathcal{O}[T_0], f_{T_0})$ are stably isomorphic π-objects in $B^{\delta}(\mathcal{O})$.

(2) This last statement implies easily $\mathcal{D}(B^{\delta}(\mathcal{O})) = \mathcal{D}(F)$ for $\mathcal{O}$ an order, K formally real and δ all (K:$\mathbb{Q}$) orderings of K, since $\underbrace{(\mathcal{O}[S_0], f_{S_0}) \perp \dots \perp (\mathcal{O}[S_0], f_{S_0})}_{k \text{ times}} \perp L$

$\cong \underbrace{(\mathcal{O}[T_0], f_{T_0}) \perp \dots \perp (\mathcal{O}[T_0], f_{T_0})}_{k \text{ times}} \perp L$

for some π-object L in $B^{\sigma}(O)$ would imply, that the π-set of elements of length 1 in the left hand side, which is just the disjoint union of k copies $\pm S_o$ and the corresponding set $L^{(1)}$ in L, must be isomorphic as a π-set to the corresponding set on the right hand side, i.e. one would have $\underbrace{S_o \dot\cup \ldots \dot\cup S_o}_{2\,k\text{ times}} \dot\cup L^{(1)} \simeq \underbrace{T_o \dot\cup \ldots \dot\cup T_o}_{2\,k\text{ times}} \dot\cup L^{(1)}$,

which readily implies $S_o \simeq T_o$ (compare for instance the orbit decomposition on both sides).

(3) Now assume, that either O is not an order, K is not formally really or σ does not contain all orderings of K. We use the following simple standard consequence of Kneser's strong approximation theorem: Under the above hypothesis there exists an inner product space (M, f) in $B^{\sigma}(O)$, such that for any two locally isomorphic π-objects (M_1, f_1) and (M_2, f_2) in $B^{\sigma}(O)$ one has

$$(M_1,f_1) \perp (M_1,f_1) \perp (M,f) \underset{O}{\otimes} (M_1,f_1) \simeq (M_2,f_2) \perp (M_2,f_2) \perp (M,f) \underset{O}{\otimes} (M_2,f_2)$$

and $(M,f) \underset{O}{\otimes} (M_1,f_1) \simeq (M,f) \underset{O}{\otimes} (M_2,f_2)$ (here (M,f) is considered as a π-object with trivial π-action), in particular any two locally isomorphic π-objects in $B^{\sigma}(O)$ are stably isomorphic.

Now assume $\pi \notin \mathcal{D}(O_{\mathfrak{p}})$ for any prime ideal $\mathfrak{p}$ in O, so we have a π-set $X_{\mathfrak{p}}$ with $(O_{\mathfrak{p}}[S_o \dot\cup X], f_{S_o \dot\cup X}) \simeq (O_{\mathfrak{p}}[T_o \dot\cup X], f_{T_o \dot\cup X})$.

Since such an isomorphism must then exist already over $O[\frac{1}{a}]$ for some $a \in O - \mathfrak{p}$, we may assume, that we have to deal only with finitely many different such π-sets $X_{\mathfrak{p}}$, which then can be replaced by their disjoint union X. Thus replacing S_o and T_o by $S_o \dot\cup X$ and $T_o \dot\cup X$ we may even assume that $(O[S_o], f_{S_o})$ and $(O[T_o], f_{T_o})$ are locally isomorphic, which then implies that they are stably isomorphic π-objects in $B^{\sigma}(O)$ by the above mentioned corollary of the strong approximation theorem.

In view of the results in [c] the following corollaries are obvious:

<u>Corollary 1:</u> Let $K(\pi, B^{\sigma}(O))$ denote the Grothendieck ring of π-objects in $B^{\sigma}(O)$. Then the induction maps $K(\gamma, B^{\sigma}(O)) \to K(\pi, B^{\sigma}(O))$ which are well defined in a canonical way for any $\gamma \leq \pi$, define surjective maps

$$\bigoplus_{\gamma \leq \pi, \gamma \in \mathcal{D}(O)} \mathbb{Q} \otimes K(\gamma, B^{\sigma}(O)) \to \mathbb{Q} \otimes K(\pi, B^{\sigma}(O))$$

and

$$\bigoplus_{\gamma \leq \pi, \gamma \in \mathfrak{h}\mathcal{D}(O)} K(\gamma, B^{\sigma}(O)) \to K(\pi, B^{\sigma}(O))$$

with $\mathfrak{h}\mathcal{D}(O) = \{\gamma \mid$ ex. $\gamma_o \trianglelefteq \gamma$

with $\gamma_o \in \mathcal{D}(O)$ and γ / γ_o a p-group$\}$, unless O is an order, K totally real and σ consists of all orderings of K.

Corollary 2: Two permutation representations $(\mathcal{O}[S], f_S)$ and $(\mathcal{O}[T], f_T)$ of π are stably isomorphic π-objects in $B^{\delta}(\mathcal{O})$ if and only if the number of γ-invariant elements in S equals the number of γ-invariant elements in T for all $\gamma \leq \pi$ with $\gamma \in \mathcal{D}(\mathcal{O})$ unless ....

Remark 1: It seems interesting to check precisely for which π-objects in $B(\mathcal{O})$ one can apply the strong approximation theorem to compute the number of elements in their genus or spinor genus.

Remark 2: In case $\mathcal{O}$ is not an order or K is not totally real one can give an elementary proof of our Theorem by extending the above mentioned weak local global principle to that case, i.e. by proving directly, that any element in the kernel of $K(\pi, B^{\delta}(\mathcal{O})) \to \prod_{\mathfrak{g}} K(\pi, B^{\delta}(\mathcal{O}_{\mathfrak{g}}))$ is nilpotent.

This follows, of course, also in the more general case from Kneser's result, which implies, that this kernel is 2-primary, together with the fact, that $K(\pi, B^{\delta}(\mathcal{O}))$ is a λ-ring, and that, by a result of Segal (cf. [c], any torsion element in a λ-ring is nilpotent.

It seems remarkable, that - vice versa - the nilpotency of this kernel implies $\mathcal{D}(B^{\delta}(\mathcal{O})) \neq \mathcal{D}(F)$ and thus all of the following statements are indeed equivalent:

(i) $\mathcal{O}$ is not an order or $|\delta| \neq (K:\mathbb{Q})$

(ii) $Ke\,(K(\pi, B^{\delta}(\mathcal{O})) \to \prod_{\mathfrak{g}} K(\pi, B^{\delta}(\mathcal{O}_{\mathfrak{g}}))$ consists of torsion elements.

(iii) $Ke\,(K(\pi, B^{\delta}(\mathcal{O})) \to \prod_{\mathfrak{g}} K(\pi, B^{\delta}(\mathcal{O}_{\mathfrak{g}})))$ is a nil ideal.

(iv) $\mathcal{D}(B^{\delta}(\mathcal{O})) = \mathcal{D}(\mathcal{O})$

(v) $\mathcal{D}(B^{\delta}(\mathcal{O})) \neq \mathcal{D}(F)$

References

c A.W.M. Dress: "Contributions to the theory of induced representations", Algebraic K-Theory II, Battelle Institute Conference 1972, Springer Lecture Notes 342(1973), 183-240

LGP A.W.M. Dress: "The weak local global principle in algebraic K-theory", Communications in algebra, 3(7), 615-661 (1975)

G J.A. Green: "Axiomatic Representation Theory for Finite Groups", J. Pure and Appl. Algebra 1 (1971), 41-77

M J. Milnor, D. Husemoller: "Symmetric bilinear forms", Ergebnisse der Math. u. i. Grenzgebiete, 73(1973), Springer-Verlag, New York - Heidelberg - Berlin

K1 M. Kneser: "Starke Approximation in Algebraischen Gruppen I", J. Reine Angew. Math. (Crelle), 218(1965), 190-203

K2 M. Kneser: "Strong Approximation" Proc. of Symp. in Pure Math. IX, AMS, Providence 1966, 187-196

The computation of surgery groups of finite groups with abelian 2-hyperelementary subgroups

by

Anthony Bak

The purpose of this note is to outline substantially the computation of the surgery obstruction groups $L_n^{s,h,p}(\pi)$ where π is a finite group such that all its 2-hyperelementary subgroup are abelian. This is equivalent to the condition that the 2-sylow subgroup of π is normal abelian. The computation of the $L_{2n+1}^{s}(\pi)$ includes an explicit basis of 4×4 matrices. Of interest to people who study $\tilde{K}_o(Z\pi)$ is a computation with an explicit basis of $H^i(\tilde{K}_o(Z\pi))$ when π is abelian of exponent 4. Of interest to those who study $SK_1(Z\pi)$ is a result which gives a bound to the number of generators of the 2-sylow subgroup of $SK_1(Z\pi)$. Computations similar to those of the surgery groups are given for the K-theory groups $KQ_i^{\pm 1}(Z\pi,\Lambda)$ $(i = 0,1)$ of nonsingular Λ-quadratic modules. We remind the reader that the groups $L_n^{s,h}(\pi)$ are the simple and homotopy surgery groups of closed orientable manifolds of C.T.C. Wall [W 1], and that the groups $L_n^{p}(\pi)$ are the surgery groups of open orientable manifolds of S. Maumary [M] and L. Taylor [T]. A thorough review of all the definitions is found in [B 9].

The paper [B 6] relates the computations here of $L_{2n}^{h,p}(\pi)$ to the problem of determining the isomorphism classes of integral representations of π which preserve a nonsingular form.

In the past half decade several authors have contributed partial reductions and answers to the current and closely related computations. To the best of my knowledge, a complete list of papers is given by [B 2-6],[BSH],[Bs 2],[BN],[C],[L],[PP],[SU],[W 3-4]. The method of proof in [B 2-3] is carried forward to the current computations. The only difference is that the current results require more arithmetic input. The details for the current computations appear in a preprint [B 5] with the same title as this paper.

The computations require the following terms. Let $Z\pi$ denote the integral group ring of π with involution which inverts each element of π. If G is on abelian group with a $Z/2Z$ - action then

$$H^i(G) = \text{i'th cohomology group of the action.}$$

A matrix group such as $K_1(Z\pi)$, or the Whitehead group $Wh(\pi) = K_1(Z\pi)/[\pm \pi]$, or $SK_1(Z\pi)$ = kernel of the reduced norm $K_1(Z\pi) \to$ units (center $Q\pi$) has the $Z/2Z$ - action which sends each matrix to its conjugate transpose. The Grothendieck group $\tilde{K}_o(Z\pi) = K_o(Z\pi)/[Z\pi]$ has the $Z/2Z$ - action which sends the class of a projective module $[P] \mapsto -[P^*]$ where $P^* = Hom_{Z\pi}(P, Z\pi)$. The relative group $K_o(Z\pi, Q\pi)$ has the $Z/2Z$ - action which sends the class of a pair of projective modules $[P,Q] \mapsto -[P^*,Q^*]$. It is worth noting that when π is abelian one can identify [B 7] $K_o(Z\pi, Q\pi)$ (resp. $\hat{K}_o(Z\pi)$)with the ideal group $I(Z\pi)$ (resp. ideal class group $Cl(Z\pi)$) of $Z\pi$. Furthermore, the identification is such that the $Z/2Z$ - action corresponds to the one induced by the involution on $Z\pi$. Let

π = finite group whose 2-sylow subgroup is normal abelian

π_p = p-sylow subgroup of π

${}_n\pi$ = elements of exponent n in π

r_∞ = number of simple factors of the real group ring $R\pi$ (every simple factor is involution invariant)

r_o = number of conjugacy classes of elements of exponent 2 in π

= number of simple factors above whose centers are isomorphic to R

= number of simple factors of $Q\pi$ whose centers are isomorphic to Q

r_2 = rank $H^1(Wh(\pi))$

= rank $H^i(SK_1(Z\pi))$ $(i > 0)$

= rank $(Z/2Z) \otimes SK_1(Z\pi)$ (providing π_2 is normal abelian)

= 0 (providing π_2 is cyclic or of order ≤ 4)

s = number of factors in a cyclic decomposition of π_2 (providing π_2 is abelian)

E = number of factors above of order 2

$\binom{E}{2}$ = binonomial coefficient

= number of subsets of order 2 in $\{1,\ldots,E\}$

$2^s = r_o$

$H(\pi)$= coker $H^2(K_o(Z\pi,Q\pi)) \to H^2(\tilde{K}_o(Z\pi))$, $[P,Q] \mapsto [P] - [Q]$

= 0 (providing π is abelian of exponent 4)

Let A be a ring with involution $a \mapsto \bar{a}$. Let $\lambda = \pm 1$. If σ is a matrix with coefficients in A let $\bar{\sigma}$ = conjugate transpose of σ.

Let

$\underline{\underline{P}}(A)$ = category with product of finitely generated projective A - modules

$\underline{\underline{Q}}^\lambda(A)$= category with product of nonsingular quadratic modules on finitely generated projective A-modules such that the associated even hermitian form is λ-symmetric (the form parameter Λ is assumed to be minimal)

$K_i(A) = K_i(\underline{\underline{P}}(A))$

$KQ_i{}^\lambda(A) = K_i(\underline{\underline{Q}}^\lambda(A))$

$\mathbb{H} : K_o(A) \to KQ_o^\lambda(A)$, $[P] \to [\mathbb{H}(P)] = [P \oplus P^*, \Psi_p]$, $\Psi_p(p,f) = f(p)$, the hyperbolic map

$\mathbb{H} : K_1(A) \to KQ_1{}^\lambda(A)$, $[a] \to [\begin{smallmatrix} a & o \\ o & \bar{a}^{-1} \end{smallmatrix}]$, the hyperbolic map

If $G'' \to G \to G'$ is any sequence of abelian groups let $H_o(G'' \to G \to G')$ denote the homology of the sequence.

If X is a $Z/2Z$ - invariant subgroup of $K_1(A)$ let

$$L_{2n+1}{}^{X}(A) = H_o\,\{X \xrightarrow{\mathbb{H}} KQ_1{}^{(-1)^n}(A) \to K_1(A)/X\}.$$

The notation $L_{2n+1}{}^{X}(A)$ is needed to introduce the group $L_{2n+1}{}^{X}(Z_2\pi_2)$ which is used in passing to describe $L_{2n+1}{}^{s,h}(\pi)$. For the reader's convenience, we recall that

$$L_{2n+1}{}^{h}(\pi) = L_{2n+1}{}^{K_1(Z\pi)}(Z\pi)/[{}^{o}_{1}{}^{(-1)^n}_{o}]$$

$$L_{2n+1}{}^{s}(\pi) = L_{2n+1}{}^{[\pm\pi]}(Z\pi)/[{}^{o}_{1}{}^{(-1)^n}_{o}]$$

$$L_{2n}{}^{P}(\pi) \quad = \text{coker}\ \mathbb{H} : K_o(Z\pi) \to KQ_o^{(-1)^n}(Z\pi)$$

$$L_{2n}{}^{h}(\pi) \quad = K_o\ \underline{\underline{Q}}^{(-1)^n}(Z\pi)_{free}/[\mathbb{H}(Z\pi)]$$

$\underline{\underline{Q}}(Z\pi)_{free}$ = full subcategory of $\underline{\underline{Q}}(Z\pi)$ of all quadratic modules whose underlying $Z\pi$-modules are free.

$$L_{2n}{}^{s}(\pi) \quad = K_o\ \underline{\underline{Q}}^{(-1)^n}(Z\pi)_{based-[\pm\pi]}/\mathbb{H}(Z\pi)$$

$\underline{\underline{Q}}(Z\pi)_{based-[\pm\pi]}$ = category with product of nonsingular quadratic modules on free modules with a prescribed basis. The only morphisms allowed are isomorphisms. Using the bases one can associate with each morphism a matrix, and one can associate with the hermitian form attached to a quadratic module a matrix. It is assumed both these matrices vanish in $K_1(Z\pi)/[\pm\pi]$.

Let

$\alpha : Z/2Z \to L_{2n}{}^{s,h,P}(\pi)$ (n odd), splitting to Kervaire-Arf invariant

$$1 \mapsto [\mathbb{Z}\pi \oplus \mathbb{Z}\pi, \begin{pmatrix} 1 & 0 \\ 1 & 1 \end{pmatrix}] - [\mathbb{H}(\mathbb{Z}\pi)]$$

sign = multisignature map on $L_{2n}^{s,h,P}(\pi)$.

Here are the results.

Theorem 1 **There is an exact sequence**

$$0 \to \begin{Bmatrix} 0 \\ \mathbb{Z}/2\mathbb{Z} \end{Bmatrix} \underset{\alpha}{\to} L_{2n}^{P}(\pi) \xrightarrow{\text{sign}} \begin{Bmatrix} \mathbb{Z}^{r_\infty} \\ \mathbb{Z}^{r_\infty - r_0} \end{Bmatrix} \to 0 \quad \begin{matrix} \text{if } n \equiv 0\ (2) \\ \text{if } n \equiv 1\ (2). \end{matrix}$$

Theorem 2 **There is an exact sequence**

$$0 \to \begin{Bmatrix} H(\pi) \oplus (\mathbb{Z}/2\mathbb{Z})^{2^s - 1 - \binom{E}{2} - s} \\ H(\pi) \oplus \mathbb{Z}/2\mathbb{Z} \end{Bmatrix} \underset{(\mathbb{H},\alpha)}{\xrightarrow{\mathbb{H}}} L_{2n}^{h}(\pi) \xrightarrow{\text{sign}} \begin{Bmatrix} \mathbb{Z}^{r_\infty} \\ \mathbb{Z}^{r_\infty - r_0} \end{Bmatrix} \to 0 \quad \begin{matrix} \text{if } n \equiv 0(2) \\ \text{if } n \equiv 1(2). \end{matrix}$$

Remark $\mathbb{H}$ is defined on $H(\pi) \oplus (\mathbb{Z}/2\mathbb{Z})^{2^s-1-\binom{E}{2}-s} =$ coker $(H^2(K_o(\mathbb{Z}\pi, \mathbb{Q}\pi)) \to H^2(\tilde{K}_o(\mathbb{Z}\pi))) \oplus (\mathbb{Z}/_{2\mathbb{Z}})^{2^s-1-\binom{E}{2}-s}$ in the following way. First of all the map $\mathbb{H} : H^2(\tilde{K}_o(\mathbb{Z}\pi)) \to L_{2n}^{h}(\pi)$ makes sense. The sequence $H^2(K_o(\mathbb{Z}\pi, \mathbb{Q}\pi)) \xrightarrow{f} H^2(\tilde{K}_o(\mathbb{Z}\pi)) \xrightarrow{\mathbb{H}} L_{2n}^{h}(\pi)$ is exact if $n \equiv 1(2)$, but if $n \equiv 0(2)$ then one only has that ker $\mathbb{H} \subset$ image f. Thus the image $\mathbb{H} \cong H(\pi) \oplus$ image $\mathbb{H}$ f. One then computes that image $\mathbb{H}$ f $\cong$ $(\mathbb{Z}/2\mathbb{Z})^{2^s-1-\binom{E}{2}-s}$.

Theorem 3 **There is an exact sequence**

$$0 \to \begin{Bmatrix} (\mathbb{Z}/2\mathbb{Z})^{r_2} \\ (\mathbb{Z}/2\mathbb{Z})^{r_2} \oplus \mathbb{Z}/2\mathbb{Z} \end{Bmatrix} \underset{(\tau,\alpha)}{\xrightarrow{\tau}} L_{2n}^{s}(\pi) \underset{\text{sign}}{\xrightarrow{(\text{sign}, f)}} \begin{Bmatrix} \mathbb{Z}^{r_\infty} \oplus (\mathbb{Z}/2\mathbb{Z})^{2^s-1-\binom{E}{2}-s} \\ \mathbb{Z}^{r_\infty - r_0} \end{Bmatrix} \to 0$$

if $n \equiv 0$ (2)

if $n \equiv 1$ (2)

τ and f are decribed in the outline below.

Outline of theorems 1-3. first one does the abelian case. Then one uses Dress induction [D 3] plus the computations in the abelian case to reduce the general case to the abelian case. The reduction is easy to make. The abelian case is handled as follows. In [BSH] the authors use the notation $W_o^{(-1)^n}(Z\pi) =$

$$\begin{cases} L_{2n}^{p}(\pi) & \text{if } n \equiv 0(2) \\ L_{2n}^{p}(\pi)/\text{image } \alpha & \text{if } n \equiv 1(2) \end{cases}$$

(The last section

§ 10 of [BSH] explains how to make the identifications above.) Theorem 1 is an immediate consequence of the computation in [BSH, 9.1] that $W_o^{(-1)^n}(Z\pi) \xrightarrow[\cong]{\text{sign}} \begin{cases} Z^{r_\infty} & n \equiv 0(2) \\ Z^{r_\infty - r_o} & n \equiv 1(2) \end{cases}$

is an isomorphism. Theorem 2 is deduced by comparing $L_{2n}^{h}(\pi)$ with $L_{2n}^{P}(\pi)$. Consider the exact sequence [R, 2.3] $H^2(\tilde{K}_o(Z\pi)) \overset{H}{\to} L_{2n}^{h}(\pi) \to L_{2n}^{P}(\pi) \xrightarrow{\text{forgetful}} H^1(\tilde{K}_o(Z\pi))$. Since both ends of the sequence are torsion it follows that the rank $(L_{2n}^{h}(\pi))$ = rank $(L_{2n}^{P}(\pi))$. It is very easy now to deduce an exact sequence

$$H^2(\tilde{K}_o(Z\pi)) \overset{H}{\to} L_{2n}^{n}(\pi) \to \begin{Bmatrix} Z^{r_\infty} & \text{if } n \equiv 0 \ (2) \\ Z^{r_\infty - r_o} \oplus Z/2Z & \text{if } n \equiv 1 \ (2) \end{Bmatrix} \to 0$$

where $Z/2Z$ is

generated by $\alpha(1)$. The sequence clearly splits. Then

one computes that the image $\mathbb{H} = \begin{cases} H(\pi)\oplus(Z/2Z)^{2^s-1-\binom{E}{2}-s} & \text{if } n \equiv 0 \ (2) \\ H(\pi) & \text{if } n \equiv 1 \ (2). \end{cases}$

(See the remark following theorem 2. The details for the case π odd order are found in [B 3].) Theorem 3 is deduced by camparing the Rothenberg exact sequence [S1,4.1] $L_{2n+1}{}^{s}(\pi) \xrightarrow{c} L_{2n+1}{}^{h}(\pi) \xrightarrow{g}$ $H^1(Wh(\pi)) \xrightarrow{T} L_{2n}{}^{s}(\pi) \to L_{2n}{}^{h}(\pi) \xrightarrow{\delta = discr} H^2(Wh(\pi))$. First one proves $g = 0$. Note that if π has odd order then this is clear because $L_{2n+1}{}^{h}(\pi) = 0$ by [B 2]. In the general case one appeals to theorems 6 and 7 which tell us the map c is surjective. Thus $g = 0$. Next one notes that by [B 8, Cor. 3] $H^1(Wh(\pi)) \cong (Z/2Z)^{r_2}$. Next one notes that since $H^1(Wh(\pi))$ is torsion it follows that the rank $(L_{2n}{}^{s}(\pi))$ = rank $(L_{2n}{}^{h}(\pi))$. Finally, one notes that if T = torsion sugroup of $L_{2n}{}^{h}(\pi)$ then the torsion subgroup of the image $(L_{2n}{}^{s}(\pi) \to L_{2n}{}^{h}(\pi))$ is the group ker (δ restricted to T) = ker $(\delta|_T)$. From all of the above observations, one deduces very easily an exact sequence

$$0 \to (Z/2Z)^{r_2} \xrightarrow{T} L_{2n}{}^{s}(\pi) \to \ker(\delta|_T) \oplus \begin{Bmatrix} Z^{r_\infty} \\ Z^{r_\infty - r_0} \end{Bmatrix} \to 0.$$

The crucial step in the argument is to

compute ker $(\delta|_T)$. This is done

as follows. Recall that $T = \begin{cases} H(\pi) \oplus (Z/2Z)^{2^s - 1 - \binom{E}{2} - s} & \text{if } n \equiv 0(2) \\ H(\pi) \oplus (Z/2Z & \text{if } n \equiv 1(2) \end{cases}$,

and $H(\pi) = \operatorname{coker} H^2(K_o(Z\pi, Q\pi)) \to H^2(\tilde{K}_o(Z\pi))$. It is clear that if $n \equiv 1(2)$ then the $Z/2Z$-summand (corresponding to $\alpha(1)$) lies in the ker δ. The rest of

T is given by the image $\mathbb{H} = \begin{cases} H(\pi) \oplus (Z/2Z)^{2^s - 1 - \binom{E}{2} - s} \\ H(\pi) \end{cases}$.

Now comes the key ingredient. The sequence $H^2(K_o(Z\pi, Q\pi)) \to$ $H^2(\tilde{K}_o(Z\pi)) \xrightarrow{\delta\, \mathbb{H}} H^2(Wh(\pi))$ is exact (theorem 4 below).

It follows very easily that $\ker(\delta|_T) = \begin{cases} (Z/2Z)^{2^s - 1 - \binom{E}{2} - s} \\ Z/2Z \end{cases}$

if $n \equiv 0\ (2)$

if $n \equiv 1\ (2)$

Thus $L_{2n}{}^{s}(\pi)$ is computed.

<u>Theorem 4</u> <u>Let</u> $\delta : L_{2n}{}^{h}(\pi) \to H^2(Wh(\pi))$ <u>denote the discriminant map above.</u> Let "$Wh(\pi)$" $=$ Nrd $K_1(Z\pi)/[\pm\pi]$, <u>and let</u> "δ" <u>denote the composite</u> $L_{2n}{}^{h}(\pi) \to H^2(Wh(\pi)) \to H^2(\text{"}Wh(\pi)\text{"})$. <u>The following sequences are exact</u>

$$H^2(K_o(Z\pi, Q\pi)) \to H^2(\widetilde{K}_o(Z\pi)) \xrightarrow{\delta\ H} H^2(Wh(\pi))$$

$$H^2(K_o(Z\pi, Q\pi)) \to H^2(\widetilde{K}_o(Z\pi)) \xrightarrow{\text{"}\delta\text{"}H} H^2(\text{"}Wh(\pi)\text{"}).$$

<u>Remark</u> There are also analogous exact sequences in which $Wh(\pi)$ is replaced by $K_1(Z\pi)$ and "$Wh(\pi)$" by "$K_1(Z\pi)$" $=$ Nrd $(K_1(Z\pi))$. In particular if π is abelian then one gets an exact sequence $H^2(I(Z\pi)) \to H^2(Cl(Z\pi)) \xrightarrow{\text{"}\delta\text{"}H} H^2(\text{units } (Z\pi))$.

<u>Proof of theorem 4</u> The main idea is to interpret δ H as a product of 2 boundry maps in certain cohomology exact sequences and then exhibit the sequence above as an exact sequence in cohomology. We fill in now the steps. First one reduces by induction [Sw § 2] to the case π is abelian. This is easy. Then consider the K-theory localization exact sequence [BSH § 3] $K_1(Z\pi) \to K_1(Q\pi) \to K_o(Z\pi, Q\pi) \to K_o(Z\pi) \to K_o(Q\pi)$. One identifies $\widetilde{K}_o(Z\pi) = \ker K_o(Z\pi) \to K_o(Q\pi)$, and deduces an exact sequence

$$0 \to K_1(Q\pi)/\text{"}K_1(Z\pi)\text{"} \to K_o(Z\pi, Q\pi) \to \widetilde{K}_o(Z\pi) \to 0$$

where "$K_1(Z\pi)$" $=$ image$(K_1(Z\pi) \to K_1(Q\pi))$($=$ Nrd $K_1(Z\pi)$).
Consider also the exact sequence

$$0 \to "K_1(Z\pi)" \to K_1(Q\pi) \to K_1(Q\pi)/"K_1(Z\pi)" \to 0.$$

Associated to these exact sequences are the following cohomology exact sequences

$$H^2(K_o(Z\pi,Q\pi)) \to H^2(\tilde{K}_o(Z\pi)) \xrightarrow{\partial_1} H^1(K_1(Q\pi)/"K_1(Z\pi)")$$

$$H^1(K_1(Q\pi)) \to H^1(K_1(Q\pi)/"K_1(Z\pi)") \xrightarrow{\partial_2} H^2("K_1(Z\pi)").$$

One identifies as in [BSH § 3] $K_o(Z\pi,Q\pi) = \coprod_p \operatorname{coker} K_1(Z_p\pi) \to K_1(Q_p\pi)$. Then using the above sequences, one computes that the sequence

$$H^2(K_o(Z\pi,Q\pi)) \to H^2(\tilde{K}_o(Z\pi)) \xrightarrow{\partial_2\partial_1} H^2("K_1(Z\pi)")$$

is exact, and that the sequence

$$H^2(K_o(Z\pi,Q\pi)) \to H^2(\tilde{K}_o(Z\pi)) \xrightarrow{p\partial_2\partial_1} H^2("Wh(\pi)")$$

is exact where p: $"K_1(Z\pi)" \to "Wh(\pi)"$ is the cononical surjection. However, by the key lemma below $"\delta"\mathbb{H} = p\partial_2\partial_1$. Thus the exactness of the second sequence in the theorem is established. The exactness of the first sequence is derived with some work from the exactness of the second.

<u>Key lemma</u> $"\delta"\mathbb{H} = p\partial_1\partial_2$.

<u>Proof</u> Let $[M] \in H^2(\tilde{K}_o(Z\pi))$. First we compute $p\partial_1\partial_2[M]$. Lift $[M]$ to a representative $[M,Z\pi^n] \in K_o(Z\pi,Q\pi)$. After stabilizing M we can assume that $M \oplus M^* = Z\pi^{2n}$. On the free hyperbolic module $\mathbb{H}(Q\pi^n)$ the two $Z\pi$-lattices $\mathbb{H}(M)$ and $\mathbb{H}(Z\pi^n)$ are free. Let $\tau(M,Z\pi^n)$ be a change of basis matrix between them. If $[\tau(M,Z\pi^n)]$ = the class of $\tau(M,Z\pi^n)$ <u>in $H^1(K_1(Q\pi)/"K_1"(Z\pi))$</u> then $\partial_1[M] = [\tau(M,Z\pi^n)]$. $\partial_2\partial_1[M] = [\tau(M,Z\pi^n)]$ $[\tau(M,Z\pi^n)]$ and $p\partial_2\partial_1[M]$ = the class of $[\tau(m,Z\pi^n)]$ $[\tau(M,Z\pi^n)]$ in $H^2("K_1"(Z\pi)/[\pm\pi]) = H^2("Wh(\pi)")$. Next we compute $"\delta"\mathbb{H}[M]$. Let

$e_1,\dots,e_n$ be a basis for $M \oplus M^* = \mathbb{H}(M)$, and let h be the hyperbolic hermitian form associated to $\mathbb{H}(M)$. If $\tau(M)$ = the matrix $(h(e_i,e_j))$ and if $[\tau(M)]$ denotes the class of $\tau(M)$ in $H^2("K_1"(Z\pi)/[\pm\pi])$ then "δ"$\mathbb{H}[M] = [\tau(M)]$. We must show that $[\tau(M)]$ = the class of $[\tau(M,Z\pi^n)]$ $[\tau(M,Z\pi^n)]$. But if one picks for the basis of $\mathbb{H}(Z\pi^n)$ the standard hyperbolic basis then $\tau(M) = \tau(M,Z\pi^n)\ (\begin{smallmatrix} o & I \\ I_n & o\end{smallmatrix} \pm n)\tau(M,Z\pi^n)$.

Next we compute the odd dimension surgery groups.

<u>Theorem 5</u> [B 2] <u>If</u> π <u>is finite odd order then</u>

$$L_{2n+1}{}^{s,h}(\pi) = 0.$$

More generally, one has the following.

<u>Theorem 6</u> <u>If</u> $n \equiv 1(2)$ <u>then</u>

$$\begin{aligned} L_{2n+1}{}^{s}(\pi) &\xrightarrow{\cong} L_{2n+1}{}^{[\pm\pi]}(Z_2\pi_2)/[\begin{smallmatrix}0&-1\\1&0\end{smallmatrix}] \\ &\cong \mathbb{H} \\ &H^2(\text{units } Z_2\pi_2)/[\pm_2\pi], [5] \\ &\cong \\ &(Z/2Z)^{2^s-1} \end{aligned}$$

$$\begin{array}{ccc} L_{2n+1}{}^{h}(\pi) & \xleftarrow{\cong} & \text{coker }(\mathbb{H}: H^2(\text{units } Z\pi) \to L_{2n+1}{}^{s}(\pi)) \\ \cong\downarrow & & \cong\downarrow \\ L_{2n+1}{}^{K_1(Z\pi_2)}(Z_2\pi_2)/[\begin{smallmatrix}0&-1\\1&0\end{smallmatrix}] & \xleftarrow{\cong} & \text{coker }(\mathbb{H}: H^2(\text{units } Z\pi_2) \to L_{2n+1}{}^{[\pm\pi_2]}(Z_2\pi_2)/[\begin{smallmatrix}0&-1\\1&0\end{smallmatrix}]) \\ & & \cong\uparrow \mathbb{H} \\ & & \text{coker }(H^2(\text{units } Z\pi_2) \to H^2(\text{units } Z_2\pi_2)/[\pm_2\pi], [5]). \end{array}$$

A basis for $L_{2n+1}^{s}(\pi)$ is constructed as follows. By theorem 5 one can assume that $\pi_2 \neq 1$. Let $\sigma_1,\ldots,\sigma_s$ be a basis for π_2. Let $\delta_i = (1-\sigma_i) + (1-\bar{\sigma}_i)$. If σ_i has order 2^{e_i}, let $\Delta_i = 1-\sigma_i^{2^{e_i-1}}$ ($\Delta_i^2 = 2\,\Delta_i$). Let T be a nonempty subset of $\{1,\ldots,s\}$ and let $t = |T|$ = order of T. Let $\Delta_T = \prod_{i\in T} \Delta_i$. Then the family of matrices below is a basis for $L_{2n+1}^{s}(\pi)$.

$$\left\{ \begin{pmatrix} (1-\delta_i) & \delta_i \\ -\delta_i & (1+\delta_i) \end{pmatrix} \;\middle|\; j = 1,\ldots,s \right\} \cup$$

$$\left\{ \begin{pmatrix} 1-\Delta_T & 2\Delta_T \\ -2^{t-1}\Delta_T & 1+\Delta_T + \Delta_T^{2} \end{pmatrix} \;\middle|\; T \subset \{1,\ldots,s\},\ t \geq 2 \right\} .$$

Theorem 7 If $n \equiv 0\ (2)$ then the commutative diagram below has exact top and bottom rows.

$$\begin{array}{ccccc}
0 \to L_{2n+1}^{s}(\pi) \to & L_{2n+1}^{[\pm\pi_2]}(Z_2\pi_2)/\left[\begin{smallmatrix}01\\10\end{smallmatrix}\right] & \to & \prod^{r_o}\left(L_{2n+1}^{[\pm 1]}(Z_2)/\left[\begin{smallmatrix}01\\10\end{smallmatrix}\right]\right) \\
 & \cong\uparrow \mathbb{H} & & \cong\uparrow \mathbb{H} \\
 & H^2(\text{units } Z_2\pi_2)/[\pm_2\pi] & \to & \prod^{r_o}\left(H^2(\text{units } Z_2)/[\pm 1]\right) \\
 & \cong\uparrow & & \cong\uparrow \\
0 \to (Z/2Z)^{2^s-s-1-\binom{E}{2}} \to & (Z/2Z)^{2^s} & \to & (Z/2Z)^{r_o = 2^s}
\end{array} .$$

Furthermore,

$$\begin{array}{ccc}
L_{2n+1}^{h}(\pi) & \xleftarrow{\ \cong\ } & \operatorname{coker}\left(\mathbb{H} : H^2(\text{units } Z\pi) \to L_{2n+1}^{s}(\pi)\right) \\
\cong\downarrow & \swarrow \cong & \\
\multicolumn{3}{l}{\ker\left\{L_{2n+1}^{K_1(Z\pi_2)}(Z_2\pi_2)/\left[\begin{smallmatrix}01\\10\end{smallmatrix}\right] \to \prod^{r_o}\left(L_{2n+1}^{K_1(Z_2)}(Z_2)/\left[\begin{smallmatrix}01\\10\end{smallmatrix}\right]\right)\right\}} \\
 & \uparrow\cong &
\end{array}$$

$$H_o\{H^2(\text{units } Z\pi_2) \to L_{2n+1}^{[\pm\pi 2]}(Z_2\pi_2)/[{}^{01}_{10}] \to \prod^{r_o}(L_{2n+1}^{[\pm 1]}(Z_2)/[{}^{01}_{10}])\}$$

$$\uparrow$$

$$H_o\{H^2(\text{units } Z\pi_2) \to H^2(\text{units } Z_2\pi_2)/[\pm_2\pi] \to \prod^{r_o}(H^2(\text{units } Z_2)/[\pm 1])\}$$

A basis for $L_{2n+1}^{s}(\pi)$ is constructed as follows. Adopt the notation in theorem 6. If $T = \{i,j\}$ and $e_i > 1$, let $\Delta'_T =$ $[(1-\sigma_i^{2^{e_i-2}}) + (1-\bar{\sigma}_i^{2^{e_i-2}}) + \Delta_i]\,\Delta_j$ $((\Delta'_T)^2 = 8\,\Delta'_T)$. For T arbitrary, let

$$\alpha_T = \begin{pmatrix} 1 - \frac{\Delta_T}{2} & \Delta_T \\ -2^{t-2}(1-2^{t-2})\Delta_T & (1-\frac{\Delta_T}{2}) + (1-2^{t-2})(1+2^{t-1})\Delta_T \end{pmatrix},$$

and for $T = \{i,j\}$ and $e_i > 1$, let

$$\alpha'_T = \begin{pmatrix} 1 - \frac{\Delta_T}{2} & \Delta'_T \\ 2\Delta'_T & (1-\frac{\Delta'_T}{2}) - 5\,\Delta'_T \end{pmatrix}.$$

Let

$$\psi = \left(\begin{array}{cc|cc} 1 & 0 & 0 & 0 \\ 0 & 0 & 0 & 1 \\ \hline 0 & 0 & 1 & 0 \\ 0 & 1 & 0 & 0 \end{array}\right).$$

The coefficients of α'_T and α_T are fixed by the involution and the det α'_T = det α_T = 1. Thus, if $\begin{pmatrix} a & b \\ c & d \end{pmatrix} = \alpha'_T$ or α_T then

$$\mathbb{H}\begin{pmatrix} a & b \\ c & d \end{pmatrix} = \left(\begin{array}{cc|cc} a & b & & \\ c & d & & \\ \hline & & d & -c \\ & & -c & a \end{array}\right).$$

<u>The family of matrices below is a basis for</u> $L_{2n+1}{}^{s}(\pi)$.

$$\{H(\alpha_T')w\ H(\alpha_T')w^{-1} \mid T = \{i,j\} \subset \{1,\dots,s\},\ e_i \text{ or } e_j > 1\} \cup$$

$$\{H(\alpha_T)w\ H(\alpha_T)w^{-1} \mid T \subset \{1,\dots,s\},\ t > 2\}.$$

<u>Proof of theorems</u> 6 <u>and</u> 7. By Dress induction [D 3] one reduces to the case π is abelian. My philosophy is to compute the group $KQ_1(Z\pi)$ and then deduce the computation of $L_{2n+1}{}^{s,h}(\pi)$ from that of $KQ_1(Z\pi)$. The group $KQ_1(Z\pi)$ is computed in theorems 9-12 below. Recall the definition of $L_{2n+1}{}^{s,h}(\pi)$ given previously in the section on notation. The groups $EKQ_1(Z\pi)$ and $SKQ_1(Z\pi)$ are defined below. From theorem 9 it follows that

$$L_{2n+1}{}^{s}(\pi) = \begin{cases} EKQ_1{}^{(-1)^n}(Z\pi)/H[\pm_2\pi],\ \left[\begin{smallmatrix}0 & -1\\ 1 & 0\end{smallmatrix}\right] & \text{if } n \equiv 1\ (2)\\ EKQ_1{}^{(-1)^n}(Z\pi)/H[\pm_2\pi] & \text{if } n \equiv 0\ (2)\end{cases}$$

$$L_{2n+1}{}^{h}(\pi) = \begin{cases} SKQ_1{}^{(-1)^n}(Z\pi)/H(H^2(K_1(Z\pi)),\ \left[\begin{smallmatrix}0 & -1\\ 1 & 0\end{smallmatrix}\right] & \text{if } n \equiv 1\ (2)\\ SKQ_1{}^{(-1)^n}(Z\pi)/H(H^2(K_1(Z\pi)) & \text{if } n \equiv 0\ (2)\end{cases}.$$

Moreover, by theorem 12 one has that $EKQ_1(Z\pi) = SKQ_1(Z\pi)$. Theorems 6 and 7 are deduced easily now from theorems 10 and 11.

In preparation for the statement of theorem 8, we need to describe a certain subgroup of the image $H^2(K_o(Z\pi,Q\pi)) \to H^2(\tilde{K}_o(Z\pi))$. The results of [BSH § 3] show that $K_o(Z\pi,Q\pi) = \coprod_p K_o(Z_p\pi,\ Q_p\pi) = \coprod_p$ coker $(K_1(Z_p\pi) \to K_1(Q_p\pi))$ = (using Nrd) $\coprod_p$ coker $(K_1(Z_p\pi) \to$ units (center $Q_p\pi$). Write center $(Q_2\pi) = K_2 \times L_2$ where $K_2 = \prod Q_2$ with trivial involution and L = product of nontrivial extensions of Q_2. Let $R_2 = \prod Z_2 \subset \prod Q_2 \subset$ center $Q_2\pi$. The units R_2 determine a subgroup of the coker $(K_1(Z_2\pi) \to \text{units (center } Q_2\pi))$; hence a subgroup of $K_o(Z\pi, Q\pi)$; and a subgroup of the image $H^2(K_o(Z\pi, Q\pi)) \to H^2(\tilde{K}_o(Z\pi))$. Denote this last subgroup by [units R_2].

<u>Theorem 8</u> <u>The following sequence is exact.</u>

$$0 \to L_{2n+1}{}^h(\pi) \to L_{2n+1}{}^P(\pi) \to \begin{cases} \text{image}\{H^2(K_o(Z\pi, Q\pi)) \to H^2(\tilde{K}_o(Z\pi))\}/[\text{units } R_2] & \text{if } n \equiv 0\ (2) \\ \text{image } H^2(K_o(Z\pi, Q\pi)) \to H^2(\tilde{K}_o(Z\pi)) & \text{if } n \equiv 1\ (2)\ . \end{cases} \to 0$$

<u>Note that if</u> π <u>has odd order then</u> $L_{2n+1}{}^h(\pi) = 0$.

<u>Proof</u> One reduces by Dress induction to the case π abelian. Consider the exact sequence [R, 2.3]

$$H^1(\tilde{K}_o(Z\pi)) \xrightarrow{\mathbb{h}} L_{2n+1}{}^h(\pi) \to L_{2n+1}{}^P(\pi) \to H^2(\tilde{K}_o(Z\pi)) \xrightarrow{\mathbb{H}} L_{2n}{}^h(\pi).$$

The map $\mathbb{h}$ is factored by the diagram

$$\begin{array}{ccc} H^1(\tilde{K}_o(Z\pi)) & \xrightarrow{\mathbb{h}} & L_{2n}{}^h(\pi) \\ \downarrow & & \downarrow f \\ H^1(\tilde{K}_o(Z_2\pi_2)) & \xrightarrow[\mathbb{h}_2]{} & L_{2n}{}^{K_1(Z_2\pi_2)}(Z_2\pi_2)/[\begin{smallmatrix} 0 & (-1)^n \\ 1 & 0 \end{smallmatrix}]\ . \end{array}$$

The group $\tilde{K}_o(Z_2\pi_2) = 0$ because $Z_2\pi_2$ is local, and the map f is injective by theorems 6 and 7. Thus $\mathbb{h} = 0$. Now using theorem 2 one shows that the ker $\mathbb{H}$ =

$$\begin{cases} \text{image}\{H^2(K_o(Z\pi, Q\pi)) \to H^2(\tilde{K}_o(Z\pi))\}/[\text{units } R_2] & \text{if } n \equiv 0\ (2) \\ \text{image}\{H^2(K_o(Z\pi, Q\pi)) \to H^2(\tilde{K}_o(Z\pi))\} & \text{if } n \equiv 1\ (2)\ . \end{cases}$$

Next we compute $KQ_1(Z\pi)$. Let

Nrd = reduced norm = the ordinary determinant if the ring is commutative

$EKQ_1(Z\pi) = \ker KQ_1(Z\pi) \to K_1(Z\pi)$

$SKQ_1(Z\pi) = \ker \mathrm{Nrd}: KQ_1(Z\pi) \to \text{units (center } Q\pi)$.

<u>Theorem 9</u> <u>There is an exact sequence</u>

$$0 \to SKQ_1^{\pm 1}(Z\pi) \to KQ_1^{\pm 1}(Z\pi) \xrightarrow{\mathrm{Nrd}} \begin{cases} (\pi/[\pi,\pi])^2 & \text{if } \lambda = -1 \\ \pm(\pi/[\pi,\pi])^2 & \text{if } \lambda = 1. \end{cases} \to 0$$

<u>Theorem 10</u> <u>Suppose</u> π <u>is abelian. Then</u>

$$\begin{array}{ccc} SKQ_1^{-1}(Z\pi) & \xrightarrow{\cong} & SKQ_1^{-1}(Z_2\pi_2) \\ & & \uparrow \cong \ (H,\beta) \\ & & (H^2(\text{units } Z_2\pi_2)/[-1],[5]) \oplus Z/4Z \\ & & \uparrow \cong \\ & & (Z/2Z)^{2^s+s-1} \oplus Z/4Z \end{array}$$

<u>where</u> $\beta[1] = \left[\begin{smallmatrix} 0 & -1 \\ 1 & 0 \end{smallmatrix}\right]$. <u>Moreover, if one recalls the notation introduced in theorem 6 then a basis for</u> $SKQ_1^{-1}(Z\pi)$ <u>is provided by the family of matrices in theorem 6 plus the matrices below</u>

$$\left\{\begin{pmatrix} 0 & -1 \\ 1 & 0 \end{pmatrix}\right\} \cup \left\{\begin{pmatrix} 1-\Delta_T & 0 \\ 0 & (1-\Delta_T)^{-1} \end{pmatrix} \;\middle|\; T \subset \{1,\dots,s\},\ t = 1\right\}.$$

<u>Theorem 11</u> <u>Suppose</u> π <u>is abelian. Then the commutative diagram below has exact top and bottom rows.</u>

$$\begin{array}{ccccc}
0 \to SKQ_1{}^1(Z\pi) & \to & SKQ_1{}^1(Z_2\pi_2) & \to & \prod^{r_0}(SKQ_1{}^1(Z_2)/\mathbb{H}[\pm 1]) \\
 & & \cong \uparrow \mathbb{H} & & \cong \uparrow \mathbb{H} \\
 & & H^2(\text{units } Z_2\pi_2) & \to & \prod^{r_0}(H^2(\text{units } Z_2)/[\pm 1]) \\
 & & \cong \uparrow & & \cong \uparrow \\
0 \to (Z/2Z)^{2^s - \binom{E}{2}} & \to & (Z/2Z)^{2^s+s+1} & \to & (Z/2Z)^{r_0 = 2^s}
\end{array} .$$

<u>Moreover, if one recalls the notation introduced in theorem 7 then a basis for</u> $SKQ_1{}^1(Z\pi)$ <u>is provided by the family of matrices in theorem 7 plus the matrices below</u>

$$\left\{\begin{pmatrix} -1 & 0 \\ 0 & -1 \end{pmatrix}\right\} \cup \left\{\begin{pmatrix} 1-\Delta_T & 0 \\ 0 & (1-\Delta_T)^{-1} \end{pmatrix} \;\middle|\; T \subset \{1,\dots,s\},\ t = 1\right\} .$$

The next result extends theorems 10 and 11 to the nonabelian case.

<u>Theorem 12</u> a) <u>If</u> π <u>is abelian then</u>

$$EKQ_1(Z\pi) = SKQ_1(Z\pi).$$

b) <u>If</u> $\pi_2 \subset \pi$ <u>is normal abelian then</u>

$$EKQ_1(Z\pi) = EKQ_1(Z\pi_2) = SKQ_1(Z\pi_2) .$$

<u>Proof of theorem 9</u>. The abelian case is found in Bass [Bs2, § 4]. The general case is obtained by copying the argument in [Bs 2, § 4], and replacing Higman's theorem [H] by Wall's theorem [W5, 6.5]. Theorem 9 is valid for any finite group π.

<u>Proof of theorems 10 and 11</u>. We handle both cases simultaneously. Before getting into the proof, we need to explain the notation $\mathbb{II}$. Let V be a set of indices. Let $\{G_v \mid v \in V\}$ and $\{H_v \mid v \in V\}$ be fami-

lies of groups such that for each $v \in V$ there is a homomorphism $f_v : H_v \to G_v$. If $I \subset J \subset V$ then there is an obvious homomorphism $\prod_{v \in I} G_v \times \prod_{v \notin I} H_v \to \prod_{v \in J} G_v \times \prod_{v \notin J} H_v$. Let $\coprod_v(G_v, H_v) = \lim_{\substack{\to \\ I}}$ $\prod_{v \in I} G_v \times \prod_{v \notin I} H_v$ where I ranges over all finite subsets of V. $\coprod$ is called the restricted direct product.

Let $\hat{Z} = \prod_p Z_p$ and $\hat{Q} = \coprod_p (Q_p, Z_p)$. To the fibred square

$$\begin{array}{ccc} Z\pi & \to & Q\pi \\ \downarrow & & \downarrow \\ \hat{Z}\pi & \to & \hat{Q}\pi \end{array}$$

one associates two exact Mayer-Vietoris sequences [B10, §15]

$$\begin{array}{ccccccc} KQ_2(\hat{Z}\pi) \oplus KQ_2(Q\pi) & \to & KQ_2(\hat{Q}\pi) & \xrightarrow{\partial} & SKQ_1(Z\pi) & \to & SKQ_1(\hat{Z}\pi) \oplus SKQ_1(Q\pi) \to SKQ_1(\hat{Q}\pi) \\ \uparrow \mathbb{H} & & \uparrow \mathbb{H} & & \uparrow \mathbb{H} & & \\ K_2(\hat{Z}\pi) \oplus K_2(Q\pi) & \to & K_2(\hat{Q}\pi) & \to & SK_1(Z\pi) & & \end{array}$$

The first key step in the proof was the exact sequences above. The second key step is the theorem below. It states that the map $\mathbb{H} : SK_1(Z\pi) \to SKQ_1(Z\pi)$ is trivial. Thus the composite mapping $K_2(\hat{Q}\pi) \xrightarrow{\mathbb{H}} KQ_2(\hat{Q}\pi) \xrightarrow{\partial} SKQ_1(Z\pi)$ is trivial. Thus the map $\partial = 0 \Leftrightarrow$ the map coker $\{K_2(\hat{Z}\pi) \xrightarrow{\mathbb{H}} KQ_2(\hat{Z}\pi)\} \oplus$ coker $\{K_2(Q\pi) \xrightarrow{\mathbb{H}} KQ_2(Q\pi)\} \xrightarrow{c}$ coker $\{K_2(\hat{Q}\pi) \xrightarrow{\mathbb{H}} KQ_2(\hat{Q}\pi)\}$ is surjective. Using Sharpe's periodicity theorem [Sh, 7.1] one interprets the cokernel groups above as certain Witt groups. The Witt groups in question are well known. One computes them (or tracks their computation down in the literature) (forthe case π odd cyclic see [B2, p273-274])and deduces that the map c is surjective. Thus $\partial = 0$. Thus there is an exact sequence

$$0 \to SKQ_1(Z\pi) \to SKQ_1(\hat{Z}\pi) \oplus SKQ_1(Q\pi) \to SKQ_1(\hat{Q}\pi).$$

Furthermore, by [B 10, § 16] the group $SKQ_1(\hat{Z}\pi) = \prod_p SKQ_1(Z_p\pi)$ and the group $SKQ_1(\hat{Q}\pi) = SKQ_1(\coprod_p(Q_p, Z_p)\pi) = \coprod_p(SKQ_1(Q_p\pi), SKQ_1(Z_p\pi))$.

The proof is completed by computing by hand the terms to the right of $SKQ_1(Z\pi)$ in the sequence above and then using the sequence to compute $SKQ_1(Z\pi)$. (In making the computations, it is helpful to keep in mind [B 2, lemma 4] and the paragraph following its proof, the kernel of the hyperbolic map $\mathbb{H} : K_1(A) \to KQ_1(A)$ as measured in [R, 3.3] (or [B 1, 6.39a)]), [BSH, 10.6], and the fact that $Z_p\pi_p$ is a complete local ring.) The computations show the following. If A is a ring let $A^{\bullet}$ = units A. If $\lambda = -1$ then one has $SKQ_1^{-1}(Z_p\pi)$ $(p \neq 2) = SKQ_1^{-1}(Q_p\pi)$(p arbitrary) $= SKQ_1^{-1}(Q\pi) = 0$, and $SKQ_1^{-1}(Z_2\pi) \underset{\approx}{\to} SKQ_1^{-1}(Z_2\pi_2) \underset{\approx}{\xleftarrow{\mathbb{H},\beta}} (H^2(Z_2\pi_2{}^{\bullet})/[-1],[5]) \oplus Z/4Z \cong$ (by Wall [W5 § 11]) (a different computation is given in [B5 § 3]) $(Z/2Z)^{2^s+s-1} \oplus Z/4Z$. The matrix $\begin{pmatrix} 0 & -1 \\ 1 & 0 \end{pmatrix}$ generates the direct summand isomorphic to $Z/4Z$ (see [B2, p272-273]). In the case $\lambda = +1$, the situation is more cluttered. For $p \neq 2$ one has that $SKQ_1{}^1(Z_p\pi) \underset{\approx}{\to} SKQ_1{}^1(Z_p\pi_2) \underset{\approx}{\to} \overset{r}{\prod}{}^{o}\, SKQ_1{}^1(Z_p) \underset{\approx}{\xleftarrow{\mathbb{H}}} \overset{r}{\prod}{}^{o}\, H^2(Z_p{}^{\bullet})$ (the map $Z_p\pi_2 \to \overset{r}{\prod}{}^{o} Z_p$ is the canonical projection of $Z_p\pi_2$ onto its direct factor $\overset{r}{\prod}{}^{o} Z_p$) and for any p $SKQ_1{}^1(Q_p\pi) \underset{\approx}{\to} SKQ_1{}^1(Q_p\pi_2) \underset{\approx}{\to} \overset{r}{\prod}{}^{o}\, SKQ_1{}^1(Q_p) \underset{\approx}{\xleftarrow{\mathbb{H}}} \overset{r}{\prod}{}^{o}\, H^2(Q_p{}^{\bullet})$. Also $SKQ_1{}^1(Q\pi) \underset{\approx}{\to} SKQ_1{}^1(Q\pi_2) \underset{\approx}{\to} \overset{r}{\prod}{}^{o}\, SKQ_1(Q) \underset{\approx}{\xleftarrow{\mathbb{H}}} \overset{r}{\prod}{}^{o}\, H^2(Q^{\bullet})$. For $p = 2$ one has $SKQ_1{}^1(Z_2\pi) \underset{\approx}{\to} SKQ_1{}^1(Z_2\pi_2) \underset{\approx}{\xleftarrow{\mathbb{H}}} H^2(Z_2\pi_2{}^{\bullet}) \cong (Z/2Z)^{2^s+s+1}$. Plugging all of this information into the exact sequence above one obtains an exact sequence

$$0 \to SKQ_1{}^1(Z\pi) \to SKQ_1{}^1(Z_2\pi_2) \oplus \overset{r}{\prod}{}^{o}\Big(\prod_{p\neq 2} SKQ_1{}^1(Z_p) \oplus SKQ_1{}^1(Q)\Big) \to$$

$$\mathbb{H}\uparrow\cong$$

$$H^2(Z_2\pi_2{}^{\bullet}) \oplus \overset{r}{\prod}{}^{o}\Big(\prod_{p\neq 2} H^2(Z_p{}^{\bullet}) \oplus H^2(Q^{\bullet})\Big) \to$$

$$\overset{r}{\prod}{}^{o}\Big(SKQ_1{}^1(Q_2) \times \prod_{p\neq 2} SKQ_1{}^1(Q_p)\Big)$$

$$\mathbb{H}\uparrow\cong$$

$$\overset{r}{\prod}{}^{o}\Big(H^2(Q_2{}^{\bullet}) \times \prod_{p\neq 2} H^2(Q_p{}^{\bullet})\Big).$$

chasing the exact sequence, one deduces easily another exact sequence

$$
\begin{array}{ccccccc}
0 \to SKQ_1{}^1(Z\pi) \to & SKQ_1{}^1(Z_2\pi_2) & \to & \prod^{r_o} \text{coker } SKQ_1{}^1(Z) \to SKQ_1{}^1(Z_2) \\
& \mathbb{H}\uparrow\cong & & \mathbb{H}\uparrow\cong \\
& H^2(Z_2\pi_2{}^{\bullet}) & \to & \prod^{r_o} H^2(Z_2{}^{\bullet})/[\pm 1]
\end{array} .
$$

The theorem follows. To check that the matrices in theorems 10 and 11 provide a basis, one computes their image in H^2(units $Z_2\pi_2$) and notes that the image consists of the appropriate elements.

Theorem 13 *If* π *is abelian then the map*

$$\mathbb{H} : SK_1(Z\pi) \to SKQ_1(Z\pi) \text{ is trivial.}$$

Proof. Let $\mathcal{O}$ = integral closure of $Z\pi$ in $Q\pi$. If $Q\pi$ has the simple decomposition $Q\pi = \prod L_i$ then $\mathcal{O}$ has a corresponding decomposition $\mathcal{O} = \prod \mathcal{O}_i$ where $\mathcal{O}_i$ is the ring of integers in the cyclotomic field $L_i = Q[\zeta_i]$. The involution on $Q\pi$ corresponds to complex conjugation in each factor L_i and thus the fixed field K_i of the involution on L_i has a real completion. This is an important fact which will be used later. Let g be any involution invariant ideal of $\mathcal{O}$ such that $g \subset Z\pi$ and $Q \cdot g = Q\pi$. (The conductor [Bs3, XI § 6] is the largest ideal with these properties.) g has a product decomposition $g = \prod g_i$ corresponding to that of $\mathcal{O}$. Now by excision [Mi 6.3] the relative group $SK_1(Z\pi, g) = \prod SK_1(\mathcal{O}_i, g_i)$ and the relative group $SKQ_1(Z\pi, g) = \prod SKQ_1(\mathcal{O}_i, g_i)$. Consider the commutative diagram

$$
\begin{array}{ccc}
\prod SKQ_1(\mathcal{O}_i, g_i) & \to & SKQ_1(Z\pi) \\
\uparrow \mathbb{H} & & \mathbb{H}\uparrow \\
\prod SK_1(\mathcal{O}_i, g_i) & \underset{f}{\to} & SK_1(Z\pi)
\end{array}
$$

The map f is surjective because of the exact sequence [Mi 4.1] $SK_1(Z\pi, g) \to SK_1(Z\pi) \to SK_1(Z\pi/g)$ and the fact [Bs3, § 9] that $SK_1(Z\pi/g) = 0$. Thus $\mathbb{H} : SK_1(Z\pi) \to SKQ_1(Z\pi)$ is trivial if each $SKQ_1(\mathcal{O}_i, g_i) = 0$. However, by the results of [B 11] $SKQ_1(\mathcal{O}_i, g_i) = 0$ providing $g_i \neq 0$ is small enough. But, of course, we can choose g_i small enough. This completes the proof. The key condition used in applying the results of [B 11] is that K_i has a real completion. Otherwise, it would follow that for g_i small enough $SKQ_1(\mathcal{O}_i, g_i) \cong$ group of roots of unity in K_i.

<u>Proof of theorem 12</u>. a) One shows easily that the basis for $SKQ_1(Z\pi)$ given in theorems 10 and 11 vanishes in $K_1(Z\pi)$. Thus $EKQ_1(Z\pi) = SKQ_1(Z\pi)$.

b) The computation of $SKQ_1(Z\pi)$ in theorems 10 and 11 shows that for π abelian $SKQ_1(Z\pi) = SKQ_1(Z\pi_2)$. By a) we know that for π abelian $SKQ_1(Z\pi_2) = EKQ_1(Z\pi_2) = EKQ_1(Z\pi)$. By [B 2, Theorem 3], $EKQ_1(Z\pi)$ has exponent a power of 2. Thus by Dress induction [D 3] $EKQ_1(Z\pi) = \lim\limits_{\overrightarrow{\pi' \subset \pi}} EKQ_1(Z\pi')$ (π' 2-hyperelementary) $= EKQ_1(Z\pi_2)$.

Next we record some interesting corollaries of our computations.

<u>Corollary 1</u> <u>The forgetful map</u>

$$L_{2n}^{P}(\pi) \to H^1(\tilde{K}_0(Z\pi))$$ <u>is surjective.</u>

<u>Proof</u> Consider the exact sequence [R 2.3] $L_{2n}^{P}(\pi) \to H^1(\tilde{K}_0(Z\pi)) \xrightarrow{\mathbb{h}} L_{2n-1}^{h}(\pi)$. The proof of theorem 8 shows that $\mathbb{h} = 0$. The corollary follows.

<u>Corollary 2</u> <u>The 2-sylow subgroup</u> $SK_1(Z\pi)_2$ <u>of</u> $SK_1(Z\pi)$ <u>is generated by</u> $r_\infty - r_0$ <u>elements</u>.

The estimate seems best when $\pi = \pi_2$.

<u>Proof</u> Reduce by induction to the case π is abelian. Consider the

composite $SK_1(Z\pi) \overset{H}{\to} SKQ_1(Z\pi) \xrightarrow{\text{forgetful}} SK_1(Z\pi)$, $[\alpha] \mapsto [\bar{\alpha}^{-1}]$. Since $H = 0$ (theorem 13), it follows that $[\alpha] = [\bar{\alpha}]$, i.e. the $Z/2Z$-action on $SK_1(Z\pi)$ is trivial (see [Bs2] or [B 8] for an alternate proof). Thus the number of generators necessary for $SK_1(Z\pi)_2$ is the same as the number of generators necessary for $H^2(SK_1(Z\pi))$. Now from the Rothenberg exact sequence [S1, 4.1]

$$L_{2n}^h(\pi) \xrightarrow{\delta} H^2(Wh(\pi)) = H^2(SK_1(Z\pi)) \times H^2(\text{units } Z\pi)/[\pm\, {}_2\pi] \overset{H}{\to} L_{2n-1}^s(\pi)$$

and theorem 13 it follows that $H^2(SK_1(Z\pi))$ is covered the image δ. $L_{2n}^h(\pi)$ is computed in theorem 2. Combining this computation with theorem 4, one deduces that there is a surjective map $Z^{r_\infty - r_o} \to H^2(SK_1(Z\pi))$.

<u>Corollary 3</u>

$$L_{2n+1}^{s,h}(\pi_2) \xrightarrow{\cong} L_{2n+1}^{s,h}(\pi)$$

$$L_{2n+1}^{s}({}_4\pi) \xrightarrow{\cong} L_{2n+1}^{s}(\pi) \quad \text{if } n \equiv 0\ (2)$$

$$L_{2n+1}^{s}(\pi) \xrightarrow{\cong} L_{2n+1}^{h}(\pi) \quad \text{if } \pi_2 \text{ has exponent } 4$$

<u>Furthermore, there are exact sequences</u>

$$0 \to (Z/2Z)^{s-E} \longrightarrow L_{2n+1}^{s}({}_2\pi) \longrightarrow L_{2n+1}^{s}(\pi) \to (Z/2Z)^{s-E} \to 0 \quad \text{if } n \equiv 1(2)$$

$$0 \longrightarrow L_{2n+1}^{s}({}_2\pi) \longrightarrow L_{2n+1}^{s}(\pi) \to (Z/2Z)^{\binom{s-E}{2}+E(s-E)} \to 0 \quad \text{if } n \equiv 0(2)$$

The computation of $L_{2n}^h(\pi)$ in theorem 2 makes it desirable to have some information on $H^2(\tilde{K}_o(Z\pi))$.

<u>Theorem 14 If π is abelian of exponent 4 then</u>

$$H^2(\tilde{K}_o(Z\pi)) \cong H^1(\tilde{K}_o(Z\pi)) \cong (Z/2Z)^{2^s-1-\binom{E}{2}-s}.$$

<u>Moreover, a basis for $H^1(\tilde{K}_o(Z\pi))$ is given by the following family</u>

of projective $Z\pi$ - submodules of $Q\pi$ (the notation is that of theorems 6 and 7)

$$\{Z\pi(1 + \frac{\Delta_T}{2}) + Z\pi(\Delta_T) \mid T \subset \{1,\dots,s\},\ t \geq 2\} \cup$$
$$\{Z\pi(1 + \frac{\Delta'_T}{2}) + Z\pi(\Delta'_T) \mid T = \{i,j\},\ e_i \text{ or } e_j \geq 1\}.$$

The isomorphism $H^2(\tilde{K}_o(Z\pi)) \cong H^1(\tilde{K}_o(Z\pi))$ is Shapiro's lemma [Se, VIII Prop.8]. The rest of the proof is quite complicated and we won't try to sketch it.

Corollary 4 If π is abelian of exponent 4 then

$$H(\pi) = 0.$$

Proof All of the generators of $H^2(\tilde{K}_o(Z\pi))$ in the theorem above are involution invariant submodules of $Q\pi$. Thus they define elements of $H^2(I(Z\pi)) = H^2(K_o(Z\pi, Q\pi))$. Thus $H(\pi) = 0$.

To end the computations of this paper, I would like to say something about even λ-hermitian forms. All the computations done so far are for K-theory groups of quadratic modules with minimum form parameter. One can define similar groups for quadratic modules with maximum form parameter (see [B 9] or [Bs 1]), and compute these groups. The case of a maximum form parameter corresponds to modules with an even λ-hermitian form. The quadratic form (μ-form) is forgotten. We describe briefly what takes place here. The proofs are similar to the ones already given.

Theorem 15

$$\left.\begin{array}{l} L_{2n,2n+1}{}^{s,h,P}(\pi,\min) \xrightarrow{\cong} L_{2n,2n+1}{}^{s,h,P}(\pi,\max) \\[2ex] KQ_1{}^{(-1)^n}(Z\pi,\min) \xrightarrow{\cong} KQ_1{}^{(-1)^n}(Z\pi,\max) \end{array}\right\} \quad \text{if } n \equiv 0(2).$$

<u>The following sequence is split exact</u>

$$0 \to \mathbb{Z}/2\mathbb{Z} \xrightarrow{\alpha} L_{2n}^{s,h,p}(\pi,\min) \to L_{2n}^{s,h,p}(\pi,\max) \to 0 \quad \underline{\text{if}}\ n \equiv 1(2).$$

$$\left.\begin{array}{r} L_{2n+1}^{s,h}(\pi,\max) = 0 \\ SKQ_1^{(-1)^n}(Z\pi,\max) = 0 \end{array}\right\} \quad \underline{\text{if}}\ n \equiv 1(2)\ .$$

In particular $L_{2n+1}^{s,h}(\pi) = L_{2n+1}^{s,h}(\pi,\min)$ and $SKQ_1^{(-1)^n}(Z\pi) = SKQ_1^{(-1)^n}(Z\pi,\min)$ $(n \equiv 1(2))$ are accounted for totally by a phenomenon one could call the K_1-Arf invariant.

References

[B 1] A. Bak, K-Theory of forms, Ann.Math. Studies, Princeton University Press, (to appear)

[B 2] ______, Odd dimension surgery groups of odd torsion groups vanish, Topology Vol. 14(1975), 367-374

[B 3] ______, The computation of even dimension surgery groups of odd torsion groups, Topology (to appear)

[B 4] ______, The computation of surgery groups of odd torsion groups, Bull. Amer. Math. Soc. 80 (1974), 1113-1116

[B 5] ______, The computation of surgery groups of finite groups with abelian 2-hyperelementary subgroups, preprint

[B 6] ______, Integral representations of a finite group which preserve a nonsingular form, Proceedings of the international conference on representations of algebras, Carleton Math. Lec Notes 9(1974)4,01-4,06

[B 7] A. Bak, Grothendieck groups of modules and hermitian forms over commutative orders, Amer. Jour. Math. (to appear)

[B 8] ______, The involution on Whitehead torsion, Proceedings of the N.S.F. regional conference in topology Knoxville Tenn., General Top. and Appl. (to appear)

[B 9] ______, Definitions and problems in surgery and related groups, Proceedings of the N.S.F. regional conference in Topology Knoxville Tenn., General Top. and Appl. (to appear)

[B 10] ______, Strong approximation for central extensions of elementary groups, preprint

[B 11] ______, Solution to the congruence subgroup problem for λ-hermitian forms, preprint (to appear in [BHM])

[BHM] A. Bak, H. Helling, J. Mennicke, The modular and related groups, Ergebnisse der Mathematik, Springer-Verlag (to appear)

[BSH] A. Bak and W. Scharlau, Grothendick and Witt groups of orders and finite groups, Inventiones Math. 23 (1974), 207-240

[Bs 1] H. Bass, Unitary algebraic K-theory, Springer Lec. Notes in Math. 343 (1973), Algebraic K-theory III, 57-265

[Bs 2] ______, L_3 of finite abelian groups, Ann. Math. 99 (1974), 118 - 153

[Bs 3] ______, Algebraic K-theory, Benjamin (1968)

[BN] I. Berstein Some algebraic calculations of Wall groups for Z_2, preprint, Cornell University

[C] F. Connolly Linking numbers and surgery, preprint, University of Notre Dame

[D 1] A. Dress Contributions to the theory of induced representations, Springer Lec. Notes in Math. 342 (1973),

[D 1] A. Dress, Algebraic K-theory II, 183-242

[D 2] ________, Induction and structure theorems for Grothendieck and Witt rings of orthogonal representations of finite groups, Bull. Am. Math. Soc. 79(1973), 741-745

[D 3] ________, Induction and structure theorems for orthogonal representations of finite groups, Ann. Math.(1975)

[F] W. Feit, Characters of finite groups, Benjamin (1967)

[H] G. Higman, The units of group rings, Proc. London Math. Soc. 46(1940), 231-248

[KM] M. Kervaire and J. Milnor, Groups of homotopy spheres, I, Ann. Math. 77 (1963), 514-537

[L] R. Lee, Computation of Wall groups, Top. 10(1971), 149-166

[M] S. Maumary, Proper surgery groups and Wall-Novikov groups, Springer Lec. Notes in Math. 343 (1973), algebraic K-theory III, 526-539

[Mi] J. Milnor, Introduction to algebraic K-theory, Ann. Math. Studies 72, Princeton University Press (1971)

[PP] D. Passman and T. Petrie, Surgery with coefficients in a field, Ann. Math. 95 (1972), 385-405

[R] A. Ranicki, Algebraic L-theory III. Twisted Laurent extensions, Springer Lec. Notes in Math. 343 (1973), Algebraic K-theory III, 412-463

[Se] J.-P. Serre, Corps locaux, Actualités scientifiques et industrielles 1296, Hermann (Paris) (1962)

[S 1] J. Shaneson, Wall's surgery obstruction groups for $G \times Z$, Ann. Math. 90 (1969), 296-334

[S 2] J. Shaneson, Hermitian K-theory in topology, Springer Lec. Notes in Math. 343 (1973), algebraic K-theory III, 1-40

[Sh] R. Sharpe, On the structure of the unitary Steinberg group, Ann. Math. 90 (1972), 444-479

[SU] M.-K. Siu, Computation of unitary Whitehead group of cyclic groups, Thesis, Columbia University (1971)

[Sw] R. Swan, K-theory of finite groups and orders, Springer Lec. Notes in Math. 149 (1970)

[T] L. Taylor, Thesis, University of California at Berkeley (1971)

[W 1] C.T.C. Wall, Surgery on compact manifolds, Academic Press (1970)

[W 2] __________, Foundations of algebraic L-theory, Springer Lec. Notes in Math. 343 (1973), Algebraic K-theory III, 266-300

[W 3] __________, Some L-groups of finite groups, Bull. Amer. Math. Soc. 79(1973), 526-529

[W 4] __________, Classification of hermitian forms, VI Group Rings, preprint, University of Liverpool

[W 5] __________, Norms of units in group rings, Proc. London Math. Soc. 29(1974), 593-632

Universität Bielefeld
Fakultät für Mathematik

4800 Bielefeld
Kurt-Schumacher-Str. 6
West-Germany

Vol. 399: Functional Analysis and its Applications. Proceedings 1973. Edited by H. G. Garnir, K. R. Unni and J. H. Williamson II, 584 pages 1974.

Vol. 400: A Crash Course on Kleinian Groups. Proceedings 1974. Edited by L. Bers and I. Kra VII, 130 pages. 1974

Vol. 401: M. F. Atiyah, Elliptic Operators and Compact Groups. V, 93 pages. 1974.

Vol 402: M. Waldschmidt, Nombres Transcendants. VIII, 277 pages 1974

Vol. 403: Combinatorial Mathematics Proceedings 1972. Edited by D. A. Holton VIII. 148 pages 1974

Vol. 404: Théorie du Potentiel et Analyse Harmonique. Edité par J. Faraut. V, 245 pages. 1974.

Vol. 405: K. J. Devlin and H. Johnsbråten, The Souslin Problem. VIII, 132 pages. 1974.

Vol. 406: Graphs and Combinatorics. Proceedings 1973 Edited by R. A. Bari and F. Harary VIII, 355 pages. 1974

Vol. 407: P Berthelot, Cohomologie Cristalline des Schémas de Caracteristique $p > o$. II, 604 pages 1974

Vol. 408: J. Wermer, Potential Theory VIII, 146 pages 1974

Vol 409: Fonctions de Plusieurs Variables Complexes, Séminaire François Norguet 1970–1973. XIII, 612 pages. 1974.

Vol. 410: Séminaire Pierre Lelong (Analyse) Année 1972–1973 VI, 181 pages 1974

Vol. 411: Hypergraph Seminar. Ohio State University, 1972. Edited by C. Berge and D. Ray-Chaudhuri. IX, 287 pages 1974.

Vol 412: Classification of Algebraic Varieties and Compact Complex Manifolds Proceedings 1974 Edited by H Popp V, 333 pages 1974.

Vol 413: M. Bruneau, Variation Totale d'une Fonction XIV, 332 pages. 1974

Vol 414: T Kambayashi, M Miyanishi and M Takeuchi, Unipotent Algebraic Groups VI, 165 pages 1974

Vol. 415: Ordinary and Partial Differential Equations. Proceedings 1974. XVII, 447 pages. 1974

Vol. 416: M. E. Taylor, Pseudo Differential Operators. IV, 155 pages. 1974.

Vol. 417: H. H. Keller, Differential Calculus in Locally Convex Spaces XVI, 131 pages 1974

Vol 418: Localization in Group Theory and Homotopy Theory and Related Topics. Battelle Seattle 1974 Seminar. Edited by P. J. Hilton. VI, 172 pages 1974.

Vol. 419: Topics in Analysis. Proceedings 1970. Edited by O. E. Lehto, I. S. Louhivaara, and R. H. Nevanlinna. XIII, 392 pages. 1974.

Vol 420: Category Seminar Proceedings 1972/73. Edited by G. M. Kelly. VI, 375 pages. 1974.

Vol. 421: V. Poénaru, Groupes Discrets VI, 216 pages. 1974

Vol. 422: J-M. Lemaire, Algèbres Connexes et Homologie des Espaces de Lacets. XIV, 133 pages. 1974.

Vol. 423: S. S. Abhyankar and A. M. Sathaye, Geometric Theory of Algebraic Space Curves. XIV, 302 pages. 1974.

Vol. 424: L Weiss and J Wolfowitz, Maximum Probability Estimators and Related Topics. V, 106 pages. 1974.

Vol. 425: P. R Chernoff and J. E. Marsden, Properties of Infinite Dimensional Hamiltonian Systems IV, 160 pages 1974.

Vol. 426: M. L. Silverstein, Symmetric Markov Processes X, 287 pages. 1974.

Vol. 427: H. Omori, Infinite Dimensional Lie Transformation Groups XII, 149 pages 1974

Vol 428: Algebraic and Geometrical Methods in Topology, Proceedings 1973 Edited by L F. McAuley XI, 280 pages. 1974.

Vol. 429: L. Cohn, Analytic Theory of the Harish-Chandra C-Function. III, 154 pages. 1974.

Vol 430: Constructive and Computational Methods for Differential and Integral Equations. Proceedings 1974 Edited by D. L. Colton and R. P. Gilbert. VII, 476 pages. 1974

Vol 431: Séminaire Bourbaki – vol. 1973/74. Exposés 436–452. IV, 347 pages. 1975

Vol. 432: R. P. Pflug, Holomorphiegebiete, pseudokonvexe Gebiete und das Levi-Problem. VI, 210 Seiten. 1975.

Vol. 433: W. G. Faris, Self-Adjoint Operators. VII, 115 pages. 1975.

Vol. 434: P. Brenner, V. Thomée, and L. B. Wahlbin, Besov Spaces and Applications to Difference Methods for Initial Value Problems. II, 154 pages 1975.

Vol. 435: C. F. Dunkl and D. E. Ramirez, Representations of Commutative Semitopological Semigroups. VI, 181 pages 1975.

Vol. 436: L. Auslander and R. Tolimieri, Abelian Harmonic Analysis, Theta Functions and Function Algebras on a Nilmanifold. V, 99 pages. 1975.

Vol. 437: D. W Masser, Elliptic Functions and Transcendence. XIV, 143 pages. 1975.

Vol. 438: Geometric Topology. Proceedings 1974. Edited by L. C. Glaser and T. B. Rushing. X, 459 pages. 1975

Vol. 439: K. Ueno, Classification Theory of Algebraic Varieties and Compact Complex Spaces. XIX, 278 pages. 1975

Vol. 440: R. K. Getoor, Markov Processes: Ray Processes and Right Processes V, 118 pages. 1975

Vol. 441: N. Jacobson, PI-Algebras. An Introduction. V, 115 pages. 1975

Vol. 442: C. H. Wilcox, Scattering Theory for the d'Alembert Equation in Exterior Domains. III, 184 pages 1975.

Vol 443: M. Lazard, Commutative Formal Groups. II, 236 pages. 1975.

Vol. 444: F. van Oystaeyen, Prime Spectra in Non-Commutative Algebra. V, 128 pages. 1975.

Vol 445: Model Theory and Topoi Edited by F. W. Lawvere, C. Maurer, and G. C. Wraith. III, 354 pages. 1975.

Vol 446: Partial Differential Equations and Related Topics. Proceedings 1974 Edited by J. A. Goldstein. IV, 389 pages. 1975.

Vol. 447: S. Toledo, Tableau Systems for First Order Number Theory and Certain Higher Order Theories. III, 339 pages 1975

Vol. 448: Spectral Theory and Differential Equations. Proceedings 1974. Edited by W. N. Everitt. XII, 321 pages. 1975

Vol. 449: Hyperfunctions and Theoretical Physics. Proceedings 1973. Edited by F. Pham IV, 218 pages. 1975.

Vol. 450: Algebra and Logic. Proceedings 1974. Edited by J. N. Crossley. VIII, 307 pages. 1975.

Vol. 451: Probabilistic Methods in Differential Equations Proceedings 1974 Edited by M. A. Pinsky. VII, 162 pages. 1975.

Vol. 452: Combinatorial Mathematics III. Proceedings 1974. Edited by Anne Penfold Street and W. D. Wallis. IX, 233 pages. 1975.

Vol. 453: Logic Colloquium. Symposium on Logic Held at Boston, 1972–73. Edited by R. Parikh. IV, 251 pages. 1975.

Vol. 454: J. Hirschfeld and W H Wheeler, Forcing, Arithmetic, Division Rings. VII, 266 pages. 1975.

Vol. 455: H. Kraft, Kommutative algebraische Gruppen und Ringe. III, 163 Seiten. 1975.

Vol. 456: R. M. Fossum, P. A. Griffith, and I. Reiten, Trivial Extensions of Abelian Categories. Homological Algebra of Trivial Extensions of Abelian Categories with Applications to Ring Theory. XI, 122 pages 1975.

Vol. 457: Fractional Calculus and Its Applications Proceedings 1974. Edited by B. Ross. VI, 381 pages. 1975

Vol. 458: P. Walters, Ergodic Theory – Introductory Lectures. VI, 198 pages. 1975.

Vol. 459: Fourier Integral Operators and Partial Differential Equations. Proceedings 1974. Edited by J. Chazarain. VI, 372 pages. 1975

Vol. 460: O. Loos, Jordan Pairs. XVI, 218 pages. 1975.

Vol. 461: Computational Mechanics. Proceedings 1974 Edited by J. T. Oden. VII, 328 pages. 1975.

Vol. 462: P. Gérardin, Construction de Séries Discrètes p-adiques. »Sur les séries discrètes non ramifiées des groupes réductifs déployés p-adiques«. III, 180 pages 1975.

Vol. 463: H.-H. Kuo, Gaussian Measures in Banach Spaces. VI, 224 pages. 1975

Vol. 464: C Rockland, Hypoellipticity and Eigenvalue Asymptotics III, 171 pages 1975

Vol. 465: Séminaire de Probabilités IX. Proceedings 1973/74. Edité par P A. Meyer. IV, 589 pages. 1975

Vol. 466: Non-Commutative Harmonic Analysis. Proceedings 1974. Edited by J. Carmona, J. Dixmier and M Vergne. VI, 231 pages. 1975

Vol. 467: M R. Essén, The Cos $\pi\lambda$ Theorem With a paper by Christer Borell VII, 112 pages. 1975.

Vol. 468: Dynamical Systems – Warwick 1974. Proceedings 1973/74. Edited by A. Manning X, 405 pages. 1975.

Vol. 469: E. Binz, Continuous Convergence on C(X). IX, 140 pages. 1975.

Vol. 470: R. Bowen, Equilibrium States and the Ergodic Theory of Anosov Diffeomorphisms. III, 108 pages 1975

Vol. 471: R. S. Hamilton, Harmonic Maps of Manifolds with Boundary. III, 168 pages. 1975.

Vol. 472: Probability-Winter School. Proceedings 1975. Edited by Z. Ciesielski, K Urbanik, and W. A. Woyczyński. VI, 283 pages 1975.

Vol. 473: D. Burghelea, R. Lashof, and M. Rothenberg, Groups of Automorphisms of Manifolds. (with an appendix by E. Pedersen) VII, 156 pages. 1975.

Vol. 474: Séminaire Pierre Lelong (Analyse) Année 1973/74. Edité par P. Lelong. VI, 182 pages. 1975.

Vol. 475: Répartition Modulo 1. Actes du Colloque de Marseille-Luminy, 4 au 7 Juin 1974. Edité par G. Rauzy. V, 258 pages. 1975 1975

Vol. 476: Modular Functions of One Variable IV. Proceedings 1972. Edited by B. J Birch and W. Kuyk. V, 151 pages. 1975.

Vol. 477: Optimization and Optimal Control. Proceedings 1974. Edited by R. Bulirsch, W. Oettli, and J. Stoer. VII, 294 pages. 1975.

Vol. 478: G. Schober, Univalent Functions – Selected Topics. V, 200 pages. 1975

Vol. 479: S. D. Fisher and J. W. Jerome, Minimum Norm Extremals in Function Spaces. With Applications to Classical and Modern Analysis. VIII, 209 pages. 1975.

Vol. 480: X. M. Fernique, J. P. Conze et J. Gani, Ecole d'Eté de Probabilités de Saint-Flour IV–1974. Edité par P.-L. Hennequin. XI, 293 pages. 1975.

Vol. 481: M. de Guzmán, Differentiation of Integrals in R^n XII, 226 pages. 1975.

Vol. 482: Fonctions de Plusieurs Variables Complexes II. Séminaire François Norguet 1974–1975. IX, 367 pages. 1975

Vol. 483: R. D M. Accola, Riemann Surfaces, Theta Functions, and Abelian Automorphisms Groups. III, 105 pages. 1975.

Vol. 484: Differential Topology and Geometry. Proceedings 1974 Edited by G. P. Joubert, R. P. Moussu, and R. H. Roussarie. IX, 287 pages. 1975.

Vol. 485: J. Diestel, Geometry of Banach Spaces – Selected Topics. XI, 282 pages. 1975.

Vol. 486: S. Stratila and D. Voiculescu, Representations of AF-Algebras and of the Group U (∞) IX, 169 pages 1975

Vol. 487: H. M. Reimann und T. Rychener, Funktionen beschränkter mittlerer Oszillation. VI, 141 Seiten. 1975.

Vol. 488: Representations of Algebras, Ottawa 1974. Proceedings 1974 Edited by V. Dlab and P. Gabriel. XII, 378 pages. 1975.

Vol. 489: J Bair and R. Fourneau, Etude Géométrique des Espaces Vectoriels. Une Introduction. VII, 185 pages. 1975.

Vol. 490: The Geometry of Metric and Linear Spaces. Proceedings 1974 Edited by L M. Kelly. X, 244 pages. 1975

Vol. 491: K A Broughan, Invariants for Real-Generated Uniform Topological and Algebraic Categories. X, 197 pages. 1975.

Vol. 492: Infinitary Logic: In Memoriam Carol Karp. Edited by D. W. Kueker. VI, 206 pages. 1975.

Vol. 493: F. W Kamber and P. Tondeur, Foliated Bundles and Characteristic Classes. XIII, 208 pages. 1975.

Vol. 494: A Cornea and G. Licea. Order and Potential Resolvent Families of Kernels IV, 154 pages 1975

Vol. 495: A. Kerber, Representations of Permutation Groups II V, 175 pages. 1975

Vol. 496: L. H. Hodgkin and V. P. Snaith, Topics in K-Theory. Two Independent Contributions. III, 294 pages 1975.

Vol. 497: Analyse Harmonique sur les Groupes de Lie. Proceedings 1973–75. Edité par P. Eymard et al VI, 710 pages. 1975.

Vol. 498: Model Theory and Algebra. A Memorial Tribute to Abraham Robinson. Edited by D H Saracino and V. B. Weispfenning. X, 463 pages. 1975.

Vol. 499: Logic Conference, Kiel 1974. Proceedings. Edited by G. H. Müller, A. Oberschelp, and K. Potthoff. V, 651 pages 1975.

Vol. 500: Proof Theory Symposion, Kiel 1974. Proceedings. Edited by J. Diller and G. H. Müller. VIII, 383 pages. 1975.

Vol. 501: Spline Functions, Karlsruhe 1975. Proceedings. Edited by K. Böhmer, G. Meinardus, and W. Schempp VI, 421 pages. 1976.

Vol. 502: János Galambos, Representations of Real Numbers by Infinite Series. VI, 146 pages. 1976.

Vol. 503: Applications of Methods of Functional Analysis to Problems in Mechanics Proceedings 1975. Edited by P. Germain and B. Nayroles. XIX, 531 pages. 1976

Vol. 504: S Lang and H. F Trotter, Frobenius Distributions in GL_2-Extensions. III, 274 pages. 1976.

Vol. 505: Advances in Complex Function Theory. Proceedings 1973/74. Edited by W. E. Kirwan and L. Zalcman. VIII, 203 pages. 1976.

Vol. 506: Numerical Analysis, Dundee 1975. Proceedings. Edited by G A. Watson. X, 201 pages. 1976.

Vol. 507: M C Reed, Abstract Non-Linear Wave Equations. VI, 128 pages 1976

Vol. 508: E. Seneta, Regularly Varying Functions. V, 112 pages. 1976

Vol. 509: D. E. Blair, Contact Manifolds in Riemannian Geometry. VI, 146 pages 1976

Vol. 510: V. Poènaru, Singularités C^∞ en Présence de Symétrie. V, 174 pages. 1976.

Vol. 511: Séminaire de Probabilités X. Proceedings 1974/75. Edité par P A. Meyer VI, 593 pages. 1976.

Vol. 512: Spaces of Analytic Functions, Kristiansand, Norway 1975, Proceedings. Edited by O. B Bekken, B. K. Øksendal, and A. Stray. VIII, 204 pages. 1976.

Vol. 513: R. B. Warfield, Jr. Nilpotent Groups VIII, 115 pages. 1976.

Vol. 514: Séminaire Bourbaki vol. 1974/75. Exposés 453 – 470. IV, 276 pages. 1976.

Vol. 515: Bäcklund Transformations. Nashville, Tennessee 1974. Proceedings. Edited by R. M. Miura. VIII, 295 pages. 1976.